Smart Nanomaterials Technology

Series Editors

Azamal Husen, Wolaita Sodo University, Wolaita, Ethiopia

Mohammad Jawaid, Laboratory of Biocomposite Technology, Universiti Putra Malaysia, INTROP, Serdang, Selangor, Malaysia

Nanotechnology is a rapidly growing scientific field and has attracted a great interest over the last few years because of its abundant applications in different fields like biology, physics and chemistry. This science deals with the production of minute particles called nanomaterials having dimensions between 1 and 100 nm which may serve as building blocks for various physical and biological systems. On the other hand, there is the class of smart materials where the material that can stimuli by external factors and results a new kind of functional properties. The combination of these two classes forms a new class of smart nanomaterials, which produces unique functional material properties and a great opportunity to larger span of application. Smart nanomaterials have been employed by researchers to use it effectively in agricultural production, soil improvement, disease management, energy and environment, medical science, pharmaceuticals, engineering, food, animal husbandry and forestry sectors.

This book series in Smart Nanomaterials Technology aims to comprehensively cover topics in the fabrication, synthesis and application of these materials for applications in the following fields:

- Energy Systems—Renewable energy, energy storage (supercapacitors and electrochemical cells), hydrogen storage, photocatalytic water splitting for hydrogen production
- Biomedical—controlled release of drugs, treatment of various diseases, biosensors,
- Agricultural—agricultural production, soil improvement, disease management, animal feed, egg, milk and meat production/processing,
- Forestry—wood preservation, protection, disease management
- Environment—wastewater treatment, separation of hazardous contaminants from wastewater, indoor air filters

Rakesh Kumar Bachheti · Archana Bachheti ·
Azamal Husen
Editors

Metal and Metal-Oxide Based Nanomaterials

Synthesis, Agricultural, Biomedical and Environmental Interventions

Springer

Editors
Rakesh Kumar Bachheti
Department of Industrial Chemistry
Addis Ababa Science and Technology
University, College of Applied Science
Addis Ababa, Ethiopia

Archana Bachheti
Department of Environment Science
Graphic Era University
Dehradun, Uttarakhand, India

Azamal Husen
Department of Biotechnology
Smt. S. S. Patel Nootan Science
and Commerce College
Sankalchand Patel University
Visnagar, Gujarat, India

ISSN 3004-8273 ISSN 3004-8281 (electronic)
Smart Nanomaterials Technology
ISBN 978-981-99-7672-0 ISBN 978-981-99-7673-7 (eBook)
https://doi.org/10.1007/978-981-99-7673-7

This Springer imprint is published by the registered company Springer Nature Singapore Pte Ltd.
The registered company address is: 152 Beach Road, #21-01/04 Gateway East, Singapore 189721, Singapore

Paper in this product is recyclable.

Preface

Nanotechnology is gaining importance in all areas of science and technology due to its numerous applications. It has been noticed that the diverse range of nanomaterials, including metal- and metal-oxide-based nanomaterials, have garnered significant attention due to their unique properties and versatile applications. This book, "Metal and Metal-Oxide Based Nanomaterials (Synthesis, Agricultural, Biomedical and Environmental Interventions)," delves into the synthesis methods, characterization techniques, and diverse interventions employing these nanomaterials in the fields of agriculture, biomedicine, and environmental remediation.

Chapter 1 explores the green synthesis of silver and silver oxide nanoparticles using plants and their subsequent characterization. The chapter discusses sustainable and eco-friendly methods for the synthesis of these nanoparticles, their properties, applications and challenges and future perspectives. Chapter 2 focuses on the plant-mediated synthesis and characterization of zinc oxide nanoparticles. The chapter discusses the potential of plants for the synthesis of zinc oxide nanoparticles and explores their unique properties and applications along with challenges. Chapter 3 delves into the plant extract-assisted green synthesis and characterization of copper nanocomposites. The chapter highlights the utilization of plant extracts in the synthesis of copper nanoparticles and examines their potential applications in diverse fields.

Chapter 4 explores the antimicrobial applications of zinc oxide nanoparticles in the food packaging industry. The chapter investigates the role of these nanoparticles in enhancing food safety and preventing microbial contamination in food packaging materials. It also focuses on the possible toxicity of zinc oxide nanoparticles in human beings. Chapter 5 explains the role of zinc nanoparticles in the management of post-harvest diseases along with the future of zinc nanoparticles in post-harvest disease management. Chapter 6 focuses on the application of zinc and zinc oxide nano-fertilizers in enhancing crop production. The chapter explores the role of these nanomaterials in improving nutrient uptake, enhancing plant growth, and increasing agricultural productivity. Chapter 7 provides recent updates on the use of nanostructures for food packaging applications. Chapter 8 examines the potential of silver and zinc nanoparticles in mediating abiotic stress tolerance in crop plants. Chapter 9 assesses the impact of silver and zinc on soil microbial structure and functionality. The chapter explores how these nanoparticles influence soil microbial communities and their crucial role in ecosystem health and functioning. Chapter 10 explores the role of gold nanoparticles in plant protection against pathogens. The chapter investigates the potential of gold nanoparticles in developing novel strategies for plant disease management.

Chapter 11 focuses on the use of silver nanoparticles in drug delivery systems. Chapter 12 delves into the role of gold nanoparticles for targeted drug delivery. The chapter explores the potential of gold nanoparticles in improving drug delivery efficiency, reducing side effects, and enhancing therapeutic outcomes. Chapter 13 explores the use of green-synthesized platinum nanoparticles for biomedical applications. The chapter investigates the unique properties of platinum nanoparticles and their potential along with challenges and future prospects. Chapter 14 examines the role of zinc oxide nanomaterials in the photocatalytic degradation of environmental pollutants. Chapter 15 focuses on the application of silver-doped nanomaterials for wastewater treatment. Chapter 16 focuses on the risks and benefits of zinc nanoparticles in aquatic ecosystems.

Taken together, this book discusses the recent risks and benefits of nanomaterials in the field of agricultural, biomedical, and environmental applications. This book provides an exciting and remarkable information to the scientists, researchers, and students working in nanotechnology, nanomedicine, environmental science, plant science, agriculture, chemistry, biotechnology, pharmacognosy, pharmaceuticals, industrial chemistry, and many other interdisciplinary subjects. It serves as a valuable resource for those interested in exploring the immense potential of these nanomaterials in addressing key challenges in agriculture, biomedicine, and environmental remediation. We also hope that the insights shared within this book will inspire further research, innovation, and the development of sustainable solutions for a better future.

Addis Ababa, Ethiopia — Rakesh Kumar Bachheti
Dehradun, India — Archana Bachheti
Visnagar, India — Azamal Husen

Contents

About the Editors

Rakesh Kumar Bachheti graduated from Hemwati Nandan Bahuguna University, Garhwal, India, in 1996. He completed his MSc in Organic Chemistry from the same university in 1998. Later, in 2001, he underwent a one-year Post Graduate Diploma in Pulp and Paper Technology from Forest Research Institute, Dehradun, India. Subsequently, he pursued his Ph.D. in Organic Chemistry from Kumaun University, Nainital, India, which he obtained in 2007. Bachheti has a total research and teaching experience of 20 Years. Before joining Addis Ababa Science and Technology University (AASTU) in Ethiopia, he worked as Dean of Project (Assistant) at Graphic Era University in Dehradun, India. It's worth noting that Graphic Era University is accredited with an 'A+' grade by the National Assessment and Accreditation Council (NAAC). Bachheti is also a Research Fellow at INTI International University Persiaran Perdana BBN, Putra Nilai, Nilai, Negeri Sembilan, Malaysia, and Adjunct Faculty at the Department of Allied Sciences, Graphic Era Hill University, Dehradun, India. Bachheti has an impressive research background and has presented papers at various international (Malaysia, Thailand, and India) and national conferences. He has also been an active member of important committees like the Internal Quality Assurance Cell (IQAC) and the Anti-ragging Committee. His research interests primarily focus on natural products for industrial applications, biofuels and bioenergy, green synthesis of nanoparticles and their applications, and pulp and paper technology. His passion for natural products is evident in all of his research endeavors. As an academic mentor, Bachheti has successfully advised 40 MSc and 6 Ph.D. students to completion. Additionally, numerous undergraduates have conducted research in his laboratory under his guidance. He actively contributes to curriculum development for BSc/MSc/Ph.D. programs. Bachheti is also recognized for his editorial work, having edited 6 books and authored 100 publications on various aspects of natural product chemistry and nanotechnology. He has contributed twenty book chapters published by prestigious publishers like Springer, Elsevier, Tylor and Francis and Nova Publisher.

Archana Bachheti completed her BSc in 1997 and MSc in 1999 from HNB Garhwal University. She earned her Ph.D. from Forest Research Institute, Dehradun, India, in 2006. Throughout her career, she has been actively involved in research projects and consultancy work in various areas, including ecorestoration and development of wasteland, physico-chemical properties of Jatropha curcus seed oil in relation to altitudinal variation, and serving as a consultant ecologist for a project funded by a government agency. Currently, Archana is a Professor at Graphic Era University in Dehradun, India. With an extensive experience of more than 17 years, she has served in different capacities within academia in India and has also provided expertise internationally. Her teaching portfolio includes subjects like Ecology and Environment, Environmental Science, Freshwater Ecology, Disaster Management, and Bryophytes and Pteridophytes. Archana's research interests cover a broad and interdisciplinary field of plant ecology. Her focus areas include ecorestoration, green chemistry, especially the synthesis of nanomaterials, and exploring the medicinal properties of plants. Her research has encompassed ecological amelioration of degraded lands, studying physical and chemical properties of plant oils, and delving into plant-based nanomaterials. Throughout her career, she has been actively involved in mentoring students and researchers. She guided one Ph.D. student and supervised three scholars. Additionally, she has provided guidance to graduate and undergraduate students in their research projects. Driven by her fascination with forest biodiversity, Archana has maintained a passion for exploring the values of biodiversity and how it can contribute to social upliftment. As an editor, she has contributed to six books and has published over 80 research articles in both international and national journals. She has also authored sixteen book chapters. Furthermore, she has organized several National seminars/conferences at Graphic Era University, India.

Azamal Husen is presently working as a Professor at Sankalchand Patel University, Visnagar, India; and Adjunct Professor at Graphic Era (Deemed to be University), Dehradun, Uttarakhand, India. He is also working as a Visiting Professor at University Putra Malaysia, Selangor, Malaysia. Previously, he served as Professor and Head of the Department of Biology, University of Gondar, Ethiopia; and worked as a Foreign Delegate at Wolaita Sodo University, Wolaita, Ethiopia. He also worked as a Visiting Faculty of the Forest Research Institute and the Doon College of Agriculture and Forest at Dehra Dun, India. His research and teaching experience of 25 years encompasses biogenic nanomaterial fabrication and application; plant responses to nanomaterials; plant adaptation to harsh environments at the physiological, biochemical, and molecular levels; herbal medicine; and clonal propagation for improvement of tree species. He has conducted research sponsored by the World Bank, the National Agricultural Technology Project, the Indian Council of Agriculture Research, the Indian Council of Forest Research Education, and the Japan Bank for International Cooperation. Husen has published extensively (over 250) and served on the Editorial Board and as reviewer of reputed journals published by Elsevier, Frontiers Media, Taylor & Francis, Springer Nature, RSC, Oxford University Press, Sciendo, the Royal Society, CSIRO, PLOS, MDPI, John Wiley & Sons, and UPM Journals. He is on the advisory board of Cambridge Scholars Publishing, UK. He is a fellow of the Plantae group of the American Society of Plant Biologists, and a member of the International Society of Root Research, Asian Council of Science Editors, and International Natural Product Sciences. He is Editor-in-Chief of the *American Journal of Plant Physiology*, and a Series Editor of *Exploring Medicinal Plants* (Taylor & Francis Group, USA); *Plant Biology,*

Sustainability, and Climate Change (Elsevier, USA); and *Smart Nanomaterials Technology* (Springer Nature, Singapore). He has achieved the distinguished honour of being recognized as one of the "World's Top 2% Scientists" for the year 2022, and again for the 2023 by Stanford University, USA. This recognition has also been prominently featured in the Elsevier Data Repository.

Green Synthesis of Silver and Silver Oxide Nanoparticles From Plants and Their Characterization

Anuj Kandwal, Shama Parveen, Rakesh Kumar Bachheti, Archana Bachheti, and Arun Kumar Khajuria

Abstract More contemporary synthetic approaches have been superseded by affordable and ecologically friendly nanoparticle (NP) production technologies. Green chemistry is environmentally safe, nontoxic, biocompatible, and financially productive. In the past, numerous research teams have produced silver and silver oxide nanoparticles from diverse plant extracts, and their findings have been widely reported. Green synthesis methods utilize microorganisms, plant extracts, or proteins as bio-capping and bio-reducing agents. They function as bio-nanofactories for material synthesis at the nanoscale size to create biogenic nanomaterials. Here, we have outlined the creation process, characterization, and composition of these plant-mediated silver oxide and silver nanoparticles made by plant-mediated synthesis. Green synthesis of silver and silver oxide nanoparticles depends on variety of parameters, such as extract concentration, exposure time, temperature, and pH. UV–visible spectroscopy, DLS, TEM, SEM, XRD, EDX, FTIR, etc. can successfully characterize plant-mediated silver and silver oxide nanoparticles. Future studies must concentrate on understanding the intricate mechanism of metal and metal oxide nanoparticle synthesis and designing nanoparticles that are less hazardous, healthier, and have precisely controlled dimensions and form. Additionally, silver release is more likely to occur in small-sized spherical silver nanoparticles. This review concludes that

A. Kandwal
Department of Chemistry, Harsh Vidya Mandir (PG) College, Raisi, Haridwar, Uttarakhand, India

S. Parveen
Department of Chemistry, Faculty of Science, Motherhood University, Haridwar, Uttarakhand, India

R. K. Bachheti
Department of Industrial Chemistry, Addis Ababa Science and Technology University, Addis Ababa, Ethiopia

A. Bachheti
Department of Environment Science, Graphic Era University, Dehradun 248002, Uttarakhand, India

A. K. Khajuria (✉)
Department of Botany, Cluster University of Jammu, Jammu 180001, India
e-mail: arun.khajuria20@gmail.com

R. K. Bachheti et al. (eds.), *Metal and Metal-Oxide Based Nanomaterials*, Smart Nanomaterials Technology, https://doi.org/10.1007/978-981-99-7673-7_1

further research into the synthesis and characterization of metallic silver and silver oxide nanoparticles made via green synthesis is necessary to bring green synthesized products to the market.

Keywords Nanoparticles · Green route synthesis · Biogenic fabrication · Mechanism · Various methods

1 Introduction

Nanoscience is one of the encouraging research fields in novel science with marvelous applications in biology, physics, chemistry, and material science. Nanoscience and nanotechnology is a branch of science and engineering defined as the design, synthesis, characterization, and application of materials whose shape and size have been manipulated at the nanoscale. Particles smaller than 100 nm in any one dimension are usually referred to as "nanoparticles" (Husen & Iqbal, 2019; Kouvaris et al., 2012). High surface area-to-volume ratio, variable pore size, and high reactivity are some of the nanostructured materials' improved or pronounced characteristics that change particles' chemical, physical, thermal, mechanical, catalytic, and electromagnetic characteristics (Bachheti et al., 2019; Iravani et al., 2014; Painuli et al., 2020; Pal et al., 2011). Metallic nanoparticles (MNPs) play a crucial role in reforming human life and its environment. This draws immense interest from researchers owing to its large variety of potential applications in optical, biomedical, water treatment, electronics, cosmetics, agriculture fields, food technology, mechanical engineering, energy, polymer science, agriculture and environmental and health science (Bachheti et al., 2020, 2020a, 2020b, 2021, 2022; Chawla et al., 2023; Fu & Fu, 2015; Husen et al. 2019; Husen & Jawaid, 2020; Husen, 2022a, 2023a, 2023b, 2023c; Husen 2023a, 2023b; Husen & Siddiqi, 2023; Mandal et al., 2023; Taghiyari et al., 2023). Today, a significant portion of the nanomaterials employed in many industries utilize chemical and physical processes. Existing synthesis techniques are known to be costly and to involve chemicals that are harmful to the environment and human health, which include solvents made from organic materials, reduction agents, and stabilizers, which limits the practical application of these nanoparticles in therapeutic and scientific sectors (Shume et al., 2020). In order to reduce or altogether avoid the use of dangerous chemicals and prevent the development of superfluous and toxic byproducts, several research organizations are now engaged in finding affordable and environmentally safe nanoparticle synthesis procedures. The green synthesis technique offers the most advantageous replacement for the current chemical and physical ways of synthesis since it is less expensive, safer to use, and environmentally benign (Abate et al., 2023; Godeto et al., 2023; Gonfa et al., 2023; Zebeaman et al., 2023). Because botanical extracts contain biomolecules including protein, carbohydrates, and coenzyme precursors that can reduce metal ions, using them as reducing substances has also started to look interesting. Additionally, the developed nanoparticles are stabilized and capped by these biomolecules. In comparison to microbe-mediated production, plant-mediated synthesis is more scalable and quicker. It has also been shown to be effective, affordable, and easily accessible. Because of the aforementioned factors, manufacturing nanoparticles from plant sources has emerged as the most desirable and popular technique (Danish et al., 2021; Elemike et al., 2017).

Metallic NPs (silver, gold, zinc oxide, silver oxide, copper oxide, and titanium oxide) have been synthesized, studied, and utilized by researchers in numerous/frequent fields (Joshi et al., 2019; Leyu et al., 2023). Silver and silver oxide nanoparticles are an excellent choicer in nanotechnology in the field of biological systems for medical imaging and for treating diseases among different metallic nanoparticles (Gonfa et al., 2022; Jeevanandam et al., 2018). Silver oxide and silver NP have been extensively used for cellular delivery because of their versatile characteristics/properties like wide availability, deep functionality, good compatibility, potential for targeted drug delivery, and moderate release of drugs. Besides medical and pharmaceutical applications, metallic NPs widely used in everyday applications (Christian et al., 2008).

Recently, scientists' attention has switched to the use of plant extracts in the production of nanoparticles. This approach is attracting more interest than other biological systems because it offers a clear, practical, economical, secure, environmentally friendly, practical, and advantageous path to synthesizing MNPs of desired shape and size. The fact that plant extract-mediated synthesis of NPs is easier to maintain, more readily available, and safe to handle than microbial synthesis of NPs is a significant advantage. Different nanoparticles (Ag, Au, silver oxide, and TiO_2) of various shapes and sizes have been produced using various plant components, including leaves, peels, roots, stems, seeds, fruits, and flowers (Jana et al., 2000). The main benefit of plant-mediated MNP synthesis over chemical synthesis is that no additional protective agent is required in the former (Shekhawat et al., 2014). Moreover, it is easy to get plant extracts to develop a successful and environmentally friendly method for manufacturing industrial quantities of uniform metallic nanoparticles. Thus, the aim of this chapter is to provide in-depth information about the plant-mediated synthesis and characterization of ZnO nanoparticles.

2 Classification of Metallic Nanoparticles

Nanoparticles can be widely classified into two categories, namely, organic nanoparticles, which comprise carbon NPs (fullerenes), and inorganic NPs, comprising magnetics NPs of noble metals (like gold, silver, platinum) and semi-conductor NPs (like titanium oxide and zinc oxide, etc.). At present, interest in inorganic metallic nanoparticles (silver, gold, platinum, titanium oxide, silver oxide, zinc oxide) is growing intensively as they provide better material properties with functional adaptability and have been tested as potential tools for medical imaging for dealing with diseases because of their size features and superiority over chemical images (Khan et al., 2019).

Metallic NPs (silver, gold, zinc oxide, silver oxide silver oxide, copper oxide, titanium oxide) have been prepared, studied, and utilized by researchers, scholars, and scientists in numerous fields. Ag and silver oxide-based nanoparticles are of excellent choice and have excellent scope in nanotechnology and biological systems for medical imaging and treating diseases among the different metallic nanoparticles (Jeevanandam et al., 2018). Silver oxide and silver NP have been extensively utilized for cellular delivery because of their adaptable characteristics/properties like widespread accessibility, deep functionality, good compatibility, the potential for targeted drug administration, and moderate absorption of drugs. Besides medical and pharmaceutical applications, metallic NPs widely used in everyday applications (Christian et al., 2008).

The number of dimensions of materials that are outside the nanoscale (100 nm) range allows for the classification of nanoparticles into three categories, as follows (Tiwari et al., 2012).

(I) **Zero-dimensional nanostructured materials (O D)**
All three dimensions are measured within the nanoscale in zero-dimensional (0 D) nanostructured materials, i.e., all dimensions are smaller than 100 nm. Zero-dimensional NSMs can be either amorphous or crystalline. It can exhibit various shapes and forms. For example, nanoparticles, quantum dots, nanolenses, hollow spheres, and uniform particle arrays.

(II) **One-dimensional nanostructured materials (1 D)**
Only one dimension is outside the nanoscale in one-dimensional nanostructured materials, which may result in needle-like nanomaterials. They substantially impact nanoelectronics, alternative energy sources, and nanocomposite materials. For example, nanotubes, nanobelts, nanoribbons, nanowires, and nanorods.

(III) **Two-dimensional nanostructured materials (2 D)**
In this class, two dimensions are outside the nanoscale. They can exhibit plate-like shapes. Examples are nanofilms, nanowalls, nanoplates, nanodisks, nanoprisms, nanolayers, nanocoatings, and graphene. They are especially interesting not only for understanding the mechanism of nanoparticle growth but also for their investigation and applications in sensors, nanoreactors, photocatalysts, nanocontainers, and templates for two-dimensional structures for other materials.

3 General Methods of Metallic and Metal Oxide Nanoparticle Synthesis

Currently, various methods, including physical, chemical, and biological, are used to synthesize metallic nanoparticles (Mittal et al., 2013). From the structural point of view, there are two basic approaches for nanoparticles synthesis which involve "top-down" and "bottom-up" (Fig. 1). "Top-down" technique is usually achieved by physical methods, in which the size of starting bulk material is reduced by removing material until the desired nanoparticles are obtained. However, the serious drawback of this approach is the imperfection of obtained structure. The other method is "bottom-up" (atom by atom or molecule by molecule), usually achieved by chemical and biological methods, in which the atoms and molecules randomly assemble themselves into the desired nanoparticles (Arole & Munde, 2014). The main advantage of the bottom-up method is that the synthesized nanoparticles are free of structural defects. Scheme 1 shows several methods and strategies used to synthesize metallic NPs (Singh et al., 2018).

The physical methods used to synthesize MNPs include evaporation–condensation, spray pyrolysis, laser ablation, pulse wire discharge, sputtering, electron beam lithography, plasma, gamma radiation, and microwave irradiation (Tran & Le, 2018). The chemical methods such as chemical reduction of metallic salt, template method, reverse micelles method, polyol process, sonochemical method, electrochemical method, hydrolysis, wet chemical method, and tollens methods are generally used to synthesize metallic NPs. A stabilizing agent must be added in chemical approaches to stabilize dispersive nanoparticles. The use of toxic chemicals and high radiation levels harmful to the environment is a severe drawback of the chemical and physical approaches (Khandel & Shahi, 2018). Therefore, researchers need to develop green routes for the synthesis of MNPs.

Green chemistry, also called sustainable chemistry, is the scheme, development, and execution of chemical reactions and products that minimizes or eradicates the

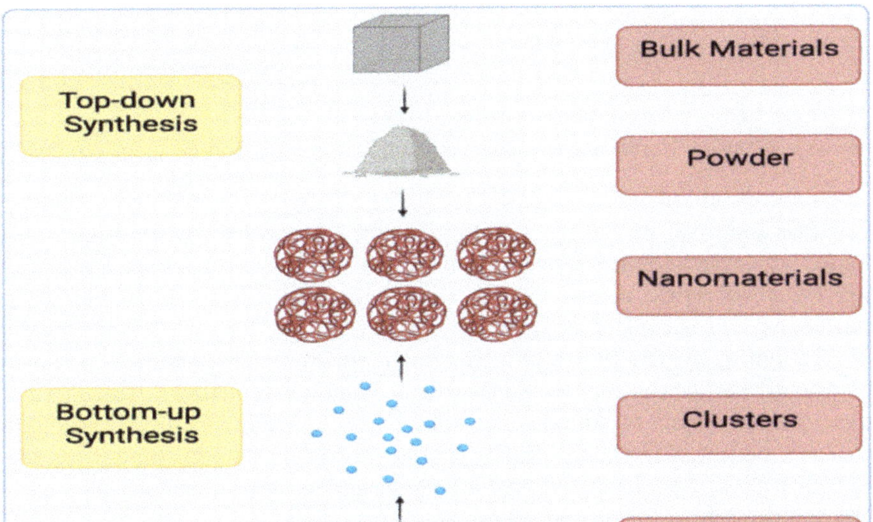

Fig. 1 Top-down and bottom-up approach for nanoparticles synthesis. Adopted from Alsaiari et al. (2023)

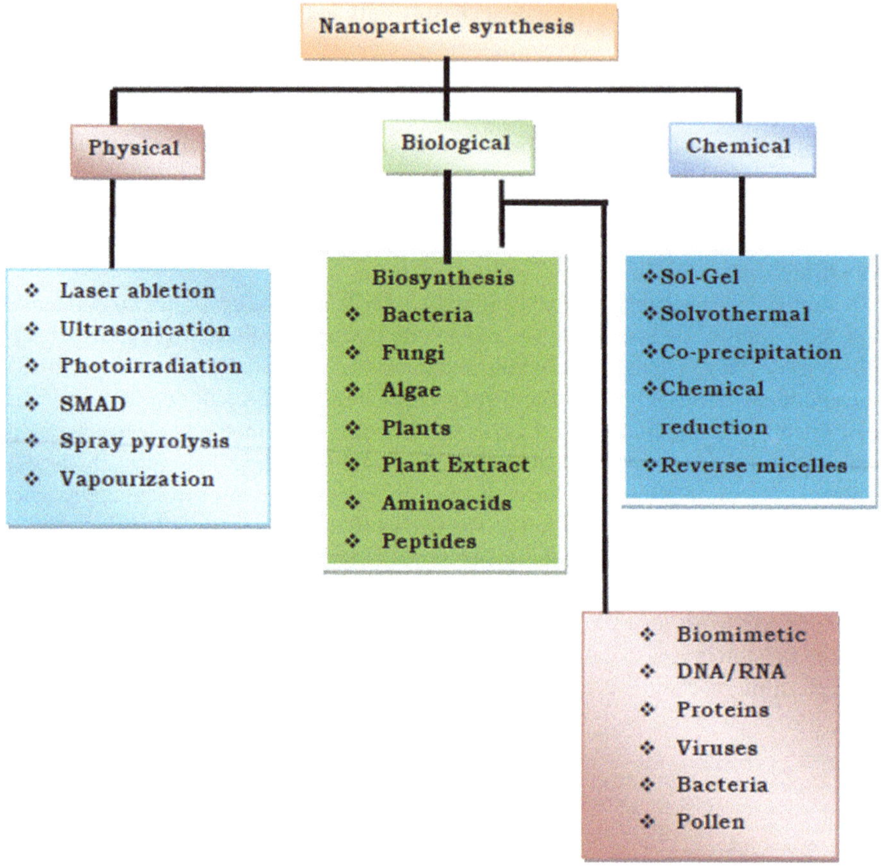

Scheme 1 Schematic representation of different methods to synthesize metallic nanoparticles

hazardous and toxic substances harmful to social life and the environment. It also focuses on energy-efficient processes. In recent years, the researcher has shifted their focus toward the biosynthesis of nanoparticles as it is green, cost-effective, feasible, rapid, sustainable, and environmentally benign. Biosynthesis of metallic NPs (bottom-up approach) is a bio-reduction method that uses naturally occurring eco-friendly substances such as plant extracts, vitamins, proteins, polysaccharides, biodegradable polymers, amino acids, microalgae, and microorganisms, including fungi and bacteria, as reducing, capping, and stabilizing agents. Bio-inspired methods are increasing steadily as they do not require hazardous toxic chemicals, high pressure, temperature, and energy (Heera & Shanmugam, 2015). The bio-enthused approach is very much close to the principles of nature, and it allows nature to synthesize stable MNPs.

4 Plant-Mediated Synthesis of Silver and Silver Oxide Nanoparticles

Recently, researchers have shifted their focus toward the biosynthesis of MNPs using plant extracts. This method is gaining more attention than other biological systems because it provides a clean, cost-effective, safe, eco-friendly, convenient, and beneficial route to synthesizing MNPs of desired shape and size. The significant advantages of plant extract-mediated synthesis of NPs over the microbial synthesis of NPs are that these are easy to maintain, readily available, and safe to handle. Different parts of plants, such as leaves, peel, roots, stem, seeds, fruits, and flowers, have been used to produce different nanoparticles (Ag, Au, silver oxide, TiO_2) of different shape and size (Jana et al., 2000). The major advantage of plant-mediated synthesis of MNPs over the chemical method is that in the former, there is no need for an additional protective agent (Shekhawat et al., 2014).

Several FTIR spectroscopic studies of plant-mediated NPs indicate that phytochemicals like flavonoids, alkaloids, terpenoids, polyphenols, polysaccharides, protein, and chlorophyll present in plant extracts act as reducing, capping, and stabilizing agents, and thus were believed to be responsible for the synthesis of Ag and silver oxide nanoparticles (Liu et al., 2017). The synthesis of bio-inspired nanoparticles results from the interaction of metallic salt with the plant extracts, which only takes place at optimum temperature, time, and pH (Kredy, 2018).

The most effective green synthesis technique for Ag and Ag_2O–NPs is using plant sources. The production of Ag and Ag_2O–NPs using plant-mediated synthesis takes less time and results in more stable nanoparticles than Ag and Ag_2O–NPs produced through microbe-mediated generation. The fact that botanical sources may be used easily, cheaply, and safely is crucial. They also include flavonoids that can stabilize the resultant nanoparticles (Husen et al. 2023d; Abbasi et al. 2020). The use of plant-derived materials has advanced to this point as more and more research teams have investigated the potential application of various plants for the manufacture of nanoparticles through the use of extracts of both aerial and beneath-the-ground parts, such as the roots, leaves, fruits, vegetables, and others (Ghotekar et al., 2020).

Silver nanoparticles were synthesized by using common and standard plant-mediated methods. At room temperature, aqueous plant extract and $AgNO_3$ solution were mixed in a fixed ratio in a 2 L Erlenmeyer conical flask. Then, the mixed solution was kept in a dark place for 72 h so that the synthesis of silver nanoparticles could take place. The solution turned a dark crimson-red color after 72 h, indicating the creation of Ag nanoparticles. To eliminate unreacted or uncoordinated components, the resulting solution was centrifuged for 20 min at 7500 rpm, rinsed with deionized distilled water, and then washed with ethanol. After being mashed in a mortar and pestle to create fine and homogeneous powdered graying black nanoparticles, the material was finally dried in an oven at 50 °C. Phytochemicals in plants are considered to mediate the synthesis of silver oxide nanoparticles. AgOH is produced initially as a result of the reaction among the hydroxyl groups (–OH) of the phytochemicals and the Ag+ from $AgNO_3$ before being changed into Ag_2O (Ravichandran et al., 2016; Sharma & Srivastava, 2020).

The roots and leaves of the plant contained various phytochemicals which plays an important role in reducing, capping, and stability of silver and silver oxide nanoparticles. For instance, *Ricinus communis* roots and leaves contained a variety of phenolic and flavonoid chemicals, including gallic acid, kaempferol-3-o-b-d-xylopyranoside, diethyl phthalate, kaempferol-3-o-b-d-xylopyranoside, 2-methyl, l-valine, triethyl citrate, quercetin-3-o-p-d-glucopyranoside, octadecanoic acid, 1-hexadecanol, and *n*-hexadecenoic acid; however, indole-3-acetic acid is the main component in root extract. Although due to the complicated chemical makeup of plants (extract), the precise method of metal nanoparticles by employing plant extract is uncertain. However, based on the above finding, a generic strategy for the creation of silver nanoparticles using three phytochemicals is given (see Fig. 2). With the release of hydrogen (reactive), keto form of the indole-3-actetic acid (present in the root extract) transforms into the enol form; the enol form, however, was unstable since it had two hydroxyl groups on the same carbon, and it changed back into keto form. Thus, silver nanoparticles are created when Ag^+ and Ag^0 combine as a result of the reactive hydrogen that has been released. Similar to this, triethyl citrate and flavonoid (quercetin-3-o-p-d-glucopyranoside) were regarded as reducing agents in the case of leaf extract. The phytochemicals (1-valine amino acid) and phenolic compounds play a significant role in the stability of metal nanoparticles, thus stabilized the silver nanoparticles (Huang et al., 2010; Rakhi & Gopal, 2012).

5 Characterization of Metallic Nanoparticles

Several techniques, such as UV–visible spectroscopy, SEM, XRD, DLS, EDX, TEM, FTIR, and TGA, can characterize metallic nanoparticles (Moond et al., 2023; Shaik et al., 2018).

In the preliminary stage, the formation of metallic nanoparticles is characterized by UV–visible spectroscopy. Metallic nanoparticles show the bright color and characteristic absorption band in the UV–visible region of the electromagnetic spectrum owing to a process known as surface plasmon resonance (SPR). Green-synthesized Ag and silver oxide nanoparticles show characteristic color and absorption UV–visible band, which is mainly dependent on their shape and size. Different surface plasmon resonance absorption bands in the range 300–700 nm are generally used to characterize different Ag (410–480 nm) and silver oxide (310–390 nm) nanoparticles of size less than 100 nm. A study revealed that the UV–visible spectrum of plant-mediated silver nanoparticles using *Allium cepa* extract has a sharp peak at around 413 nm. In another study, SPR peaks at around 422 and 447 nm were reported for the plant-mediated silver oxide NPs using *Vitex negundo* extract (Zargar et al., 2011).

X-ray diffraction (XRD) is a widely used technique for characterizing and identifying metallic nanoparticles' phase, crystalline nature, and crystalline structure).

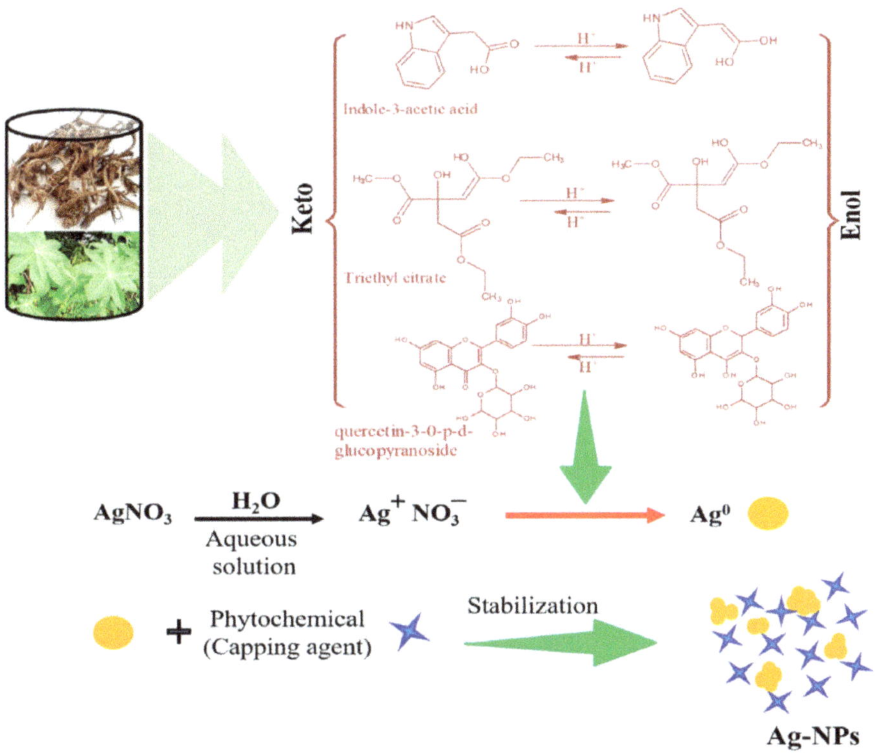

Fig. 2 Probable mechanism for plant-mediated synthesis of silver nanoparticles. Adopted from Gul et al. (2021)

In X-ray diffraction, X-rays enter the powdered sample, resulting in a diffraction pattern according to Bragg's equation n $\lambda = 2dSin\theta$. This diffraction pattern is compared with the standard data to obtain the crystal structures. In XRD pattern consists of diffraction peaks as a function of the diffraction angle 2 θ. The size of metallic nanoparticles is determined by the Debye–Scherrer equation:

$$D = K\lambda / \beta \cos\theta$$

where D is the average crystallite size, λ is the X-ray wavelength, β is the full width at half maxima (FWHM), K is Scherrer's constant, and θ is Bragg's diffraction angle (Fig. 3).

Scanning electron microscopy (SEM) is a type of electron microscope that is used to produce the image of a sample and thus gives the morphology of metallic nanoparticles (Arjunan et al., 2012). In SEM, electron beams are generated from a filament (tungsten) and then accelerated toward a condenser lens by anode, where these are converged and then projected onto the specimen by the objective lens. These beams scan the surface sample in a raster pattern. The electron beam interacts with

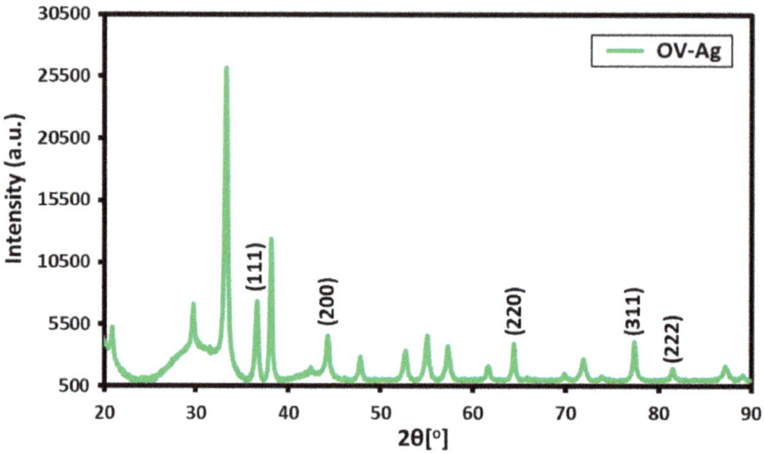

Fig. 3: XRD spectra of silver NP using *Origanum vulgare* leaf extract. Adopted from Shaik et al. (2018)

the atom in the sample, producing various signals and giving information about the surface morphology and composition of samples (Pradeep, 2007).

Energy-dispersive X-ray (EDX) analysis is a chemical microanalysis technique used to characterize the elemental composition of nanoparticles. SEM coupled with energy-dispersive X-ray spectroscopy (SEM–EDX) is widely used for surface analysis and elemental composition. EDX spectra measure the relative abundance of emitted X-rays from samples against their energy (Fig. 4). EDX spectrum shows the peaks at around 0.5 keV for oxygen and 3 keV for the elemental silver (Alaraidh et al., 2014).

Transmission electron microscopy (TEM) is a valuable and prominent microscopy approach that involves passing an electron beam through a specimen to create an image. Transmission electron microscopes can image at a substantially higher resolution than other microscopes due to the smaller de Broglie wavelength of electrons (Fig. 5). TEM can reveal the finest details of internal structure. High-resolution TEM can be used to analyze the structure and constitutional analysis of Ag and silver oxide nanoparticles (Ankamwar et al., 2005).

Fourier-transform infrared (FTIR) spectroscopy is used to identify the phytochemicals responsible for the capping, reducing, and effective stabilization of metallic nanoparticles. It is a technique that produces an IR spectrum of absorption or emission of a solid, gas, and liquid over a wide spectral range of 4000–400 cm^{-1} (Fig. 6).

For instance, silver nanoparticles were synthesized from the leaf extract of *Moringa oleifera,* and their antimicrobial and catalytic properties were evaluated (Sathyavathi et al., 2011). The absorption peak of silver nanoparticles at λ_{max} = 460 nm was recorded by UV–visible spectrophotometer, XRD and FRSEM results confirmed the average size of NPs was less than 40 nm. FTIR study showed several stretching peaks at 3308, 1629, 1069, and 1029 cm^{-1} indicating the presence of

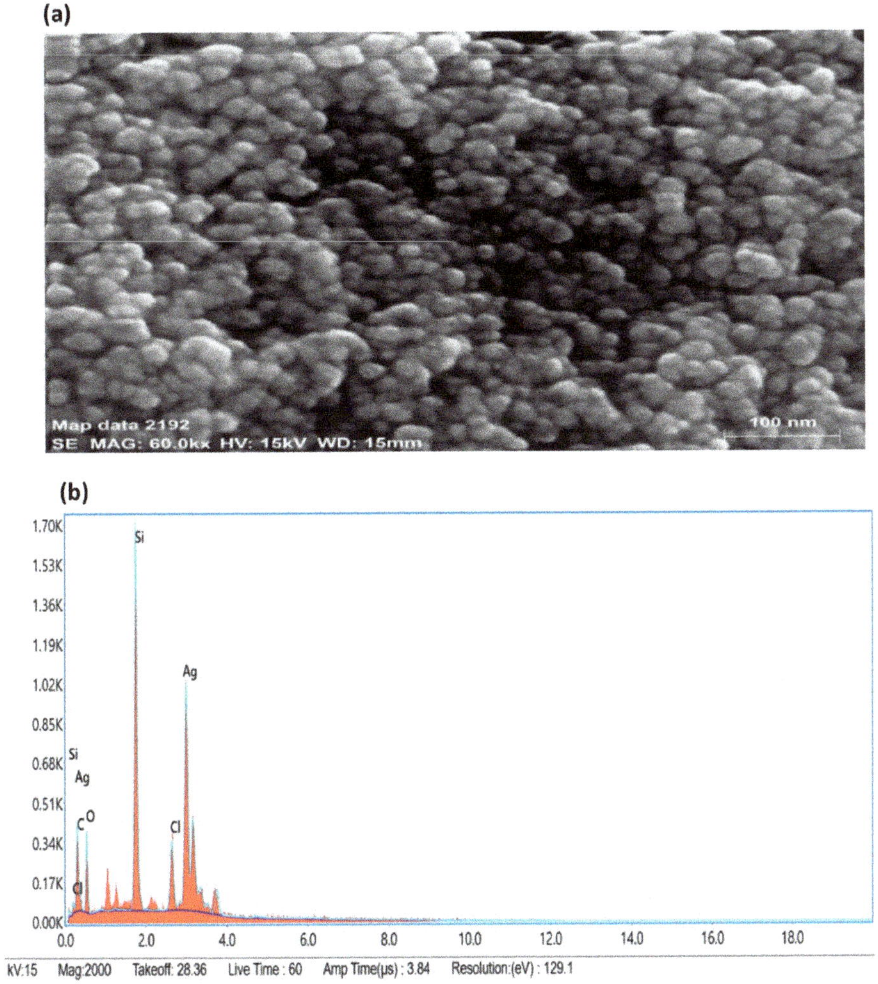

Fig. 4 **a** SEM image of Ag nanoparticles **b** EDX image of Ag nanoparticles using *Trigonella foenum-graceum* leaf extract. Adopted from Moond et al. (2023)

several functional groups acting as capping and stabilizing agents. In another study, Roy et al. (2017) reported the synthesis of silver NPs from the leaf extract of *Azadirachta indica* and the screening of these NPs for antioxidant and antibacterial activities. SPR peaks of these nanoparticles at $\lambda_{max} = 430$ nm were recorded by a spectrophotometer, particle size analyzer (PSA) showed the average size range of nanoparticles was 21.07 nm, and FTIR study confirmed the presence of phenolic compounds. In another report, spherical silver NPs were synthesized from *Abutilon indicum* leaf extract, and their antioxidant, antibacterial, and anticancer activity was investigated (Mata et al., 2015). TEM imaging showed the formation of spherical NPs with an average size of 70–75 nm, and XRD results revealed wurtzite hexagonal

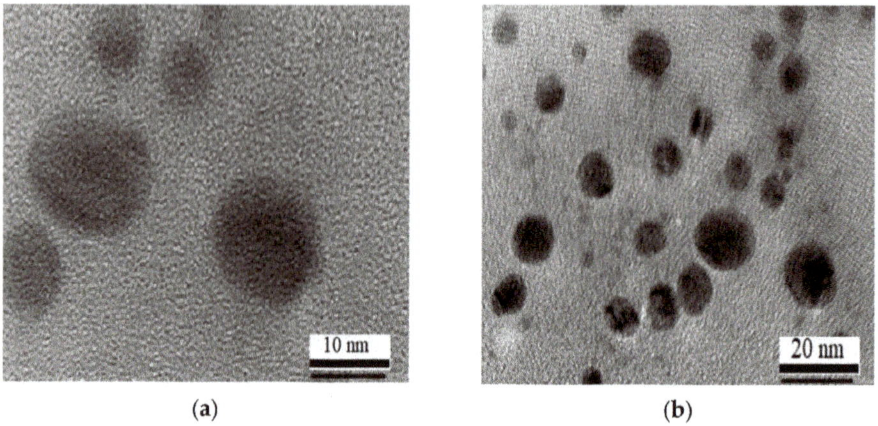

Fig. 5 TEM images of Ag nanoparticles using *Trigonella foenum-graceum* leaf extract. Adopted from Moond et al. (2023)

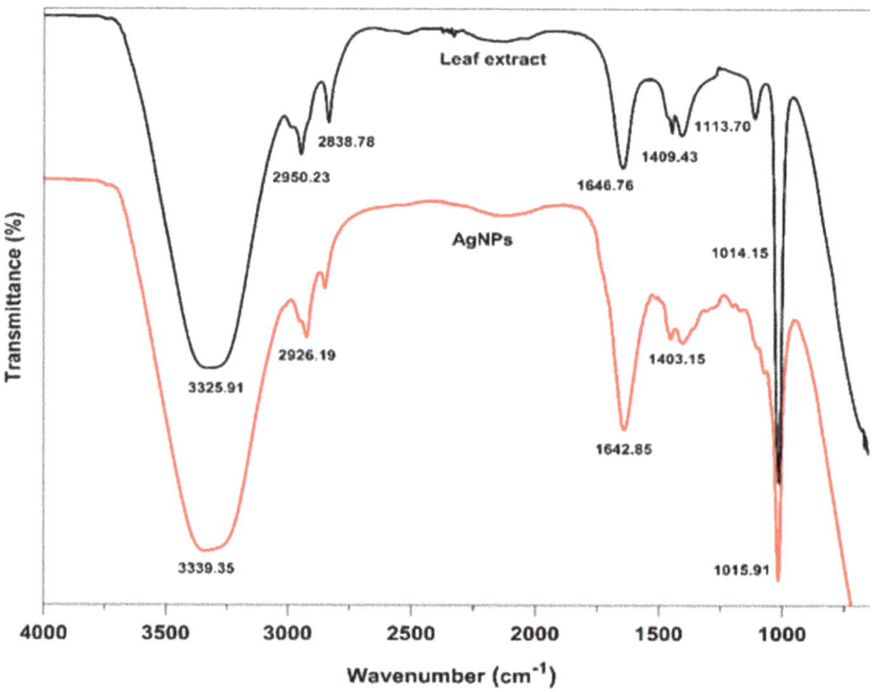

Fig. 6 FTIR spectrum of Ag nanoparticles using *Trigonella foenum-graceum* leaf extract. Adopted from Moond et al. (2023)

silver oxide NPs. FTIR spectrum showed a peak at 3291 cm^{-1} due to the presence of hydroxyl functional groups, while other peaks at 2930, 1663, 1530, 1239, and 459 cm^{-1} confirmed the presence of C-H, amide I and II groups, C-O bond, and metal–oxygen groups.

In order to create Ag_2O–NPs, Rashmi et al. (2020) combined separately milled plant powders of *Centella Asiatica* and *Tridax procumbens* using $AgNO_3$. Gotu kola, or *Centella Asiatica*, is part of the Apiaceae family. It is important in medicine and cosmetics because of the pentacyclic triterpenoids, which are asiaticoside, brahmoside, Asiatic acid, and Brahmic acid that it is said to possess. These substances can stabilize the resultant nanoparticles as well as decrease $AgNO_3$ to Ag_2O. The synthesized Ag_2O nanoparticles appear to have outstanding electrochemical capabilities, good photocatalytic activity for AO-8 under UV light, and antimicrobial properties. The generated nanoparticles, with a diameter of particles of 11–12 nm, are suggested by the results to have these qualities.

Using the hydroethanolic leaf extracts of the Ranunculaceae family's medicinal plant, *Helleborus odors*, Sutan et al. (2021) created Ag_2O–NPs in a bottom-up manner. Using a 1:1 mixture of crude hydroethanolic leaf extracts and $AgNO_3$, the Ag_2O–NPs were created. The reaction was allowed to run its course for 24 h in the dark at room temperature.

The gathered nanoparticles were identified as Cytogen toxic leaf extracts of papaya (*Carica papaya*) and fenugreek (*Trigonella foenum-graecum*). Separate mixtures of fenugreek leaf and papaya leaf extracts were combined with $AgNO_3$, which was then agitated at room temperature for 8 h to produce Ag_2O–NPs with a particle size range of 19.4–30.4 nm. These Ag_2O-NPs were then tested for hemolytic activity.

Shah et al. (2019) used *Paeonia emodi* leaf extracts combined with $AgNO_3$ while constantly stirring at 55 °C for 1 h 30 min. The resultant nanomaterials were investigated for their potential as photocatalytic and antibacterial agents. According to Suresh et al. (2018), the plant Pavetta indica Linn was employed to create Ag_2O–NPs. In order to create distorted cubic-shaped nanomaterials with a size less than 50.5 nm, leaf extracts were combined with an $AgNO_3$ solution and agitated continuously for 24 h at 100–120 °C. These particles were then examined for their ability to reduce inflammation and for their phytochemical content.

As a reaction mediator and capping agent, Muthukumar et al. (2022) used Amaranthus sp. leaf extract to create Ag_2O–NPs in the shape of husks. Aqueous leaf extract of Amaranthus sp. was gradually added to $AgNO_3$ solution with constant magnetic stirring at 35 °C for 15–20 min to produce Ag_2O–NPs. According to Muthukumar et al. (2022), Amaranthus sp. leaf extract was utilized as a capping agent and reaction mediator in forming Ag_2O nanoparticles in the shape of husks. Other reports of green synthesis of Ag and silver oxide NPs from the various parts of plant extract are illustrated in Table 1.

Table 1 Plant-mediated synthesis of silver and silver oxide nanoparticles of some important medicinal plants

Plant used	Characterization technique	Nanoparticle synthesized	Key references
Acalypha indica	XRD, EDS, SEM, HRTEM	Ag	Krishnaraj et al. (2010)
Aerva lanata	TEM, EDS, FTIR, XRD	Ag	Balashanmugam et al. (2014)
Allium cepa	UV–VIS, XRD, SEM,	Ag	Bouqellah et al. (2019)
Allium sativum	UV–VIS, TEM,	Ag	El-Refai et al. (2018)
Aloe vera	HRTEM	Ag	Vélez et al. (2018)
Alstonia scholaris	DLS, FTIR, XRD, UV–VIS,	Ag	Shetty et al. (2014)
Azadirachta indica	TEM, DLS, FTIR, UV–VIS	Ag	Ahmed et al. (2016)
Berberis vulgaris	XRD, UV–VIS, FTIR, DLS	Ag	Behravan et al. (2019)
Carica papaya	TEM, XRD, FTIR, SEM, UV–VIS	Ag	Balavijayalakshmi and Ramalakshmi (2017)
Cinnamon zeylanicum	UV–VIS, HRTEM, EDX	Ag	Anjum et al. (2019)
Clerodendrum serratum	UV–VIS, TEM,	Ag	Patil and Mengane (2016)
Datura stramonium	FTIR, EDAX, TEM, XRD, UV–VIS,	Ag	Gomathi et al. (2017)
Desmodium gageticum	FTIR, EDAX, TEM, UV–VIS, SEM	Ag	Thirunavoukkarasu et al. (2013)
Diospyros montana	SEM, XRD, EDAX, FTIR, TEM, DLS, UV–VIS	Ag	Bharathi et al. (2018)
Eugenia stipitata	TEM, XRD, SEM, UV–VIS	Ag	Kumar et al. (2016a)
Ficus carica	XRD, TEM, DLS, UV–VIS	Ag	Kumar et al. (2016b)
Justicia adhatoda	UV–VIS, TEM,	Ag	Bose and Chatterjee (2015)
Kalanchoe pinnata	TGA, FTIR, DSC, XRD, EDX, SEM, PSA, UV–VIS	Ag	Phatak and Hendre (2015)
Petroselinum crispum	TGA, FTIR, DSC, XRD, EDX, SEM, TGA, XPS, DLS, UV–VIS	Ag	Talabani et al. (2021)
Parthenium hysterophorus	FTIR, TGA, DSC, TEM	Ag	Ahsan et al. (2020)

(continued)

Table 1 (continued)

Plant used	Characterization technique	Nanoparticle synthesized	Key references
Mimosa pudica	TEM, XRD, EDX, SEM, UV–VIS, FTIR	Ag	Fatimah and Mutiara (2016)
Mulberry	TEM, FTIR, XRD, UV–VIS	Ag	Chunfa et al. (2018)
Musa paradisiaca	TEM, FTIR, XRD, UV–VIS	Ag	Anbazhagan et al. (2017)
Sambucus nigra	TEM, XRD, UV–VIS	Ag	Moldovan et al. (2016)
Sida cordifolia	UV–VIS, ZETA POTENTIAL, EDAX, SEM, FTIR, TEM	Ag	Srinithya et al. (2016)
Thevetia peruviana	EDX, TEM, SEM, FTIR, XRD, UV–VIS,	Ag	Oluwaniyi et al. (2016)
Tinospora cordifolia	EDX, XRD, SEM, FTIR, UV–VIS	Ag	Selvam et al. (2017)
Zingiber officinale	TEM, FTIR, XRD, UV–VIS	Ag	Yang et al. (2017)
Ziziphus oenoplia	EDAX, XRD, DLS, FTIR, TEM, UV–VIS	Ag	Soman and Ray (2016)
Panicum miliaceum	UV–VIS, FTIR, TEM, TGA, XPS, EDX,	Silver oxide	Velsankar et al. (2022)
Centella Asiatica	XRD, UV–VIS, FTIR, SEM, EDX,	Silver oxide	Rashmi et al. (2020)
Cleome gynandra	UV–VIS, ZETA POTENTIAL, SEM, FTIR, TEM, DLS,	Silver oxide	Mani et al. (2021)
Helleborus odorus	UV–VIS, XRD, TEM, EDX	Silver oxide	Şuţan et al. (2021)
Herniaria hirsuta	UV–VIS, FTIR, XRD, SEM,	Silver oxide	El-Ghmari et al. (2021)
Trigonella foenum-graecum	UV–VIS, FTIR, TEM, SAED	Silver oxide	Ashokraja et al. (2017)
Amaranthus sp.	UV–VIS, XRD, SEM, DLS	Silver oxide	Muthukumar et al. (2022)
Diospyros montana	UV–VIS, ZETA POTENTIAL, EDAX, SEM, FTIR, TEM,	Silver oxide	Sujatha et al. (2023)
Pavetta indica Linn	XRD, SEM, EDX	Silver oxide	Suresh et al. (2018)
Artocarpus heterophyllus	UV–VIS, XRD, FTIR, DLS, TEM	Silver oxide	Sharma and Srivastava (2020)
Lawsonia inermis	UV–VIS, EDAX, XRD, SEM, FTIR	Silver oxide	Fayyadh and Jaduaa Alzubaidy (2021)

(continued)

Table 1 (continued)

Plant used	Characterization technique	Nanoparticle synthesized	Key references
Daphne alpina	XRD, SEM, EDX, FTIR	Silver oxide	Haq et al. (2021)
Thunbergia mysorensis	UV–VIS, FTIR, XRD, SEM, TEM,	Silver oxide	Kokila et al. (2022)
Cyathea nilgirensis	FTIR, XRD, SEM, EDX	Silver oxide	Pradheesh et al. (2020)
Salix integra	UV–VIS, XRD, XPS, TEM, FTIR	Silver oxide	Patel and Joshi (2023)
Zephyranthes rosea	FTIR, TEM, SAED, EDX, XPS, XRD	Silver oxide	Maheshwaran et al. (2020)
Rhamnus virgata	FTIR, XRD, SEM, EDX, DLS, UV–VIS	Silver oxide	El-Ghmari et al. (2021)
Eupatorium odoratum	UV–VIS, XRD, SEM, TEM	Silver oxide	Elemike et al. (2017)

6 Challenges and Future Perspectives

The most frequently studied and utilized nanomaterials are undoubtedly silver nanoparticles. Green synthesis approaches have garnered a lot of interest recently among the various ways to produce silver nanoparticles since they are easy, effective, economical, and ecologically safe. In order to effectively use the biosynthesized silver nanoparticles for potential applications, it is necessary to obtain insight into the various inevitable challenges that must be addressed in the future. Metal and metal oxide nanoparticles have various properties that can be greatly influenced by their physical and chemical properties, especially their size, shape, and composition. This, in turn, can have a positive impact on their health benefits. Additionally, the metallic nanoparticle's typical size has a direct impact on both their catalytic and antibacterial properties (Chi et al., 2019; Ferdous & Nemmar, 2020; Hong et al., 2016; Yin et al., 2020). Although plant-mediated synthesis has received greater attention recently because it is a green strategy, it is still a novel concept that is in the early stages of research. In order to effectively use the plant-mediated silver and silver oxide nanoparticles for potential applications, it is necessary to obtain insight into the following inescapable issues that must be addressed in the future.

The shape and size of the silver nanoparticles produced by plants have varied greatly. To effectively manage the shape, size, and monodispersity of nanoparticles, it is necessary to experiment with and validate a variety of parameters, such as extract concentration, exposure time, temperature, and pH. Additionally, silver release is more likely to occur in small-sized spherical silver nanoparticles. Silver nanoparticles less than 10 nm are easily able to infiltrate bacterial cells, alter cell permeability, and cause cell damage. A significant issue in addition to this is the toxicity of metal nanoparticles. According to reports, the use of metal NPs in various industries has had extremely damaging impacts on both human health and the environment. For instance, prolonged exposure to silver can cause eye and skin discoloration and bluish-gray skin. Additionally, recent studies appear to support the idea that some bacteria are less susceptible to silver nanoparticles or are resistant to them. The mechanism behind the mechanism of silver nanoparticles is still poorly understood and now in a dormant state. In the future, attention should be paid to thorough investigations that identify the biochemistry of the bioactive molecules involved in capping and reducing during the creation of nanoparticles. In plant-mediated biosynthesis, different outcomes from the extract of the same plant species gathered from different parts of the world are expected due to differences in their makeup. This problem can be solved by correctly identifying the variety of phytochemicals present in plant extract and then researching how they affect capping and decreasing. Studies on the production of silver nanoparticles to date have solely used laboratory-scale experiments. For mass or industrial production to be possible, more optimization effort is required. The potential of biosynthesized silver nanoparticles for various applications has been shown in several research. Future research is necessary to assess the toxicity of biosynthesized silver nanoparticles, their biocompatibility (with cells and tissues), their safety for the environment (soil, water, and air), their accumulation in plants and animals, and their long-term effects on human and agricultural crop health. In comparison to conventional techniques, biogenic production of nanoparticles is more affordable. However, there is still room to cut costs, which are mostly caused by metal salts, microbial media, and downstream processing in biosynthesis. This can be resolved in the future by taking waste recycling into account as a different cost-saving measure.

7 Conclusion

Silver and silver oxide nanoparticles are the most desirable nanomaterials for various applications, including electrical, catalytic, antibacterial, and other biological activities. The literature for understanding the production and characterization of silver and silver oxide nanoparticles utilizing plant extracts is summarized in the current review. The use of plant extract in the synthesis of silver and silver oxide nanoparticles is advantageous due to its lower environmental impact and ability to create vast numbers of nanoparticles. Plant extracts may function in creating silver and silver oxide nanoparticles as both reducing and stabilizing agents. The use of plant extract in synthesizing silver oxide nanoparticles has advantages over other physical and chemical approaches since it is secure, environmentally benign, and easy to implement. UV–visible spectroscopy, DLS, TEM, SEM, XRD, EDX, and FTIR can successfully characterize these nanoparticles. Plants have a huge potential for producing silver and silver oxide nanoparticles with the desired shape and size that have a vast potential for use. In a nutshell, future studies must concentrate on understanding the intricate mechanism of metal and metal oxide nanoparticle synthesis and designing nanoparticles that are less hazardous, healthier, and have precisely controlled dimensions and form. Additionally, silver release is more likely to occur in small-sized spherical silver nanoparticles. Furthermore, the necessity to use plant-based metallic NPs more frequently in related disciplines must also be taken into consideration.

References

Abate, L., Megra, M. B., Nigussie, Y., Zebeaman, M., Tadesse, M. G., Bachheti, A., Bachheti, R. K., & Husen, A. (2023). Medicinally important seed extract and seed oil-mediated nanoparticles synthesis and their role in drug delivery and other applications. In R. K. Bachheti & A. Bachheti (Eds.), *Secondary metabolites from medicinal plants* (pp. 199–212). CRC Press.

Abbasi, B. A., Iqbal, J., Nasir, J. A., Zahra, S. A., Shahbaz, A., Uddin, S., Hameed, S., Gul, F., Kanwal, S., & Mahmood, T. (2020). Environmentally friendly green approach for the fabrication of silver oxide nanoparticles: Characterization and diverse biomedical applications. *Microscopy Research and Technique, 83*(11), 1308–1320.

Ahmed, S., Saifullah, A. M., Swami, B. L., & Ikram, S. (2016). Green synthesis of silver nanoparticles using *Azadirachta indica* aqueous leaf extract. *Journal of Radiation Research and Applied Sciences, 9*(1), 1–7.

Ahsan, A., Farooq, M. A., Ahsan Bajwa, A., & Parveen, A. (2020). Green synthesis of silver nanoparticles using Parthenium hysterophorus: Optimization, characterization and in vitro therapeutic evaluation. *Molecules, 25*(15), 3324.

Alaraidh, I. A., Ibrahim, M. M., & El-Gaaly, G. A. (2014). Evaluation of green synthesis of Ag nanoparticles using *Eruca sativa* and *Spinacia oleracea* leaf extracts and their antimicrobial activity. *Iranian Journal of Biotechnology, 12*(1), 50–55.

Alsaiari, N. S., Alzahrani, F. M., Amari, A., Osman, H., Harharah, H. N., Elboughdiri, N., & Tahoon, M. A. (2023). Plant and microbial approaches as green methods for the synthesis of nanomaterials: Synthesis, applications, and future perspectives. *Molecules, 28*(1), 1–38.

Anbazhagan, P., Murugan, K., Jaganathan, A., Sujitha, V., Samidoss, C. M., Jayashanthani, S., Amuthavalli, P., Higuchi, A., Kumar, S., Wei, H., Nicolett, M., Canale, A., & Benelli, G. (2017). Mosquitocidal, antimalarial and antidiabetic potential of *Musa paradisiaca*-synthesized silver nanoparticles: In vivo and in vitro approaches. *Journal of Cluster Science, 28*(1), 91–107.

Anjum, S., Jacob, G., & Gupta, B. (2019). Investigation of the herbal synthesis of silver nanoparticles using *Cinnamon zeylanicum* extract. *Emergent Materials, 2*(1), 113–122.

Ankamwar, B., Damle, C., Ahmad, A., & Sastry, M. (2005). Biosynthesis of gold and silver nanoparticles using *Emblica officinalis* fruit extract, their phase transfer, and transmetallation in an organic solution. *Journal of Nanoscience and Nanotechnology, 5*(10), 1665–1671.

Arjunan, N. K., Murugan, K., Rejeeth, C., Madhiyazhagan, P., & Barnard, D. R. (2012). Green synthesis of silver nanoparticles for the control of mosquito vectors of malaria, filariasis, and dengue. *Vector-Borne and Zoonotic Diseases, 12*(3), 262–268.

Arole, V. M., & Munde, S. V. (2014). Fabrication of nanomaterials by top-down and bottom-up approaches-an overview. *Journal of Materials Science, 1*, 89–93.

Ashokraja, C., Sakar, M., & Balakumar, S. (2017). A perspective on the hemolytic activity of chemical and green-synthesized silver and silver oxide nanoparticles. *Materials Research Express, 4*(10), 105406.

Bachheti, A., Bachheti, R. K., Abate, L., & Husen, A. (2022). Current status of Aloe-based nanoparticle fabrication, characterization and their application in some cutting-edge areas. *South African Journal of Botany, 147*, 1058–1069.

Bachheti, R. K., Fikadu, A., Bachheti, A., & Husen, A. (2020). Biogenic fabrication of nanomaterials from flower-based chemical compounds, characterization and their various applications: A review. *Saudi Journal of Biological Sciences, 27*(10), 2551–2562.

Bachheti, R. K., Sharma, A., Bachheti, A., Husen, A., Shanka, G. M., & Pandey, D. P. (2020b). Nanomaterials from various forest tree species and their biomedical applications. In A. Husen, M. Jawaid (Eds.), *Nanomaterials for agriculture and forestry applications* (pp. 81–106). Elsevier. https://doi.org/10.1016/B978-0-12-817852-2.00004-4

Bachheti, R. K., Abate, L., Bachheti, A., Madhusudhan, A., & Husen, A. (2021). Algae-, fungi-, and yeast-mediated biological synthesis of nanoparticles and their various biomedical applications. In *Handbook of greener synthesis of nanomaterials and compounds* (pp. 701–734). Elsevier. https://doi.org/10.1016/B978-0-12-821938-6.00022-0

Bachheti, R. K., Konwarh, R., Gupta, V., Husen, A., & Joshi, A. (2019). Green synthesis of iron oxide nanoparticles: Cutting edge technology and multifaceted applications. In *Nanomaterials and plant potential* (pp. 239–259), Springer. https://doi.org/10.1007/978-3-030-05569-1_9

Bachheti, R. K., Godebo, Y., Bachheti, A., Yassin, M. O., & Husen, A. (2020a). Root-based fabrication of metal/metal-oxide nanomaterials and their various applications. In A. Husen, M. Jawaid (Eds.), *Nanomaterials for agriculture and forestry applications* (pp.135–166). Elsevier. https://doi.org/10.1016/B978-0-12-817852-2.00006-8

Balashanmugam, P., Santhosh, S., Mukesh Kumar, D. J., Anbazhakan, S., & Kalaichelvan, P. T. (2014). Synthesis of plant-mediated silver nanoparticles using *Aerva lanata* leaf aqueous extract and evaluation of its antibacterial activities. *Indo-American Journal of Pharmaceutical Research, 4*, 475–480.

Balavijayalakshmi, J., & Ramalakshmi, V. (2017). Carica papaya peel mediated synthesis of silver nanoparticles and its antibacterial activity against human pathogens. *Journal of Applied Research and Technology, 15*(5), 413–422.

Behravan, M., Panahi, A. H., Naghizadeh, A., Ziaee, M., Mahdavi, R., & Mirzapour, A. (2019). Facile green synthesis of silver nanoparticles using *Berberis vulgaris* leaf and root aqueous extract and its antibacterial activity. *International Journal of Biological Macromolecules, 124*, 148–154.

Bharathi, D., Josebin, M. D., Vasantharaj, S., & Bhuvaneshwari, V. (2018). Biosynthesis of silver nanoparticles using stem bark extracts of *Diospyros montana* and their antioxidant and antibacterial activities. *Journal of Nanostructure in Chemistry, 8*(1), 83–92.

Bose, D., & Chatterjee, S. (2015). Antibacterial activity of green synthesized silver nanoparticles using Vasaka (*Justicia adhatoda* L.) leaf extract. *Indian Journal of Microbiology, 55*(2), 163–167.

Bouqellah, N. A., Mohamed, M. M., & Ibrahim, Y. (2019). Synthesis of eco-friendly silver nanoparticles using *Allium sp.* and their antimicrobial potential on selected vaginal bacteria. *Saudi Journal of Biological Sciences, 26*(7), 1789–1794.

Chawla, S., Singh, S., & Husen, A. (2023). *Smart nanomaterials targeting pathological hypoxia.* Springer Nature Singapore Pte Ltd. ISBN: 978–981–99–1717–4. https://doi.org/10.1007/978-981-99-1718-1

Chi, M., Qi, M., Wang, P., Weir, M. D., Melo, M. A., Sun, X., Dong, B., Li, C., Wu, J., Wang, L., & Xu, H. H. (2019). Novel bioactive and therapeutic dental polymeric materials to inhibit periodontal pathogens and biofilms. *International Journal of Molecular Sciences, 20*(2), 1–29.

Christian, P., Von der Kammer, F., Baalousha, M., & Hofmann, T. (2008). Nanoparticles: Structure, properties, preparation, and behavior in environmental media. *Ecotoxicology, 17*(5), 326–343.

Chunfa, D., Fei, C., Xianglin, Z., Xiangjie, W., Xiuzhi, Y., & Bin, Y. (2018). Rapid and green synthesis of monodisperse silver nanoparticles using mulberry leaf extract. *Rare Metal Materials and Engineering, 47*, 1089–1095.

Danish, M. S. S., Estrella, L. L., Alemaida, I. M. A., Lisin, A., Moiseev, N., Ahmadi, M., Nazari, M., Wali, M., Zaheb, H., & Senjyu, T. (2021). Photocatalytic applications of metal oxides for sustainable environmental remediation. *Metals, 11*(1), 1–25.

El-Ghmari, B., Farah, H., & Ech-Chahad, A. (2021). A new approach for the green biosynthesis of Silver Oxide nanoparticles Ag_2O, characterization and catalytic application. *Bulletin of Chemical Reaction Engineering and Catalysis, 16*(3), 651–660.

El-Refai, A. A., Ghoniem, G. A., El-Khateeb, A. Y., & Hassaan, M. M. (2018). Eco-friendly synthesis of metal nanoparticles using ginger and garlic extracts as biocompatible novel antioxidant and antimicrobial agents. *Journal of Nanostructure in Chemistry, 8*(1), 71–81.

Elemike, E. E., Onwudiwe, D. C., Ekennia, A. C., Sonde, C. U., & Ehiri, R. C. (2017). Green synthesis of Ag/Ag_2O nanoparticles using aqueous leaf extract of Eupatorium odoratum and its antimicrobial and mosquito larvicidal activities. *Molecules, 22*(5), 1–15.

Fatimah, I., & Mutiara, N. A. L. (2016). Biosynthesis of silver nanoparticles using Putri Malu (Mimosa pudica) leaves extract and microwave irradiation method. *Molekul, 11*(2), 288–298.

Fayyadh, A. A., & Jaduaa Alzubaidy, M. H. (2021). Green synthesis of Ag_2O nanoparticles for antimicrobial assays. *Journal of the Mechanical Behavior of Materials, 30*(1), 228–236.

Ferdous, Z., & Nemmar, A. (2020). Health impact of silver nanoparticles: A review of the biodistribution and toxicity following various routes of exposure. *International Journal of Molecular Sciences, 21*(7), 1–31.

Fu, L., & Fu, Z. (2015). *Plectranthus amboinicus* leaf extract–assisted biosynthesis of ZnO nanoparticles and their photocatalytic activity. *Ceramics International, 41*(2), 2492–2496.

Ghotekar, S., Dabhane, H., Pansambal, S., Oza, R., Tambade, P., & Medhane, V. (2020). A review of biomimetic synthesis of Ag_2O nanoparticles using plant extract, characterization, and its recent applications. *Advanced Journal of Chemistry-Section B, 2*(3), 102–111.

Godeto, Y. G., Ayele, A., Ahmed, I. N., Husen, A., & Bachheti, R. K. (2023). Medicinal plant-based metabolites in nanoparticles synthesis and their cutting-edge applications: An overview. In: R. K. Bachheti, A. Bachheti (Eds.), *Secondary metabolites from medicinal plants* (pp. 1–34). CRC Press.

Gomathi, M., Rajkumar, P. V., Prakasam, A., & Ravichandran, K. (2017). Green synthesis of silver nanoparticles using *Datura stramonium* leaf extract and assessment of their antibacterial activity. *Resource-Efficient Technologies, 3*(3), 280–284.

Gonfa, Y. H., Tessema, F. B., Tadesse, M. G., Bachheti, A., & Bachheti, R. K. (2023). Medicinally important plant roots and their role in nanoparticles synthesis and applications. In R. K. Bachheti & A. Bachheti (Eds.), *Secondary metabolites from medicinal plants* (pp. 227–242). CRC Press.

Gonfa, Y. H., Tessema, F. B., Bachheti, A., Tadesse, M. G., Eid, E. M., Abou Fayssal, S., Adelodun, B., Choi, K. S., Širić, I., Kumar, P., & Bachheti, R. K. (2022). Essential oil composition of aerial part of *Pluchea ovalis* (Pers.) DC., Silver nanoparticles synthesis, and larvicidal activities against Fall Armyworm. *Sustainability, 14*(23),15785.

Gul, A., Fozia, Shaheen, A., Ahmad, I., Khattak, B., Ahmad, M., Ullah, R., Bari, A., Ali, S.S., Alobaid, A., Asmari, M.M. & Mahmood, H. M. (2021). Green synthesis, characterization, enzyme inhibition, antimicrobial potential, and cytotoxic activity of plant mediated silver nanoparticle using Ricinus communis leaf and root extracts. *Biomolecules, 11*(2), 1–15.

Haq, S., Yasin, K. A., Rehman, W., Waseem, M., Ahmed, M. N., Shahzad, M. I., Shahzad, N., Shah, A., Rehman, M. U., & Khan, B. (2021). Green synthesis of silver oxide nanostructures and investigation of their synergistic effect with moxifloxacin against selected microorganisms. *Journal of Inorganic and Organometallic Polymers and Materials, 31*, 1134–1142.

Heera, P., & Shanmugam, S. (2015). Nanoparticle characterization and application: An overview. *International Journal of Current Microbiology & Applied Science, 4*(8), 379–386.

Hong, X., Wen, J., Xiong, X., & Hu, Y. (2016). Shape effect on the antibacterial activity of silver nanoparticles synthesized via a microwave-assisted method. *Environmental Science and Pollution Research, 23*, 4489–4497.

Huang, X., Wu, H., Liao, X., & Shi, B. (2010). One-step, size-controlled synthesis of gold nanoparticles at room temperature using plant tannin. *Green Chemistry, 12*(3), 395–399.

Husen, A. (2022). *Engineered nanomaterials for sustainable agricultural production, soil improvement and stress management*. Elsevier Inc https://doi.org/10.1016/C2021-0-00054-7

Husen, A. (2023a). *Secondary metabolites based green synthesis of nanomaterials and their applications*. Springer Nature Singapore Pte Ltd. https://doi.org/10.1007/978-981-99-0927-8

Husen, A. (2023b). *Nanomaterials from agricultural and horticultural products*. Springer Nature. https://link.springer.com/book/9789819934348

Husen, A. (2023c). *Nanomaterials and nanocomposites exposures to plants (Response, Interaction, Phytotoxicity and Defense Mechanisms)*. Springer Nature. https://doi.org/10.1007/978-981-99-2419-6

Husen, A., Bachheti, R. K., & Bachheti, A. (2023d). *Current trends in green nano-emulsions (Food, Agriculture and Biomedical Sectors)*. Springer Nature Singapore Pte Ltd.

Husen, A., & Iqbal, M. (2019). *Nanomaterials and plant potential*. Springer International Publishing AG. https://doi.org/10.1007/978-3-030-05569-1

Husen, A., & Jawaid, M. (2020). *Nanomaterials for agriculture and forestry applications*. Elsevier Inc. https://doi.org/10.1016/C2018-0-02349-X

Husen, A., & Siddiqi, K. S. (2023). *Advances in smart nanomaterials and their applications*. Elsevier Inc. https://doi.org/10.1016/C2021-0-02202-1

Husen, A., Rahman, Q. I., Iqbal, M., Yassin, M. O., & Bachheti, R. K. (2019). Plant-mediated fabrication of gold nanoparticles and their applications. In Nanomaterials and plant potential (pp. 71–110). Springer. https://doi.org/10.1007/978-3-030-05569-1_3

Iravani, S., Korbekandi, H., Mirmohammadi, S. V., & Zolfaghari, B. (2014). Synthesis of silver nanoparticles: Chemical, physical and biological methods. *Research in Pharmaceutical Sciences, 9*(6), 385–406.

Jana, N. R., Wang, Z. L., Sau, T. K., & Pal, T. (2000). Seed-mediated growth method to prepare cubic copper nanoparticles. *Current Science-Bangalore, 79*(9), 1367–1369.

Jeevanandam, J., Barhoum, A., Chan, Y. S., Dufresne, A., & Danquah, M. K. (2018). Review on nanoparticles and nanostructured materials: History, sources, toxicity, and regulations. *Beilstein Journal of Nanotechnology, 9*(1), 1050–1074.

Joshi, A., Sharma, A., Bachheti, R. K., Husen, A., & Mishra, V. K. (2019). Plant-mediated synthesis of copper oxide nanoparticles and their biological applications. In *Nanomaterials and plant potential* (pp. 221–237). Springer. https://doi.org/10.1007/978-3-030-05569-1_8

Khan, I., Saeed, K., & Khan, I. (2019). Nanoparticles: Properties, applications, and toxicities. *Arabian Journal of Chemistry, 12*(7), 908–931.

Khandel, P., & Shahi, S. K. (2018). Mycogenic nanoparticles and their bio-prospective applications: Current status and future challenges. *Journal of Nanostructure in Chemistry, 8*(4), 369–391.

Kokila, N. R., Mahesh, B., Roopa, K. P., Prasad, B. D., Raj, K., Manjula, S. N., Mruthunjaya, K., & Ramu, R. (2022). Thunbergia mysorensis mediated nano silver oxide for enhanced antibacterial, antioxidant, anticancer potential and in vitro hemolysis evaluation. *Journal of Molecular Structure, 1255*, 132455.

Kouvaris, P., Delimitis, A., Zaspalis, V., Papadopoulos, D., Tsipas, S. A., & Michailidis, N. (2012). Green synthesis and characterization of silver nanoparticles produced using *Arbutus unedo* leaf extract. *Materials Letters, 76*, 18–20.

Kredy, H. M. (2018). The effect of pH, Temperature on the green synthesis and biochemical activities of silver nanoparticles from *Lawsonia inermis* extract. *Journal of Pharmaceutical Sciences and Research, 10*(8), 2022–2026.

Krishnaraj, C., Jagan, E. G., Rajasekar, S., Selvakumar, P., Kalaichelvan, P. T., & Mohan, N. J. C. S. B. B. (2010). Synthesis of silver nanoparticles using *Acalypha indica* leaf extracts and its antibacterial activity against water-borne pathogens. *Colloids and Surfaces b: Biointerfaces, 76*(1), 50–56.

Kumar, B., Smita, K., Cumbal, L., & Debut, A. (2016b). *Ficus carica* (Fig) fruit mediated green synthesis of silver nanoparticles and its antioxidant activity: a comparison of thermal and ultrasonication approach. *BioNanoScience, 6*(1), 15–21.

Kumar, B., Smita, K., Debut, A., & Cumbal, L. (2016a). Extracellular green synthesis of silver nanoparticles using Amazonian fruit Araza (*Eugenia stipitata* McVaugh). *Transactions of Nonferrous Metals Society of China, 26*(9), 2363–2371.

Leyu, A. M., Debebe, S. E., Bachheti, A., Rawat, Y. S., & Bachheti, R. K. (2023). Green synthesis of gold and silver nanoparticles using invasive alien plant *Parthenium hysterophorus* and their antimicrobial and antioxidant activities. *Sustainability, 15*(12), 9456.

Liu, H., Zhang, H., Wang, J., & Wei, J. (2017). Effect of temperature on the size of biosynthesized silver nanoparticle: Deep insight into microscopic kinetics analysis. *Arabian Journal of Chemistry, 13*, 1011–1019.

Maheshwaran, G., Bharathi, A. N., Selvi, M. M., Kumar, M. K., Kumar, R. M., & Sudhahar, S. (2020). Green synthesis of Silver oxide nanoparticles using Zephyranthes Rosea flower extract and evaluation of biological activities. *Journal of Environmental Chemical Engineering, 8*(5), 104137.45.

Mandal, A. K., Ghorai, S., & Husen, A. (2023). *Functionalized smart nanomaterials for point-of-care testing*. Springer Nature Singapore Pte Ltd.

Mani, M., Harikrishnan, R., Purushothaman, P., Pavithra, S., Rajkumar, P., Kumaresan, S., Farraj, D. A. A., Elshikh, M. S., Balasubramanian, B., & Kaviyarasu, K. (2021). Systematic green synthesis of silver oxide nanoparticles for antimicrobial activity. *Environmental Research, 202*, 111627.

Mata, R., Nakkala, J. R., & Sadras, S. R. (2015). Biogenic silver nanoparticles from *Abutilon indicum*: Their antioxidant, antibacterial and cytotoxic effects in vitro. *Colloids and Surfaces b: Biointerfaces, 128*, 276–286.

Mittal, A. K., Chisti, Y., & Banerjee, U. C. (2013). Synthesis of metallic nanoparticles using plant extracts. *Biotechnology Advances, 31*(2), 346–356.

Moldovan, B., David, L., Achim, M., Clichici, S., & Filip, G. A. (2016). A green approach to phytomediated synthesis of silver nanoparticles using *Sambucus nigra* L. fruits extract and their antioxidant activity. *Journal of Molecular Liquids, 221*, 271–278.

Moond, M., Singh, S., Sangwan, S., Rani, S., Beniwal, A., Rani, J., Kumari, A., Rani, I., & Devi, P. (2023). Phytofabrication of silver nanoparticles using trigonella foenum-graceum L. Leaf and evaluation of its antimicrobial and antioxidant activities. *International Journal of Molecular Sciences, 24*(4), 3480.

Muthukumar, H., Palanirajan, S. K., Shanmugam, M. K., Arivalagan, P., & Gummadi, S. N. (2022). Photocatalytic degradation of caffeine and *E. coli* inactivation using silver oxide nanoparticles obtained by a facile green co-reduction method. *Clean Technologies and Environmental Policy, 24*(4), 1087–1098.

Oluwaniyi, O. O., Adegoke, H. I., Adesuji, E. T., Alabi, A. B., Bodede, S. O., Labulo, A. H., & Oseghale, C. O. (2016). Biosynthesis of silver nanoparticles using aqueous leaf extract of *Thevetia peruviana* Juss and its antimicrobial activities. *Applied Nanoscience, 6*(6), 903–912.

Painuli, S., Semwal, P., Bacheti, A., Bachheti, R. K., & Husen, A. (2020). Nanomaterials from nonwood forest products and their applications. In: A. Husen, M. Jawaid (Eds.), *Nanomaterials for agriculture and forestry applications* (pp. 15–40). Elsevier. https://doi.org/10.1016/B978-0-12-817852-2.00002-0

Pal, S. L., Jana, U., Manna, P. K., Mohanta, G. P., & Manavalan, R. (2011). Nanoparticle: An overview of preparation and characterization. *Journal of Applied Pharmaceutical Science, 1*(6), 228–234.

Patel, H., & Joshi, J. (2023). Green and chemical approach for synthesis of Ag_2O nanoparticles and their antimicrobial activity. *Journal of Sol-Gel Science and Technology, 105*, 814–826.

Patil, A. B., & Mengane, S. K. (2016). Green synthesis of silver nanoparticles from *Clerodendrum serratum* and antimicrobial activity against human pathogens. *Journal of Bionanoscience, 10*(6), 491–494.

Phatak, R. S., & Hendre, A. S. (2015). Sunlight-induced green synthesis of silver nanoparticles using sundried leaves extract of Kalanchoe pinnata and evaluation of its photocatalytic potential. *Der Pharmacia Lettre, 7*(5), 313–324.

Pradeep, T. (2007). *Nano: the essentials*. Tata McGraw-Hill Education.

Pradheesh, G., Suresh, S., Suresh, J., & Alexramani, V. (2020). Antimicrobial and anticancer activity studies on green synthesized silver oxide nanoparticles from the medicinal plant *Cyathea nilgiriensis* Holttum. *International Journal of Pharmaceutical Investigation, 10*(2), 146–150.
Rakhi, M., & Gopal, B. B. (2012). Terminalia arjuna bark extract mediated size controlled synthesis of polyshaped gold nanoparticles and its application in catalysis. *International Journal of Research in Chemistry and Environment, 2*(4), 338–344.
Rashmi, B. N., Harlapur, S. F., Avinash, B., Ravikumar, C. R., Nagaswarupa, H. P., Kumar, M. A., Gurushantha, K., & Santosh, M. S. (2020). Facile green synthesis of silver oxide nanoparticles and their electrochemical, photocatalytic, and biological studies. *Inorganic Chemistry Communications, 111*, 107580.
Ravichandran, S., Paluri, V., Kumar, G., Loganathan, K., & Kokati Venkata, B. R. (2016). A novel approach for the biosynthesis of silver oxide nanoparticles using aqueous leaf extract of Callistemon lanceolatus (Myrtaceae) and their therapeutic potential. *Journal of Experimental Nanoscience, 11*(6), 445–458.
Roy, P., Das, B., Mohanty, A., & Mohapatra, S. (2017). Green synthesis of silver nanoparticles using *Azadirachta indica* leaf extract and its antimicrobial study. *Applied Nanoscience, 7*(8), 843–850.
Sathyavathi, R., Krishna, M., & Rao, D. N. (2011). Biosynthesis of silver nanoparticles using *Moringa oleifera* leaf extract and its application to optical limiting. *Journal of Nanoscience and Nanotechnology, 11*(3), 2031–2035.
Selvam, K., Sudhakar, C., Govarthanan, M., Thiyagarajan, P., Sengottaiyan, A., Senthilkumar, B., & Selvankumar, T. (2017). Eco-friendly biosynthesis and characterization of silver nanoparticles using *Tinospora cordifolia* (Thunb.) Miers and evaluate its antibacterial, antioxidant potential. *Journal of Radiation Research and Applied Sciences, 10*(1), 6–12.
Shah, A., Haq, S., Rehman, W., Waseem, M., Shoukat, S., & Rehman, M. U. (2019). Photocatalytic and antibacterial activities of Paeonia emoji mediated silver oxide nanoparticles. *Materials Research Express, 6*(4), 045045.
Shaik, M. R., Khan, M., Kuniyil, M., Al-Warthan, A., Alkhathlan, H. Z., Siddiqui, M. R. H., Shaik J. P., & Adil, S. F. (2018). Plant-extract-assisted green synthesis of silver nanoparticles using Origanum vulgare L. extract and their microbicidal activities. *Sustainability, 10*(4), 913.
Sharma, S. N., & Srivastava, R. (2020). Silver oxide nanoparticles synthesized by green method from Artocarpus Hetrophyllus for antibacterial and antimicrobial applications. *Materials Today: Proceedings, 28*, 332–336.
Shekhawat, M. S., Ravindran, C. P., & Manokari, M. (2014). Biosynthesis of zinc oxide nanoparticles from *Passiflora foetida* L. extracts and their characterization. *International Journal of Green and Herbal Chemistry, 3*, 518–523.
Shetty, P., Supraja, N., Garud, M., & Prasad, T. N. V. K. V. (2014). Synthesis, characterization, and antimicrobial activity of *Alstonia scholaris* bark-extract-mediated silver nanoparticles. *Journal of Nanostructure in Chemistry, 4*(4), 161–170.
Shume, W. M., Murthy, H. A., & Zereffa, E. A. (2020). A review on synthesis and characterization of Ag_2O nanoparticles for photocatalytic applications. *Journal of Chemistry, 2020*, 1–15.
Singh, J., Dutta, T., Kim, K. H., Rawat, M., Samddar, P., & Kumar, P. (2018). 'Green'synthesis of metals and their oxide nanoparticles: Applications for environmental remediation. *Journal of Nanobiotechnology, 16*(1), 1–24.
Soman, S., & Ray, J. G. (2016). Silver nanoparticles synthesized using aqueous leaf extract of *Ziziphus oenoplia* (L.) Mill: Characterization and assessment of antibacterial activity. *Journal of Photochemistry and Photobiology b: Biology, 163*, 391–402.
Srinithya, B., Kumar, V. V., Vadivel, V., Pemaiah, B., Anthony, S. P., & Muthuraman, M. S. (2016). Synthesis of biofunctionalized AgNPs using medicinally important *Sida cordifolia* leaf extract for enhanced antioxidant and anticancer activities. *Materials Letters, 170*, 101–104.
Sujatha, V., Kaviyasri, G., Venkatesan, A., Thirunavukkarasu, C., Acharya, S., Dayel, S. B., Ghamdi, S.A., Abdelzaher, M.H., Shahid, M., & Ramesh, T. (2023). Biomimetic formation of silver oxide nanoparticles through Diospyros montana bark extract: Its application in dye degradation,

antibacterial and anticancer effect in human hepatocellular carcinoma cells. *Journal of King Saud University-Science*, 102563.
Suresh, S., Pradheesh, G., & Ramani, V. A. (2018). Biosynthesis and characterization of CuO, MgO, and Ag_2O nanoparticles, anti-inflammatory activity and phytochemical screening of the ethanolic extract of the medicinal plant Pavetta indica Linn. *Journal of Pharmacognosy and Phytochemistry, 7*(4), 1984–1990.
Şuţan, N. A., Fierăscu, I., Şuţan, C., Soare, L. C., Neblea, A. M., Somoghi, R., & Fierăscu, R. C. (2021). In vitro mitodepressive activity of phytofabricated silver oxide nanoparticles (Ag_2O-NPs) by leaves extract of Helleborus odorus Waldst. & Kit. ex Willd. *Materials Letters, 286*, 129194.
Taghiyari, H. R., Morrell, J. F., Husen, A. (2023). *Emerging nanomaterials* (*Opportunities and Challenges in Forestry Sectors*) (Vol. 11, pp. 6330). Springer Nature. https://doi.org/10.1007/978-3-031-17378-3
Talabani, R. F., Hamad, S. M., Barzinjy, A. A., & Demir, U. (2021). Biosynthesis of silver nanoparticles and their applications in harvesting sunlight for solar thermal generation. *Nanomaterials, 11*(9), 2421.
Thirunavoukkarasu, M., Balaji, U., Behera, S., Panda, P. K., & Mishra, B. K. (2013). Biosynthesis of silver nanoparticle from leaf extract of *Desmodium gangeticum* (L.) DC. and its biomedical potential. *Spectrochimica Acta Part a: Molecular and Biomolecular Spectroscopy, 116*, 424–427.
Tiwari, J. N., Tiwari, R. N., & Kim, K. S. (2012). Zero-dimensional, one-dimensional, two-dimensional, and three-dimensional nanostructured materials for advanced electrochemical energy devices. *Progress in Materials Science, 57*(4), 724–803.
Tran, Q. H., & Le, A. T. (2018). Corrigendum: Silver nanoparticles: Synthesis, properties, toxicology, applications and perspectives. *Advances in Natural Sciences: Nanoscience and Nanotechnology, 9*(4), 1–20.
Velsankar, K., Parvathy, G., Sankaranarayanan, K., Mohandoss, S., & Sudhahar, S. (2022). Green synthesis of silver oxide nanoparticles using Panicum miliaceum grains extract for biological applications. *Advanced Powder Technology, 33*(7), 103645.
Vélez, E., Campillo, G., Morales, G., Hincapié, C., Osorio, J., & Arnache, O. (2018). Silver nanoparticles obtained by aqueous or ethanolic aloe Vera extract an assessment of the antibacterial activity and mercury removal capability. *Journal of Nanomaterials, 7215210*, 1–7.
Yang, N., Li, F., Jian, T., Liu, C., Sun, H., Wang, L., & Xu, H. (2017). Biogenic synthesis of silver nanoparticles using ginger (*Zingiber officinale*) extract and their antibacterial properties against aquatic pathogens. *Acta Oceanologica Sinica, 36*(12), 95–100.
Yin, I. X., Zhang, J., Zhao, I. S., Mei, M. L., Li, Q., & Chu, C. H. (2020). The antibacterial mechanism of silver nanoparticles and its application in dentistry. *International Journal of Nanomedicine*, 2555–2562.
Zargar, M., Hamid, A. A., Bakar, F. A., Shamsudin, M. N., Shameli, K., Jahanshiri, F., & Farahani, F. (2011). Green synthesis and antibacterial effect of silver nanoparticles using *Vitex negundo* L. *Molecules, 16*(8), 6667–6676.
Zebeaman, M., Bachheti, R. K., Bachheti, A., Pandey, D. P., & Husen, A (2023). Green and cost-effective nanoparticles synthesis from some frequently used medicinal plants and their various applications. In *Secondary metabolites from medicinal plants* (pp. 287–304). CRC Press.

Plant-Mediated Synthesis and Characterization of Zinc Oxide Nanoparticles

Arun Kumar Khajuria, Rakesh Kumar Bachheti, Archana Bachheti, and Anuj Kandwal

Abstract Nanotechnology, one of the fastest growing branches of science, revolutionized traditional science in every aspect. The world is witnessing the swing of every aspect of science toward nanotechnology. This science provides more surface functionality in the given volume, but sometimes other characters are too improved compared to the original or bulky materials. Zinc and zinc-based products are in high demand due to their use in almost all fields that service humanity. Zinc at its nano-range proved more potent than the normal one, and zinc oxide nanoparticles are presently used in every industry (paints, rubber, cosmetics, food, bio-imaging, pharmacognosy, and pharmacology). Further, zinc oxide nanoparticles synthesized by the green route promote their use frequently due to their nontoxic, more stable and easier to synthesize, and excellent biocompatibility compared to the other methods. This chapter attempts to summarize the importance of zinc oxide nanoparticles, their synthesis, and the different tools used to characterize the synthesized nanoparticles.

Keywords Nanofabrication · Plant extract · Phytochemicals · Procedure · Mechanism

A. K. Khajuria
Department of Botany, Cluster University of Jammu, Jammu 180001, India

R. K. Bachheti
Department of Industrial Chemistry, Addis Ababa Science and Technology University, Addis Ababa, Ethiopia

A. Bachheti
Department of Environment Science, Graphic Era University, Dehradun 248002, Uttarakhand, India

A. Kandwal (✉)
Department of Chemistry, Harsh Vidya Mandir (PG) College, Raisi, Haridwar, Uttarakhand, India
e-mail: anujkandwal4@gmail.com

R. K. Bachheti et al. (eds.), *Metal and Metal-Oxide Based Nanomaterials*, Smart Nanomaterials Technology, https://doi.org/10.1007/978-981-99-7673-7_2

1 Introduction

Nanotechnology, the hot spot of material science, has revolutionized every aspect of today's science. The basic idea of nanoscience was spread by one of the great scientists, Richard Feynman, in 1959. Nanotechnology is basically based on manipulating the material or molecules to have new particles of the level of nanoscale, i.e., 1–100 nm, for functioning below the sub-microscopic level. Due to their nanosizes, the newly synthesized particles exhibit new and improved properties compared to their bulk form, i.e., changed morphology, size, distribution and this is further coupled with the enhanced or superior physiochemical properties such as their non-linear optical performance, catalytic activity, thermal conductivity, and superior biological properties (Bachheti et al., 2019; Husen et al., 2019; Painuli et al., 2020). Nanotechnology, one of the thrust areas of research, revolutionizes physical, chemical, biological, agricultural, environmental and engineering science, and other fields of science by synthesizing material at the nano-level, which behaves as a complete unit in terms of their characteristics. Further, with improved properties such as large surface area relative to their volume and improved quantum size effect, they can exhibit atom-like characteristics, and exhibited several applications (Bachheti et al., 2020a, 2020b, 2020c, 2021, 2022, 2023; Chawla et al., 2023; Husen & Iqbal, 2019a; Husen & Jawaid, 2020; Husen 2022, 2023a, 2023b, 2023c; Husen et al., 2023; Husen & Siddiqi, 2023; Mandal et al., 2023; Taghiyari et al., 2023; Worku et al., 2023). Nanoparticles are generally synthesized by two different approaches, i.e., first includes the synthesis via "bottom-up" method and second approach includes "top-down" method (Husen, 2019, 2020; Husen & Iqbal, 2019b; Husen & Siddiqi, 2014). In the first approach, the synthesis of nanoparticles was carried out by manipulating or reducing or size reduction of the starting material to its nano-size by applying physical or some chemical tools. While in the second method, the nanoparticles are synthesized by joining smaller molecules and particles, mostly using biological and chemical methods (Belay et al., 2023; Husen, 2019, 2020; Husen & Iqbal, 2019b; Husen & Siddiqi, 2014). The main advantage of the bottom-up method is that the synthesized nanoparticles are free of structural defects.

Nanoparticles are broadly divided into two parts, i.e., organic nanoparticles and inorganic nanoparticles and are traditionally synthesized by physical or chemical methods; in physical methods, a large area for setting up costly machines and equipment is required, which increases the cost of a new level and makes the whole process more complicated, while in chemical method use of toxic chemicals limits their application. Thus both methods are generally costly, and their use sometimes becomes hazardous for the person handling or involved in the synthesis and hazardous for the environment also. Thus need for an alternative method that is safe and cheaper than the physical and chemical methods arose. Biogenic synthesis (use of plants, microbes, and biomolecules) included living organisms and was thus used as an alternative approach for nanoparticle synthesis. In biogenic synthesis, plants always looked at for their medicinal purposes gained a special place in the synthesis. Green nanotechnology is a groundbreaking approach that offers an additional advantage of

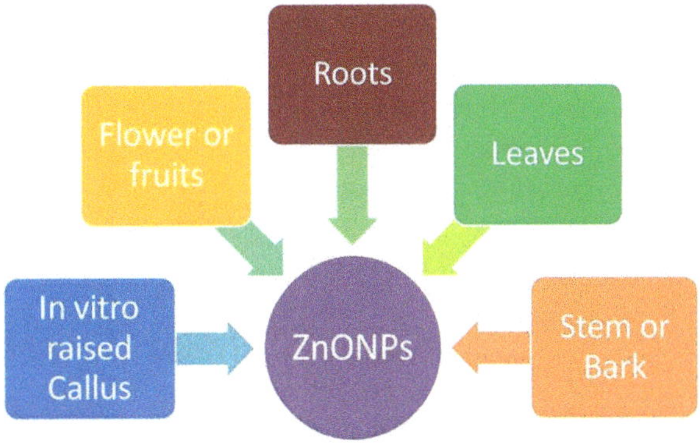

Fig. 1 Biosynthesis of ZnONPs using plant extract

the extended life span to the synthesized nanoparticles when compared to synthesize via other routes (chemical or physical) (Fig. 1).

This synthesized nanoscale material is called the "Wonder of Modern Medicine" (Abbasi et al., 2017). Owing to their use in the pharma sector as an antimicrobial agent, nanoscale material is also known as nano-antibiotics because of their antimicrobial activities (Agarwal et al., 2017). Metal-based synthesis of the nanoparticles is one of the most attractive and active branches of nanoparticles and among the various metal oxide nanoparticles, i.e., iron oxide (Fe_3O_4), copper oxide (CuO), silver oxide (Ag_2O), cerium oxide, zinc oxide, and titanium dioxide nanoparticles have gained wide attention due to their long shelf life, biocompatibility, and cost-effective nature.

Plant-based metal nanoparticles (MNPs) are becoming the core of research due to their high pharmacological potential and safe to use in pharmacognosy and the food industry. Table 1 shows the synthesis of ZnO from some important medicinal plants. Zinc oxide is of great industrial and economic value due to its broad range of properties that allow its use in many different areas, such as agriculture, pharmacology, food, paint, the rubber industry, and the biomedical field (Husen, 2019; Siddiqi et al., 2018). In nature, ZnO is present as zincite in the earth's crust, but for commercial use, ZnO is synthesized as a white powder that is non-water soluble.

Zinc oxide is also called as II-IV binary compound, because oxygen and zinc belong to the second and sixth groups of the periodic table. ZnO crystallizes in two forms, i.e., either cubic zinc blend structures or hexagonal wurtzite. The wurtzite structure is regarded as more stable in ambient situations. ZnO semiconductors are unique in that they are on the cusp of being ionic and covalent semiconductors, besides this, they have several unique properties, i.e., large electron mobility, wide bandgap (3.34–3.46 eV) with large exciton binding energy (55–65 meV), high catalytic energy, UV filtering characteristics, anti-inflammatory activity, good transparency, and strong luminescence at room temperature (Agarwal et al., 2017; Mirzaei & Darroudi, 2017; Vaseem et al., 2010).

Green synthesis of ZnO nanoparticles has attracted the interest of workers from the globe in past majorly in the last 2–3 years, owing to their facilitating properties some of which includes their nontoxic and eco-friendly nature and their biocompatible and

Table 1 Green synthesis of ZnO nanoparticles of some important medicinal plants

Plant	Family	Material	Size in nm	Key references
Aloe vera	Liliaceae	Leaves	22.18	Varghese and George (2015)
Azadirachta indica	Meliaceae	Leaves	18	Elumalia and Velmurugan (2015)
Brassica oleracea	Brassicaceae	Aerial part	14–17	Osuntokun et al. (2019)
Calotropis gigantea	Asclepiadaceae	Leaves	30-35 nm	Vidya et al. (2013)
Catharanthus roseus	Apocynaceae	Leaves	23–57	Bhumi and Savithramma (2014)
Coptidis rhizoma	Ranunculaceae	Rhizome	2.9–25.2	Nagajyothi et al. (2015)
Hibiscus sabdariffa	Malvaceae	Flower	8–30	Soto-Robles et al. (2019)
Moringa oleifera	Moringaceae	Leaves	24	Elumalai et al. (2015)
Ocimum basilicum	Lamiaceae	Leaves	50	Salam et al. (2014)
Ocimum tenuiflorum	Lamiaceae	Leaves	13.68	Raut et al. (2015)
Rosa canina	Rosaceae	Fruit	32–36	Jafariradet al. (2016)
Solanum nigrum	Solanaceae	Leaves	20–30	Ramesh et al. (2015)
Solanum rantonnetii	Solanaceae	Leaves	12	Salem and Awwad (2022)
Syzygium Cumini	Myrtaceae	Leaves	11.35	Sadiq et al. (2021)
Viola canescens	Violaceae	Leaves	26	Khajuria et al. (2017)
Vitex negundo	Vitaceae	Leaves	38.17	Ambika and Sundrarajan (2015)

biodegradable nature. Besides all these they have a wide range of applicability in different fields of the science, i.e., optic, medical, electronic, agriculture, and various material sciences.

Further, ZnO nanoparticles have a broad range of biomedical activity because many reports suggest that ZnO nanoparticles are essential trace elements in the human physiological system and are nontoxic to human cells. Further, the presence of the –OH group allows it to dissolve slowly in acidic and basic microenvironments. Thus, these nanoparticles are often used in drug delivery, anticancer, antimicrobial, antioxidant, and other pharmacological activities. Besides biomedical use, they are also employed for manufacturing rubber, paints, removing sulfur and arsenic from water, batteries, fire retardants, and dental application (Fig. 2).

The green route for the synthesis of nanoparticles reduced the work on physical and chemical methods. This method is gaining more attention than other biological systems because it provides a clean, cost-effective, safe, eco-friendly, convenient, and beneficial route to synthesizing MNPs of desired shape and size. Among the different nanomaterials, metal-based nanoparticle is the most dynamic area of research due to their unique physicochemical properties. Plant material as a reducing agent in nanoparticles has gained importance. Different parts of plants, such as leaves, peel,

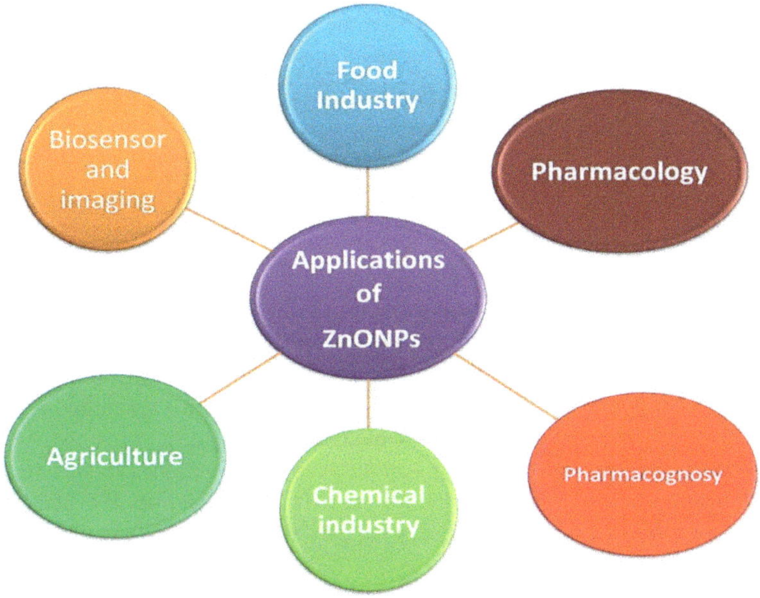

Fig. 2 Applications of ZnONPs in various sectors

roots, stem, seeds, fruits, and flowers, have been used to produce different nanoparticles (Ag, Au, silver oxide, TiO_2) of different shape and size. The major advantage of plant-mediated synthesis of MNPs over the chemical method is that in the former, there is no need for an additional protective agent (Kumar & Yadav, 2009; Lee et al., 2011; Park et al., 2011). Zinc oxide nanoparticles were synthesized using leaf extract of *Ocimum basilicum,* and synthesized nanoparticles were effective against both gram positive and negative bacteria (Abdelsattar et al., 2022). The antimicrobial activity of synthesized nanoparticles of palladium using grape polyphenols was reported by Amarnath et al. (2012). This chapter aims to provide in-depth information about the possible mechanism involved in the plant-mediated synthesis of zinc oxide nanoparticles and various techniques employed in the characterization of ZnO nanoparticles.

2 Plant-Mediated Synthesis of Zinc Oxide Nanoparticles

Plant-mediated synthesis of nanoparticles includes the use of plant extract as a reducing and capping agent in nanoparticle synthesis. Since no external stabilizers or capping agents are required in the chemical method, this method has gained popularity over the other methods. Synthesis of nanoparticles by this method has increased considerably in the last decade and has become the core of the research for nanomaterial formation. The popularity of this method is not only due to the absence of toxic chemicals (chemical preparations) but also the absence of applying a high amount of energy (physical methods) which increases the cost of synthesis of nanoparticles. Thus this technique is a more cost-effective, eco-friendly, and accessible approach for nanoparticle synthesis.

2.1 Preparation of Plant Extract

Each plant part may be used as an extract-preparing component (root, leaves, bark, flowers, and complete plant) because of their diverse range of phytochemicals. Plant extract can be prepared by collecting the desired plants from the wild or from some agri-farm or sometimes by another means. The collected plants were then kept under the tap water for at least 10 min to remove attached dust, insects, soil, and then finally washed with sterilized water. If found, any infected plant or plant part was discarded, and only healthy plant parts were used for the extract preparation. Washed plants or plant parts are then allowed to shade dry for 20 days or until the constant weight of dry components is achieved.

The dried plant material was then homogenized separately using mortar and pestle. A required amount of dry powder is weighed and used in preparing the plant extract. In a very basic protocol, the plant extract is generally made by dissolving the required amount of dry powder of plant extract into the required amount of distilled autoclaved water in a flask and boiling at 65 °C for 20 min, followed by allowing the extract cool and filtered using muslin cloth followed by Whatman filter paper.

2.2 Preparation of Zinc Nitrate Solution

Zinc nitrate hexahydrate, zinc acetate, or zinc chloride may serve as a precursor of zinc ions for the synthesis of ZnO nanoparticles. Using a magnetic stirrer, a required molar concentration of zinc precursor was made by dissolving zinc salt in 100 ml of distilled autoclaved water. After which complete dilution solution was filtered by using Whatman filter paper.

2.3 Preparation of ZnO Nanoparticles

The synthesis was performed by mixing an equal amount of zinc (precursor) nitrate solution and plant extract (1:1) in a 500 mL flask and warmed at 65 °C for 15 min under constant stirring using a magnetic stirrer. Mixing the zinc nitrate solution to plant extract in the flask is a very continuous and slow process. The prepared solution was then held at 60 °C with vigorous stirring for the required time, typically 3–4 h. The interaction between the zinc salt and the plant extract resulted in a color

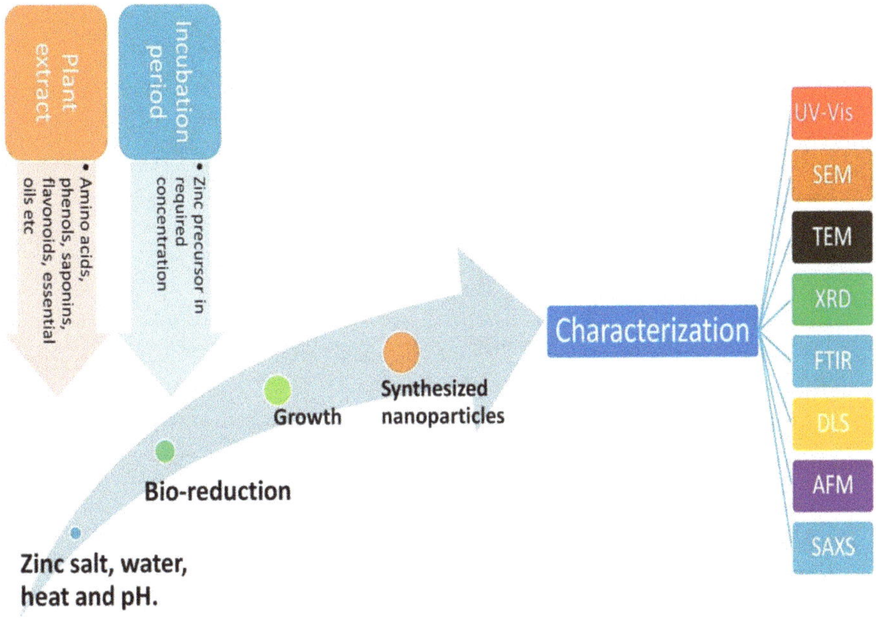

Fig. 3 Biosynthesis and characterization of plant-based ZnONPs

change of the solution. This color change is generally considered as a successful synthesis of nanoparticles and is regarded as the first preliminary observation of nanoparticles synthesis. After 3 h of vigorous stirring, precipitates began to form and these formed precipitates were allowed to stand at room temperature for 1–2 days. The supernatant of the solution was discarded and the precipitates were centrifuged at 5,000–10,000 rpm for 20 min. This process of centrifuging was repeated thrice and care is always taken to wash the collected pellets at least three times with double distilled water. In last, the formed pellets were dry for 10 h at 70 °C in an oven. These pellets were then ground with a mortar and pestle to obtain a fine powder and stored in an airtight bottle for further characterization (Fig. 3).

3 Characterization of Synthesized ZnO Nanoparticles

The synthesized nanoparticles are characterized by various analytical techniques, which include UV–Vis, SEM (scanning electron microscopy), TEM (transmission electron microscopy), XRD (X-ray diffraction), FTIR (Fourier transform infrared spectroscopy), atomic force microscopy (AFM), small-angle X-ray scattering (SAXS), and dynamic light scattering (DLS) (Fig. 4). Depending on the requirement of the exercise, each of these techniques has its peculiar advantage.

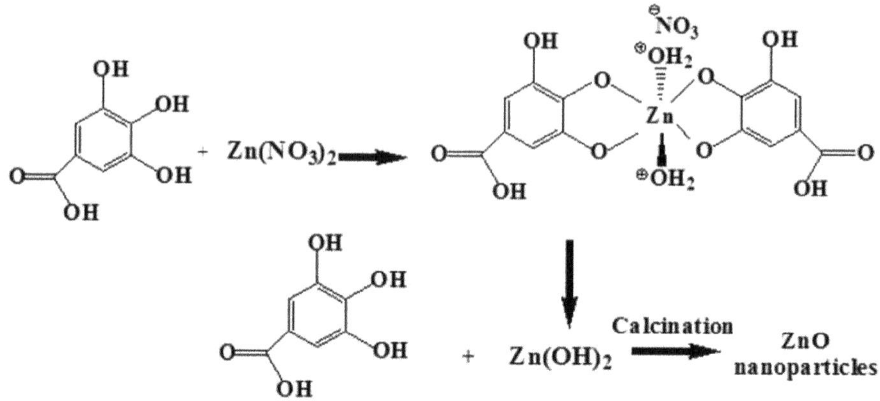

Fig. 4 Possible mechanism of zinc oxide nanoparticle synthesis by phytochemicals (phenols) from the plant extract (Safavinia et al., 2021)

3.1 UV–Visible Spectroscopy

One of the preliminary characterization techniques for the nanoparticle is based on exposing the sample to UV–Vis spectroscopy. Both (UV and Vis) lights are energetic that can excite electrons to higher energy levels. This technique measures the extinction of light passing through the sample (scatter + absorption). The sharp peaks in the 340–390 nm region, with their energy gap in the range of 3.10–3.72 eV, confirm the formation of ZnO nanoparticles.

3.2 EDX (Energy-Dispersive X-ray)

Energy-dispersive X-ray (EDX) analysis is a chemical microanalysis technique used to characterize the elemental composition of nanoparticles. SEM coupled with energy-dispersive X-ray spectroscopy (SEM–EDX) is widely used for surface analysis and elemental composition. This spectroscopy method is based on the idea that an electron beam passing through a sample causes X-rays to be produced. The characteristics and composition of the elements present in the sample determine how X-rays are produced, and the X-ray detector detects the X-rays emitted from the specimen. EDX spectra measure the relative abundance of emitted X-rays from samples against their energy peaks at around 1 keV are characteristics for zinc and 0.5 keV for oxygen.

3.3 TEM (Transmission Electron Microscopy)

Transmission electron microscopy, another one of the most important utilized approaches for the characterization of nanoparticles, was invented by E. Ruska and M. Knoll in 1933. This microscopy is utilized to analyze the surface structure (morphology), surface imperfection (defects), crystal structure, and size of the particles. This technique employs a focused electron beam on a prepared sample to record and produce the image of nanomaterials. Further, by spatially concentrating and restricting the beam, one can study the crystalline structure of the selected region in the sample and further facilitate the study of shape, size, and particle mean distribution size in a sample.

3.4 SEM (Scanning Electron Microscopy)

A scanning electron microscope is a type of microscope that uses a focused beam of high-energy electrons to create high-resolution images of a sample surface. In SEM, electron beams are generated from a filament (tungsten) and then accelerated toward a condenser lens by anode, where these are converged and then projected onto the specimen by the objective lens. These beams scan the surface sample in a raster pattern and a three-dimensional image is produced by combining the detected signals along with the beam position. During the study, SEM was used to observe the morphology of the synthesized ZnO nanoparticles.

3.5 FTIR (Fourier Transform Infrared Spectroscopy)

FTIR is the most important nondestructive technique for material analysis. It is primarily used to identify the bond type, structure, and functional group in compounds. Each molecule has specific frequencies that is directly associated with its rotational and vibrational motions. FTIR's primary objective is to identify the numerous functional groups that are present in the materials. This is due to the fact that different functional groups absorb infrared natural frequencies, which affects the rotational and vibrational states of molecular bonds. Thus, the spectrum with different peaks generated corresponds to the vibrational frequencies between the bonds of the atoms that make up the molecule, providing the intrinsic properties of the functional groups that make up the molecule and the overall composition of the molecule. This makes it possible to identify substances based on their infrared absorption spectra. Since no two compounds produce the same infrared spectrum (only optical isomers have that ability), the size of the peaks in the spectrum provides a unique fingerprint which helps in identifying the phytochemicals responsible for the capping, reducing, and effective stabilization of metallic nanoparticles.

3.6 XRD (X-Ray Diffraction)

X-ray diffraction is another important and versatile nondestructive analytical method for identifying and quantifying various crystal forms in solids, including their lattice parameter and geometry. Unknown compounds are identified by comparing the diffraction data with standard data from the International Center for Diffraction. Crystalline solids are composed of regularly arranged particles (atoms, ions, or molecules). The interaction of certain crystalline solids with X-rays helps study their structure.

X-ray diffraction (XRD) is a popular spectroscopic technique utilized for determining the phase, crystalline structure, and crystalline nature of metallic nanomaterials. A crystalline sample and monochromatic X-rays' constructive interference form the basis of most X-ray diffraction experiments. When the circumstances are in accordance with Bragg's law, the interaction of the incident rays with the sample results in constructive interference.

Here Bragg law is

$$2d\ \sin\theta = n\lambda$$

where Greek letter lambda (λ) was used as a symbol of the wavelength of incident X-ray radiations; θ = diffraction angle; n = order of diffraction; and d = interplanar distance.

The size of metallic nanoparticles is determined by the Debye–Scherrer equation

$$\mathrm{D} = \mathrm{K}\lambda / \beta \cos\theta$$

where D is the average size of crystal, β is the full width at half maxima (FWHM), K is Scherrer's constant, and θ is Bragg's diffraction angle.

3.7 Atomic Force Microscopy (AFM)

It was G. Binnig, in 1986, who first demonstrated the ideas of AFM. This technique uses probes and visualizes the nanomaterial objects. In general, this technique can be explained as scanning probe microscopy in which an image can be recorded by measuring the forces or interactions between the probe and the surface. The tip of the probe gently touches the sample surface. It records the topographical image with the help of a laser beam reflected off the cantilever tip onto the photodiode array. The probe, made up of silicon, does not move while the piezoelectric material moves the sample in the three dimensions.

Hook's law can be used to describe the forces

$$\mathrm{F} = -\mathrm{k} \cdot \mathrm{x}$$

where x = cantilever deflection and k = spring constant.

3.8 *Small-Angle X-Ray Scattering (SAXS)*

SAXS is one of the versatile nondestructive methods of nanoparticle characterization. In this technique, a detector positioned at modest angles, typically between 0.1 and 5°, can detect the dispersed X-ray radiation after the sample has been exposed to it. The technique can be performed on any samples such as solids, powders, colloidal suspensions, thin films, or dispersions of nanomaterials in a liquid phase. The intensity distribution of the scattered X-ray photons traveling through the sample can be used to gather data regarding agglomeration, molecular weight, crystallinity, particle size, morphology, and size distribution. This technique has several advantages over other electron microscopic methods because this technique provides a more statistically reliable estimation of nanoparticle sizes. After all, the particle size distribution obtained by this method is typically estimated over many nanoparticles. Compared to other microscopic methods, they are generally based on the measurements of up to few thousand nanoparticles.

4 Possible Mechanism of ZnO Nanoparticles

A lot of attention, to using green routes for the synthesis of metallic nanoparticles, has been worked by a number of researchers because of their inexpensive and environmentally safe nature (Agarwal et al., 2020; Bala et al., 2015; Umamaheswari et al., 2021). Among the different sources, i.e., bacteria, fungi, algae, and plants, plants are considered to be the first choice owing to their easy availability and easy handling (Gunalan et al., 2012; Khajuria et al., 2019; Ramesh et al., 2014; Sharmila et al., 2018). Further, plant phytochemicals, i.e., alkaloids, amino acids, flavonoids, polysaccharides, and proteins are used as stabilizing agents in the green route of nanoparticle synthesis (Murali et al., 2021; Sabir et al., 2014).

Zinc oxide nanoparticles are of high demand in approximately every sphere of science and can be generally synthesized by using any common and standard plant-mediated methods, this includes the reaction of fixed ratio of zinc precursor with the aqueous plant extract of the desired plants. But the number of studies suggested that plant type, source, plant species, and plant part affect the size of the synthesized nanoparticles because different plants have different phytochemical compositions (Naseer et al., 2020; Ramesh et al., 2014; Sabir et al., 2014). Then the question arises, how these nanoparticles are synthesized using plant extract and why they are of different size? The probable answer for the synthesis of zinc oxide nanoparticles by green route (plant-mediated) may be attributed to the presence of phytochemicals and the amount of the particular phytochemicals. These phytochemicals react with the zinc precursor to reduce them to their nano-size. Polyphenolics are the major compounds in plant extract and are the main constituents present in plants and it seems to be majorly responsible for the creation and stability of zinc oxide nanoparticles (Sharmila et al., 2018; Umamaheshwari et al., 2021), furthermore the polyphenolic plant is more the stable nanoparticles are supposed to synthesize. In a hypothetical mechanism, these phenols combine with zinc precursor (zinc nitrate or acetate) and form a Zn-OR complex via weak hydrogen bonds and attract Zn to its phenolic component and in the chemical reaction is converted into Zn $(OH)_2$. Finally, these formed zinc hydroxide complexes undergo calcination to obtain zinc oxide nanoparticles.

In order to synthesis ZnO nanoparticles, the flower extract of *Cassia auriculata* was mixed with 1 mM aqueous of zinc acetate by Ramesh et al. (2014). Similarly, Divya et al. (2013) use the leaf extract of *Hibiscus rosasinensi* and mix with zinc sulfate as zinc source for the synthesis of ZnO nanoparticles. Biogenic synthesis of ZnO nanomaterials using the leaf extract of *Plectranthus amboinicus* displayed good antimicrobial activity (Vijayakumar et al., 2015). In another research, Upadhyay et al. (2020) worked out a comparative study on synthesis of ZnO by conventional chemical method and green method using leaves of *Ocimum tenuiflorum* for extract preparation and reported that the ZnO nanoparticles synthesized via green route have better structural, morphological, and optical properties, when compared to the nanoparticles synthesized by chemical methods. Gunalan et al. (2012) synthesized ZnO nanoparticles using *Aloe vera* extract and reported that nanoparticles synthesized via this method had higher antibacterial activity when compared with the nanoparticles synthesized via the chemical route. These studies not only provide the evidential role of different phytochemicals in the synthesis and capping of the ZnO nanoparticles but also suggested that the green route or plant-mediated nanoparticles are cost-effective methods, easy, environmentally stable, and exhibit greater biological activities than the chemical route synthesized nanoparticles (Fig. 4).

5 Challenges and Future Perspectives

Zinc NPs have diverse uses and are widely studied in the agricultural, biomedical, pharmaceutical, food and herbal industry. The synthesis of the ZnO nanoparticles has some challenges. One of the common problems is the monodispersion of nanoparticles. Besides this, ZnONPs tend to agglomerate; sometimes, ZnO nanoparticles may change their size and surface chemistry and reduces biological activity with time due to change in nanoparticle chemistry. To overcome this problem, many workers use silica gel as an amorphous matrix in the green synthesis of ZnONPs. Besides these challenges, plant-mediated synthesized nanoparticles are considered a promising alternative to dealing with resistant pathogens or superbugs and solving the increasing problem of drug resistance in them. However, their toxicity evaluation, understanding of transport, distribution, and interactions within the biological systems are crucial and are the prime concerns ahead of their use as therapeutic agents. Future studies must concentrate on unraveling the intricate mechanisms underlying metal nanomaterial synthesis and designing nanoparticles (NPs) that are less hazardous, healthier, and have precisely controlled dimensions and form.

6 Conclusion

Synthesis of nanoparticles via green method has been the area of hot research for the last decade. A number of protocols were published using different biological materials to increase the in-depth knowledge of using the biological material as the fuel for the synthesis, stabilizing, and capping agent. Numerous reports by different workers on the ZnO nanoparticles have been on record. The green synthesized nanoparticles are considered a reliable source for use in the food and pharma sectors. Further, the citing literature in this chapter also suggests that the zinc oxide nanoparticles can be successfully synthesized using biological extracts (plant, fungi, bacteria, and algae) and showed enhanced properties than the bulky material. However, the change in the conditions (temperature, pH, source of zinc) and concentrations (extract from different plant parts) has several effects on nanoparticle stability and size. Hence, it indicates more scope is open for a better understanding of the interaction of plant extract and zinc source and the open field for the research. The current book chapter aims to discuss the synthesis and characterization of ZnO nanoparticles, and it's anticipated that the chapter will enhance the knowledge regarding green synthesis, characterization, and their correlations.

References

Abbasi, B. H., Anjum, S., & Hano, C. (2017). Differential effect of In vitro cultures of *Linum usitatissimum* L. on biosynthesis, stability, antibacterial and antileishmanial activites of Zinc oxide nanoparticles: A mechanistic approach. *Royal Society of Chemistry, 7*, 15931–15943.

Abdelsattar, A. S., Farouk, W. M., Gouda, S. M., Safwat, A., Hakim, T. A., & El-Shibiny, A. (2022). Utilization of *Ocimum basilicum* extracts for zinc oxide nanoparticles synthesis and their antibacterial activity after a novel combination with phage. *Materials Letters, 309*, 131344.

Agarwal, H., Menon, S., & Shanmugam, V. K. (2020). Functionalization of zinc oxide nanoparticles using Mucuna pruriens and its antibacterial activity. *Surfaces and Interfaces, 19*, 100521.

Aggarwal, H., Kumar, S. V., & Kumar, R. S. (2017). A review on green synthesis of Zinc oxide nanoparticles-an eco-friendly approach. *Resource-Efficient Technologies*. https://doi.org/10.1016/j.reffit.2017.03.002

Alhujaily, M., Albukhaty, S., Yusuf, M., Mohammed, M. K., Sulaiman, G. M., Al-Karagoly, H., & AlMalki, F. A. (2022). Recent advances in plant-mediated zinc oxide nanoparticles with their significant biomedical properties. *Bioengineering, 9*(10), 1–19.

Amarnath, K., Kumar, J., Reddy, T., Mahesh, V., Ayyappan, S. R., & Nellore, J. (2012). Synthesis and characterization of chitosan and grape polyphenols stabilized palladium nanoparticles and their antibacterial activity. *Colloid Surface, 92*, 254–261.

Ambika, S., & Sundrarajan, M. (2015). Antibacterial behavior of Vitex negundo extract assistedZnO nanoparticles against pathogenic bacteria. *Journal of Photochemistry and Photobiology b: Biology, 146*, 52–57.

Bachheti, R. K., Abate, L., Hunde, Y., Tadesse, M. G., Bachheti, A., Pandey, D. P., Sharma, A., Zebeaman, M., & Husen, A. (2023). Essential oils from medicinal plants and their role in nanoparticles synthesis, characterization, and applications. In R. K. Bachheti & A. Bachheti (Eds.), *Secondary metabolites from medicinal plants* (pp. 177–198). CRC Press.

Bachheti, A., Bachheti, R. K., Abate, L., & Husen, A. (2022). Current status of Aloe-based nanoparticle fabrication, characterization and their application in some cutting-edge areas. *South African Journal of Botany., 147*, 1058–1069.

Bachheti, R. K., Fikadu, A., Bachheti, A., & Husen, A. (2020a). Biogenic fabrication of nanomaterials from flower-based chemical compounds, characterization and their various applications: A review. *Saudi Journal of Biological Sciences., 27*(10), 2551–2562.

Bachheti, R. K., Sharma, A., Bachheti, A., Husen, A., Shanka, G. M., & Pandey, D. P. (2020b). Nanomaterials from various forest tree species and their biomedical applications. In A. Husen & M. Jawaid (Eds.), *Nanomaterials for agriculture and forestry applications* (pp. 81–106). Elsevier. https://doi.org/10.1016/B978-0-12-817852-2.00004-4

Bachheti, R. K., Abate, L., Bachheti, A., Madhusudhan, A., & Husen, A. (2021). Algae-, fungi-, and yeast-mediated biological synthesis of nanoparticles and their various biomedical applications. In *Handbook of greener synthesis of nanomaterials and compounds* (pp. 701–734). Elsevier. https://doi.org/10.1016/B978-0-12-821938-6.00022-0

Bachheti, R. K., Konwarh, R., Gupta, V., Husen, A., & Joshi, A. (2019). Green synthesis of iron oxide nanoparticles: Cutting edge technology and multifaceted applications. In *Nanomaterials and plant potential* (pp. 239–259). Springer. https://doi.org/10.1007/978-3-030-05569-1_9

Bachheti, R. K., Godebo, Y., Bachheti, A., Yassin, M. O., & Husen, A. (2020c). Root-based fabrication of metal/metal-oxide nanomaterials and their various applications. In A. Husen, & M. Jawaid (Eds.), *Nanomaterials for agriculture and forestry applications* (pp.135–166). Elsevier. https://doi.org/10.1016/B978-0-12-817852-2.00006-8

Bala, N., Saha, S., Chakraborty, M., Maiti, M., Das, S., Basu, R., & Nandy, P. (2015). Green synthesis of zinc oxide nanoparticles using Hibiscus subdariffa leaf extract: Effect of temperature on synthesis, anti-bacterial activity and anti-diabetic activity. *RSC Advances, 5*(7), 4993–5003.

Belay, T., Worku, L. A., Bachheti, R. K., Bachheti, A., & Husen, A. (2023). Nanomaterials: Introduction, synthesis, characterization, and applications. In A. Husen & K. S. Siddiqi (Eds.), *Advances in smart nanomaterials and their applications* (pp. 1–21). Elsevier.

Bhumi, G., & Savithramma, N. (2014). Biological synthesis of zinc oxide nanoparticles from Catharanthus roseus (L.) G. Don. Leaf extract and validation for antibacterial activity. *International Journal of Drug Development and Research, 6*(1), 208–214.

Chawla, S., Singh, S., & Husen, A. (2023). *Smart nanomaterials targeting pathological hypoxia.* Springer Nature Singapore Pte Ltd. ISBN: 978-981-99-1717-4. https://doi.org/10.1007/978-981-99-1718-1

Divya, M. J., Sowmia, C., Joona, K., & Dhanya, K. P. (2013). Synthesis of zinc oxide nanoparticle from hibiscus rosa-sinensis leaf extract and investigation of its antimicrobial activity. *Research Journal of Pharmaceutical, Biological and Chemical Sciences, 4*, 1137–1142.

Elumalai, K., & Velmurugan, S. (2015). Green synthesis, characterization and antimicrobial activities of zinc oxide nanoparticles from the leaf extract of Azadirachta indica (L.). *Applied Surface Science, 345*, 329–336.

Elumalai, K., Velmurugan, S., Ravi, S., Kathiravan, V., & Ashokkumar, S. (2015). Green synthesis of zinc oxide nanoparticles using Moringa oleifera leaf extract and evaluation of its antimicrobial activity. *Spectrochimica Acta Part a: Molecular and Biomolecular Spectroscopy, 143*, 158–164.

Gunalan, S., Sivaraj, R., & Rajendran, V. (2012). Green synthesized ZnO nanoparticles against bacterial and fungal pathogens. *Progress in Natural Science: Materials International, 22*(6), 693–700. https://doi.org/10.1016/j.matlet.2021.131344

Husen, A., & Siddiqi, K. S. (2014). Phytosynthesis of nanoparticles: Concept, controversy and application. *Nanoscale Research Letters, 9*(229), 1–24. https://doi.org/10.1186/1556-276X-9-229

Husen, A. (2019). Natural product-based fabrication of zinc oxide nanoparticles and their application. In A. Husen & M. Iqbal (Eds.), *Nanomaterials and plant potential* (pp. 193–291). Springer International Publishing AG. https://doi.org/10.1007/978-3-030-05569-1_7

Husen, A. (2020). Introduction and techniques in nanomaterials formulation. In A Husen & M. Jawaid (Eds.), *Nanomaterials for agriculture and forestry applications* (pp. 1–14). Elsevier Inc. https://doi.org/10.1016/B978-0-12-817852-2.00001-9

Husen, A. (2022). *Engineered nanomaterials for sustainable agricultural production, soil improvement and stress management*. Elsevier Inc. https://doi.org/10.1016/C2021-0-00054-7

Husen, A. (2023a). *Secondary metabolites based green synthesis of nanomaterials and their applications*. Springer Nature Singapore Pte Ltd. https://doi.org/10.1007/978-981-99-0927-8

Husen, A. (2023b). *Nanomaterials from agricultural and horticultural products*. Springer Nature Singapore Pte Ltd. https://link.springer.com/book/9789819934348

Husen, A. (2023c). *Nanomaterials and nanocomposites exposures to plants (Response, Interaction, Phytotoxicity and Defense Mechanisms)*. Springer Nature Singapore Pte Ltd. https://doi.org/10.1007/978-981-99-2419-6

Husen, A., Bachheti, R. K., & Bachheti, A. (2023). *Current trends in green nano-emulsions (Food, Agriculture and Biomedical Sectors)*. Springer Nature Singapore Pte Ltd.

Husen, A., Iqbal, M. (2019a). *Nanomaterials and plant potential*. Springer International Publishing AG. https://doi.org/10.1007/978-3-030-05569-1

Husen, A., Iqbal, M. (2019b). N*anomaterials* and plant potential: An overview. In A. Husen, & M. Iqbal (Eds.), *Nanomaterials and plant potential* (pp. 3–29). Springer International Publishing AG. https://doi.org/10.1007/978-3-030-05569-1_1

Husen, A., & Jawaid, M. (2020). *Nanomaterials for agriculture and forestry applications*. Elsevier Inc. https://doi.org/10.1016/C2018-0-02349-X

Husen, A., & Siddiqi, K. S. (2023). *Advances in smart nanomaterials and their applications*. Elsevier Inc. https://doi.org/10.1016/C2021-0-02202-1

Husen, A., Rahman, Q. I., Iqbal, M., Yassin, M. O., Bachheti, R. K. (2019). Plant-mediated fabrication of gold nanoparticles and their applications. In *Nanomaterials and plant potential* (pp. 71–110). Springer. https://doi.org/10.1007/978-3-030-05569-1_3

Jafarirad, S., Mehrabi, M., Divband, B., & Kosari-Nasab, M. (2016). Biofabrication of zinc oxide nanoparticles using fruit extract of Rosa canina and their toxic potential against bacteria: A mechanistic approach. *Materials Science and Engineering: C, 59*, 296–302.

Khajuria, A. K., Bisht, N. S., Manhas, R. K., & Kumar, G. (2019). Callus mediated biosynthesis and antibacterial activities of zinc oxide nanoparticles from Viola canescens: An important Himalayan medicinal herb. *SN Applied Sciences, 1*, 1–13.

Khajuria, A. K., Bisht, N. S., & Kumar, G. (2017). Synthesis of zinc oxide nanoparticles using leaf extract of Viola canescens Wall. ex, Roxb. and their antimicrobial activity. *Journal of Pharmacognosy and Phytochemistry, 6*(5), 1301–1304.

Kumar, V., & Yadav, S. K. (2009). Plant-mediated synthesis of silver and gold nanoparticles and their applications. *Journal of Chemical Technology and Biotechnology, 84*, 151–157.

Lee, H. J., Lee, G., Jang, N. R., Yun, J. H., Song, J. Y., & Kim, B. S. (2011). Biological synthesis of copper nanoparticles using plant extract. *Nanotechnology, 1*, 371–374.

Mandal, A. K., Ghorai, S., & Husen, A. (2023). *Functionalized smart nanomaterials for point-of-care testing*. Springer Nature

Mirzaei, H., & Darroudi, M. (2017). Zinc oxide nanoparticles: Biological synthesis and biomedical applications. *Ceramics International, 43*(1), 907–914.

Murali, M., Kalegowda, N., Gowtham, H. G., Ansari, M. A., Alomary, M. N., Alghamdi, S., & Amruthesh, K. N. (2021). Plant-mediated zinc oxide nanoparticles: Advances in the new millennium towards understanding their therapeutic role in biomedical applications. *Pharmaceutics, 13*(10), 1–44.

Nagajyothi, P. C., Cha, S. J., Yang, I. J., Sreekanth, T. V. M., Kim, K. J., & Shin, H. M. (2015). Antioxidant and anti-inflammatory activities of zinc oxide nanoparticles synthesized using Polygala tenuifolia root extract. *Journal of Photochemistry and Photobiology b: Biology, 146*, 10–17.

Naseer, M., Aslam, U., Khalid, B., & Chen, B. (2020). Green route to synthesize Zinc Oxide Nanoparticles using leaf extracts of Cassia fistula and Melia azadarach and their antibacterial potential. *Scientific Reports, 10*(1), 1–10.

Osuntokun, J., Onwudiwe, D. C., & Ebenso, E. E. (2019). Green synthesis of ZnO nanoparticles using aqueous Brassica oleracea L. var. italica and the photocatalytic activity. *Green Chemistry Letters and Reviews, 12*(4), 444–457.

Painuli, S., Semwal, P., Bacheti, A., Bachheti, R. K., Husen, A. (2020). Nanomaterials from nonwood forest products and their applications. In: A. Husen, M. Jawaid (Eds.), Nanomaterials for agriculture and forestry applications (pp. 15–40). Elsevier. https://doi.org/10.1016/B978-0-12-817852-2.00002-0

Park, Y., Hong, Y. N., Weyers, A., Kim, Y. S., & Linhardt, R. J. (2011). Polysaccharides and phytochemicals: A natural reservoir for the green synthesis of gold and silver nanoparticles. *IET Nanobiotechnology, 5*, 69–78.

Ramesh, M., Anbuvannan, M., & Viruthagiri, G. (2015). Green synthesis of ZnO nanoparticles using *Solanum nigrum* leaf extract and their antibacterial activity. *Spectrochimica Acta Part a: Molecular and Biomolecular Spectroscopy, 136*, 864–870.

Ramesh, P., Rajendran, A., & Meenakshisundaram, M. (2014). Green syntheis of zinc oxide nanoparticles using flower extract cassia auriculata. *Journal of NanoScience and NanoTechnology, 2*(1), 41–45.

Raut, S., Thorat, P. V., & Thakre, R. (2015). Green synthesis of Zinc oxide nanoparticles using Ocimumteniflorum leaves. *International Journal of Science and Research, 4*(5), 1225–1228.

Sabir, S., Arshad, M., & Chaudhari, S. K. (2014). Zinc oxide nanoparticles for revolutionizing agriculture: Synthesis and applications. *The Scientific World Journal, 2014*, 1–8.

Sadiq, H., Sher, F., Sehar, S., Lima, E. C., Zhang, S., Iqbal, H. M., et al. (2021). Green synthesis of ZnO nanoparticles from Syzygium Cumini leaves extract with robust photocatalysis applications. *Journal of Molecular Liquids, 335*, 116567.

Safavinia, L., Akhgar, M. R., Tahamipour, B., & Ahmadi, S. A. (2021). Green synthesis of highly dispersed zinc oxide nanoparticles supported on silica gel matrix by Daphne oleoides extract and their antibacterial activity. *Iranian Journal of Biotechnology, 19*(1), 86–95.

Salam, H. A., Sivaraj, R., & Venckatesh, R. (2014). Green synthesis and characterization of zinc oxide nanoparticles from Ocimumbasilicum L. var. purpurascens Benth.-Lamiaceae leaf extract. *Materials Letters, 131*, 16–18.

Salem, N. M., & Awwad, A. M. (2022). Green synthesis and characterization of ZnO nanoparticles using Solanum rantonnetii leaves aqueous extract and antifungal activity evaluation. *Chemistry International, 8*(1), 12–17.

Sharmila, G., Muthukumaran, C., Sandiya, K., Santhiya, S., Pradeep, R. S., Kumar, N. M., & Thirumarimurugan, M. (2018). Biosynthesis, characterization, and antibacterial activity of zinc oxide nanoparticles derived from Bauhinia tomentosa leaf extract. *Journal of Nanostructure in Chemistry, 8*, 293–299.

Siddiqi, K. S., Rahman, A., & Tajuddin, H. A. (2018). Properties of zinc oxide nanoparticles and their activity against microbes. *Nanoscale Research Letters, 13*(141), 1–13. https://doi.org/10.1186/s11671-018-2532-3

Soto-Robles, C. A., Luque, P. A., Gómez-Gutiérrez, C. M., Nava, O., Vilchis-Nestor, A. R., Lugo-Medina, E., & Castro-Beltrán, A. (2019). Study on the effect of the concentration of Hibiscus sabdariffa extract on the green synthesis of ZnO nanoparticles. *Results in Physics, 15*, 102807.

Taghiyari, H. R., Morrell, J. F., & Husen, A. (2023). *Emerging nanomaterials* (*Opportunities and Challenges in Forestry Sectors*). Springer Nature. https://doi.org/10.1007/978-3-031-17378-3

Umamaheswari, A., Prabu, S. L., John, S. A., & Puratchikody, A. (2021). Green synthesis of zinc oxide nanoparticles using leaf extracts of Raphanus sativus var. Longipinnatus and evaluation of their anticancer property in A549 cell lines. *Biotechnology Reports, 29*, 1–9.

Upadhyay, P. K., Jain, V. K., Sharma, S., Shrivastav, A. K., & Sharma, R. (2020) Green and chemically synthesized ZnO nanoparticles: A comparative study. *IOP Conference Series: Materials Science and Engineering, 798*(1), 012025. IOP Publishing.

Varghese, E., & George, M. (2015). Green synthesis of Zinc oxide nanoparticles. *International Journal of Advanced Research in Science and Engineering, 4*(1), 307–314.

Vaseem, M., Umar, A., & Hahn, Y. B. (2010). ZnO nanoparticles: Growth, properties, and applications. *Metal Oxide Nanostructures and Their Applications, 5*, 1–36.

Vidya, C., Hiremath, S., Chandraprabha, M. N., Antonyraj, M. L., Gopal, I. V., Jain, A., & Bansal, K. (2013). Green synthesis of ZnO nanoparticles by Calotropis gigantea. *International Journal of Current Engineering and Technology, 1*, 118–120.

Vijayakumar, S., Vinoj, G., Malaikozhundan, B., Shanthi, S., & Vaseeharan, B. (2015). Plectranthusamboinicus leaf extract mediated synthesis of zinc oxide nanoparticles and its control of methicillin resistant Staphylococcus aureus biofilm and blood sucking mosquito larvae. *Spectrochimica Acta Part a: Molecular and Biomolecular Spectroscopy, 137*, 886–891.

Worku, L. A., Nigussie, Y., Bachheti, A., Bachheti, R. K., & Husen, A. (2023). Antimicrobial activities of nanomaterials. In A. Husen & K. S. Siddiqi (Eds.), *Advances in smart nanomaterials and their applications* (pp. 127–148). Elsevier.

Plant Extract-Assisted Green Synthesis and Characterization of Copper Nanocomposite

Addisu Tamir Wassie, Ibrahim Nasser Ahmed, Archana Bachheti, Azamal Husen, and Rakesh Kumar Bachheti

Abstract The era of nanomaterials is here. Science has undergone a revolution thanks to nanotechnology. The use of nanoparticles has greatly helped every area of technology. Nanomaterials were created through different chemical and physical methods. But green synthesis has received a lot of attention recently, especially when it involves the usage of extracts of plants or microbes. The promotion of environmental sustainability can benefit from this. Because green synthesis is simpler, cheaper, and more reproducible than other methods, it is also a more advanced way to make nanomaterials. Compared to other methods, plants create more stable metal nanocomposites and/or nanoparticles, and scaling up is easy. Additionally, there is less chance of contamination. The most prevalent element among the numerous

A. T. Wassie · I. N. Ahmed · R. K. Bachheti (✉)
Department of Industrial Chemistry, College of Applied Sciences, Addis Ababa Science and Technology University, P.O. Box: 16417, Addis Ababa, Ethiopia
e-mail: rkbacheti@gmail.com

A. T. Wassie
Department of Chemistry, College of Natural Sciences, Wollo University, P.O. Box: 1145, Dessie, Ethiopia

I. N. Ahmed · R. K. Bachheti
Nanotechnology Center of Excellence, Department of Industrial Chemistry, Addis Ababa Science and Technology University, Addis Ababa, Ethiopia

A. Bachheti
Department of Environment Science, Graphic Era (Deemed to Be University), Dehradun, India

A. Husen
Department of Biotechnology, Smt. S.S. Patel Nootan Science & Commerce College, Sankalchand Patel University, Visnagar 384315, Gujarat, India

Department of Biotechnology, Graphic Era (Deemed to be University), Dehradun 248002, Uttarakhand, India

Laboratory of Bio Resource Management, Institute of Tropical Forestry and Forest Product, University Putra Malaysia, 43400 UPM Serdang, Selangor, Malaysia

R. K. Bachheti
Department of Allied Sciences, Graphic Era Hill University (G.E.H.U), Society Area, Clement Town, Dehradun 248002, India

R. K. Bachheti et al. (eds.), *Metal and Metal-Oxide Based Nanomaterials*, Smart Nanomaterials Technology, https://doi.org/10.1007/978-981-99-7673-7_3

metals is copper, which is essential to an organism's ability to operate normally. When compared to modern antibiotics, copper nanoparticles and/or nanocomposites show stronger antibacterial activity. Due to a lack of compiled study about copper-based green synthesis, this article addressed an inclusive assembly of investigations concentrated mostly on the plant-based green synthesis of copper nanoparticles and/or nanocomposites. We also discuss the various advantages and disadvantages of copper nanocomposites and/or nanoparticles, in addition to the limitations and advantages of physical and chemical characterization techniques for copper nanocomposites. Lastly, the latent uses of green synthesized copper nanocomposites and nanoparticles, such as antimicrobial activity, catalysis, biosensing, drug delivery, and so forth have been reviewed and discussed.

Keywords Copper nanocomposites · Green synthesis · Applications · Plants extracts · Copper nanoparticles

1 Introduction

Materials with solid structures, known as nanocomposites, have spacing between phases that is minimal on the order of nanometers and generally take the form of an organic matrix in an inorganic phase or an inorganic matrix in an organic phase. These new materials, known as nanocomposites, are created when metal nanoparticles are incorporated into a polymer matrix. Nanocomposites are a subset of composite materials that are distinguished from conventional composite materials by having superior mechanical properties (Bhat et al., 2021). With numerous theoretical and experimental procedures developing and redefining the methods of analysis, synthesis, and cost control of nanocomposites, the study of nanocomposites has grown rapidly in its numerous scientific fields over the years. It is possible to create a variety of nanocomposites, including metal–metal oxide-polymer, cellulose-metal oxide-polymer, metal–metal oxides, metal-oxide polymer, metal-polymer, metal-graphene oxide-polymer, metal–carbon dots, and metal-graphene oxide-polymer in which the wastes contribute to at least one of the components. As a general qualitative assessment, it is suggested that the variety of processes used to create and analyze nanocomposites have varying effects on them, affecting their physical makeup, optical properties, chemical behavior, production costs, and biological interactions, which could limit how they are used (Bachheti & Bachheti, 2023; Husen & Iqbal, 2019; Husen & Siddiqi, 2023; Husen et al., 2023a, 2023b; Husen, 2022, 2023a, 2023b; Vinyas et al., 2019). Various electrical, mechanical, optical, and antibacterial properties of nanoparticles and polymers can be combined to create nanocomposites with better properties. The multifunctional features of the nanocomposites include the high surface area-to-volume ratios for loading biomolecules like enzymes, catalytic activity, redox reactivity, thermal stability, and also have high durability, electrical conductivity, flame retardancy, chemical resistance, and high mechanical strength (Bhat et al., 2021; Tamayo et al., 2016).

Nowadays, different types of copper nanocomposites are synthesized with various techniques under appropriate conditions and used for different applications in various fields. In most cases, the preparation of nanocomposites has focused on explorations in the field of reduction and inclusive studies have been carried out using high-cost and infrequent materials, like silver and gold. Similarly, nanocomposites take much time to manufacture nanocatalysts and the materials used to prepare them are not economically friendly. For instance, Naghash-Hamed and co-workers synthesized the cost-effective, toxicity-free, and appropriate magnetic nanocomposite of $CuFe_2O_4$ NPs/carbon quantum dots (CQDs) of citric acid by using the co-precipitation technique. After achieving magnetic nanocomposite, they used it as a nanocatalyst to reduce the *o*-NA and *p*-NA using sodium borohydride as a reducing agent. As a result, $CuFe_2O_4$@CQD nanocomposite was successfully synthesized for the first time from citric acid for catalytic usage to cut down on o-nitroaniline and p-nitroaniline organic pollutants (Naghash-Hamed et al., 2023). Revathi and co-workers also created a simple and inexpensive Cu–CuO nanocomposite to minimize the nitroarenes and examined the performance of nanocomposite materials. Instead of Cu and CuO, they found that the produced Cu–CuO nanocomposite showed better catalytic improvement to reduce the *p*-NA into para-phenylenediamine (*p*-PDA) (Revathi et al., 2019).

Formation of NPs and their composites has recently become a widespread practice known as "green synthesis." In this process, nanoparticles are created using organisms such as algae, fungi, bacteria, and plant components. These biomolecules which are naturally occurring have the potential to actively contribute to the synthesis of NPs and/or nanocomposites (NCs), which can lead to the development of safer, more sanitary, and environmentally friendly methods. But chemicals that can sometimes be poisonous and dangerous are used in the method used to create metallic nanoparticles. As a result, an environmentally friendly process is needed for their synthesis. Due to easy availability, relatively low production cost, less toxic nature, and eco-friendly in most studies; plants are used as the production assembly in the synthesis of nanoparticles and their composites, which has garnered interest. These processes are quick, efficient, environmentally benign, nonpathogenic, and convergent. In addition, plant extracts can make more stable NPs and/or NCs (Ameta et al., 2021; Jain & Banik, 2020). These all listed above things push us to compile recent articles to increase the attention of researchers fatherly toward the green synthesis of copper nanocomposites. This review also presents a comprehensive assembly of investigations (Table 1) focused mainly on the plant-based green synthesis of copper nanoparticles and/or nanocomposites. We also discuss various green synthesis advantages and disadvantages for copper nanoparticles and/or their nanocomposites and also the advantages and limitations of physical and chemical characterization techniques of copper nanocomposites. In addition, the potential applications of green synthesized copper NPs as catalysis, antimicrobial activity, biosensing, and drug delivery are reviewed and discussed in this chapter.

Table 1 Green synthesized copper-based nanocomposites and copper nanoparticles from plants and their application/s

Copper nanoparticles/ nanocomposite	Plant name	Family name	The plant part used to extract	Major phytochemicals	Type of extracts	Role of plant extract	Uses of products	References
CuO/CNC	Adhatoda vasica Nees	Acanthaceae	Leaves	Alkaloids, tannins, saponins, phenolics, and flavonoids	Crude extract	Reducing, capping agent and source of carbon	Antimicrobial activities	Bhavyasree and Xavier (2020)
CuFO/RGO NC	*Azadirachta indica*	Meliaceae	Leaves	Flavonoids phenols and tannins	Crude extract	Reduce the metal ions stabilized Metal NPs	Photoluminescence, the activity of anticancer and antimicrobial	Reddy et al. (2022)
Cu/MgO NC	*Cassytha filiformis*	Lauraceae	Fruit	Tannin, flavonoid, saponin, and alkaloid	Crude extract	Stabilizing and reducing agent	Catalyst activity for reduction of dyes	Nasrollahzadeh et al. (2018)
Cu/MnO_2 NC	*Centella asiatica*	Apiaceae	Leaves	Flavonoids, terpenoids, and saponins	Crude extract	Reducing agent	The catalyst for the reduction dyes as well as nitro compounds with $NaBH_4$	Nasrollahzadeh et al. (2018)
Cu/GO/ MnO_2NC	*Cuscuta reflexa*	Convolvulaceae	Leaves	Terpenoids and alkaloids	Crude extract	Biosynthesis of CuNPs and the functionalizing surface of graphene	As a catalyst to reduce nitroarenes and organic dyes in $NaBH_4$	Naghdi et al. (2018)
Fe, Cu oxide NC	*Eriobotrya japonica*	Rosaceae	Leaves	Phenolic acids and flavonoids	Crude extract	Reducing agent	Removal of Norfloxacin and Ciprofloxacin	Liu et al. (2020)

(continued)

Table 1 (continued)

Copper nanoparticles/ nanocomposite	Plant name	Family name	The plant part used to extract	Major phytochemicals	Type of extracts	Role of plant extract	Uses of products	References
Cu/Fe_3O_4 NC	*Eriobotrya japonica*	Rosaceae	seed	Carotenoids, vitamins, polyphenolic compounds,	Crude extract	Reducing, stabilizing agents, and reducing agglomeration	Antibacterial effects	Joharian et al. (2022)
Cu-Hibiscus @Fe_3O_4 NC	*Hibiscus sabdariffa*	Malvaceae	Flower	Polyphenols, anthocyanins, and polysaccharides	Crude extract	Reducing agent and stabilizer of NPs	Catalytic activity and anti-colon cancer	Wang et al. (2022)
CuO/ZnO NC	*Melissa Officinalis*	Lamiaceae	Leaves	Polyphenolic compounds, flavonoids, tannins, and terpenoid	Crude extract	Reducing agent and efficient stabilizer	Reduction of 4-nitrophenol and Rhodamine B	Bordbar et al. (2018)
Fe_3O_4@CNC/ CuNC	*Petasites hybridus*	Asteraceae	Leaves	Alkaloids and sesquiterpene esters	Crude extract	Stabilizing and reducing agent for Cu	For the sensitive detection of venlafaxine	Khalilzadeh et al. (2020)
MoO_3/Cu complex bio-NC	*Sesbania sesban*	Legume	Leaves	Glycoside, steroids, flavonoids, saponins, Alkaloids	Crude extract	NA	Oxidize alcohols	Naeimi et al. (2019)
CuO-CS NC	*Sida cordifolia*	Malvaceae	Leaves	Glycosides, alkaloids, carbohydrates, and coumarins	Crude extract	NA	Antibacterial and against lung and breast cancer cell	Sathiyavimal et al. (2023)

(continued)

Table 1 (continued)

Copper nanoparticles/ nanocomposite	Plant name	Family name	The plant part used to extract	Major phytochemicals	Type of extracts	Role of plant extract	Uses of products	References
CuO/C NC	*Vitex negundo*	Lamiaceae	Leaves	Terpenes, phenolic compounds, flavonoids, and phytosteroids	Crude extract	As origination of carbon and reducing and capping agents	Degradation of MB, CBB, and CR dyes	Bhavyasree and Xavier (2021)
CuNPs	*Azadirachta indica*	Meliaceae	Flower	Flavonoids, limonoids, catechins, phenols, polyphenols, nimbins, and gallic acid	Crude extract	Reduction of Cu^{2+} to CuNPs, used as stabilizing and capping agent	Excellent antibacterial activity	Gopalakrishnan Muniraj (2021)
CuNPs	*Celastrus paniculatus* willd	Celastraceae	Leaves	Flavonoids, alkaloids, diterpenoids, phenylpropanoids, tetraterpenes, and triterpenoids	Crude extract	Stabilizing, reducing, and capping agents	As photocatalytic and antifungal agents	Mali et al. (2020)
CuNPs	*Cinnamomum verum*	Lauraceae	Bark	Glycosides, saponin, phenol, and terpenoids	Crude extract	Bio-reductant and stabilizer in citric acid mediation	Antibacterial and disinfectant on textile fabrics	Sarwar et al. (2021)
CuNPs	*Cissus arnotiana*	Vitaceae	Leaves	Alkaloids, phenolic, terpenoids, and flavonoids	Crude extract	Stabilizing and reducing agent	Antibacterial and antioxidant	Rajeshkumar et al. (2019)

(continued)

Table 1 (continued)

Copper nanoparticles/ nanocomposite	Plant name	Family name	The plant part used to extract	Major phytochemicals	Type of extracts	Role of plant extract	Uses of products	References
CuNPs	*Cissus vitiginea*	Vitaceae	Leaves	Polyphenols, steroids, carbohydrates, saponins, proteins, tannins alkaloids, flavonoids, anthraquinone, and terpenoids	Crude extract	As reducing agent	Antioxidant and antibacterial activity against DPPH and urinary tract infection pathogens	Wu et al. (2020)
CuNPs	Citrus medica *Linn*	Rutaceae	Fruits	Phenolic acids, flavonoids, and essential oils	Juice	NP	Antimicrobial activity	Shende et al. (2015)
CuNPs	*Eclipta prostrata*	Asteraceae	Leaves	Phenolic acid derivatives, flavonoids, triterpenoids, and steroid saponins	Crude extract	Reducing agent	Antioxidant and cytotoxic activities	Chung et al. (2017)
CuNPs	*Fortunella margarita*	Rutaceae	Leaves	Carotenoids, flavonoids, minerals, and vitamins	Crude extract	Reducing agent	Drug formulation, drug delivery systems, biomedical applications	Amjad et al. (2021)
CuNPs	Green tea	Theaceae	Leaves	Tannins, flavonoids, alkaloids, and steroids	Crude extract	Stabilize NPs from oxidation & aggregation	Anticodon cancer activity, catalytic and antioxidant activity	Xu et al. (2022)

(continued)

Table 1 (continued)

Copper nanoparticles/ nanocomposite	Plant name	Family name	The plant part used to extract	Major phytochemicals	Type of extracts	Role of plant extract	Uses of products	References
CuNPs	*Jatropha curcas*	Euphorbiaceae	Leaves	Alkaloids, tannins, phenols, saponins flavonoids, and glycosides	Crude extract	Reducing agent	Photocatalytic activity	Ghosh et al. (2020)
CuNPs	*Nerium oleander*	Apocynaceae	Leaves	Flavonoids, cardiac glycosides, carbohydrates, steroids, alkaloids, and tannins	Crude extract	Reducing agent for $CuSO_4$ ions into CuO NPs	Nano-adsorbents of MB and crystal violet from H_2O	Sebeia et al. (2019)
CuNPs	*Piper nigrum*	Piperaceae	Seeds	Tannins, alkaloids, phytosterols, phenols, flavonoids, and saponins	Crude extract	Reducing agent	–	Sirisha and Asthana (2018)
CuNPs	*Plantago asiatica*	Plantaginaceae	Leaves	Phenolic acids, flavonoids, and triterpenoids	Crude extract	Reducing agent of copper ions	Cyanation of aldehydes using $K_4Fe(CN)_6$	Nasrollahzadeh et al. (2017)
CuNPs	*Prunus mahaleb*	Rosaceae	Leave, stems, and fruit	Flavonoids, cyanogenic glycosides, and alkaloids	Crude extract	Capping agents and prevented agglomeration of NPs	High-potential cytotoxic and anticancer activities	Dashtizadeh et al. (2021)
CuNPs	*Prunus nepalensis*	Rosaceae	Fruits	Flavonoids, tannins, and alkaloids	Crude extract	Capping agent	Breast cancer cell (MCF7) treatments	Biresaw and Taneja (2022)

(continued)

Table 1 (continued)

Copper nanoparticles/nanocomposite	Plant name	Family name	The plant part used to extract	Major phytochemicals	Type of extracts	Role of plant extract	Uses of products	References
CuNPs	*Punica granatum*	Punicaceae	Fruit	Tannins, flavonoids, carbohydrates, saponins, phenols, terpenoids, glycosides, and curcumins	Crude extract	Reducing and capping agent	Antimicrobial agent and development of nanobandages	Padma et al. (2018)
CuNPs	*Tinospora cordifolia*	Menispermaceae	Leaves	Flavonoids, alkaloids, glycosides, and terpenes	Crude extract	Capping and reducing agent	Antimicrobial activity of fabrics	Sharma et al. (2019)
CuNPs	*Ziziphus spina-christi*	Rhamnaceae	Fruits	Alkaloids, saponins, and flavonoids	Crude extract	Novel reducing agent	Antibacterial activity and adsorbent for dye removal	Khani et al. (2018)

CR: Congo red; 4-NP: 4-nitrophenol; MB: Methylene blue; 2,4-DNPH: 2,4-dinitrophenylhydrazine; CBB: Coomassie Brilliant Blue

2 Green Synthesis (GS) of Copper Nanocomposites From Plant Materials and Its Applications

Considering the electron source, we can classify the synthesis methods of copper nanocomposites into three categories. first, the nature of the synthesis process is a chemical method if the reducing agent (electron source) is a chemical compound. Second, if the synthesis uses biological organisms (such as plants, algae, molds, and bacteria) as reducing agents, it is a green synthesis method. Third, when the sources are physical, like electricity, it is classified as a physical method of synthesis. The order of these synthesis methods based on the use of safe solvents for human beings during synthesis is GS, which is better than physical methods, and the physical method is also better than chemical methods (Crisan & Lucian, 2021). The GS referred to the synthesis of numerous MNPs (metal nanoparticles) using bioactive agents like extracts of plant parts, microorganisms and different biowastes such as vegetable, fruit peel, eggshell, agricultural waste, and so forth (Awwad, 2020; Kalantari et al., 2020). In comparison to physical and chemical techniques of synthesis of metal nanoparticles that are used for the manufacture of nanocomposites, GS, which uses plant extracts to synthesize metal nanoparticles used to prepare nanocomposites, has many benefits. These benefits include the use of non-hazardous solvents; a straightforward work-up procedure; cleaner reaction outlines; very safe reaction conditions; the abolition of noxious and risky materials; the removal of high temperatures, energy, and pressure; fewer costs; diminutions contamination; and enhance ecological and human health protection (Pozdnyakov et al., 2021). However, the main disadvantages of the GS method of NPs and nanocomposites are the fabrication of them with imperfect surface structures; higher costs related to research, development, and production; the inability to modify some processes; and the time-consuming approach (Garibo et al., 2020; Parveen et al., 2016). But to our best knowledge, there are no investigated articles that show significant disadvantages of copper nanoparticle and/or nanocomposite synthesis methods to date. Hereinafter, the recent articles that are concerned with the only GS of copper nanocomposites via different plant materials and their applications are presented.

Naghdi and co-workers prepared copper/graphene oxide/manganese oxide nanocomposite for the first time by using the leaves extract of *Cuscuta reflexa*. They investigate the use of *this plant* leaf's extract for the biosynthesis of the copper/graphene oxide/manganese oxide nanocomposite as a restorable and solid catalyst to catalyze the reduction of organic dyes with $NaBH_4$ in an aqueous medium at 25 °C. The common organic dyes are, for instance, methylene blue, methyl orange, Congo red, and Rhodamine B, and also nitro compounds like 2,4-DNPH and 4-NP. First, they created GO/MnO_2NC with a one-step hydrothermal technique devoid of hazardous solvents. Next, utilizing a brand-new, eco-friendly, quick, easy, and cost-effective process, CuNPs were created and immobilized on the surface of the GO/MnO_2NC by reducing Cu^{2+} to Cu(0) with *C. reflexa* leaf's extract as a reducing and stabilizing agent under organic solvent-free conditions (Naghdi et al., 2018).

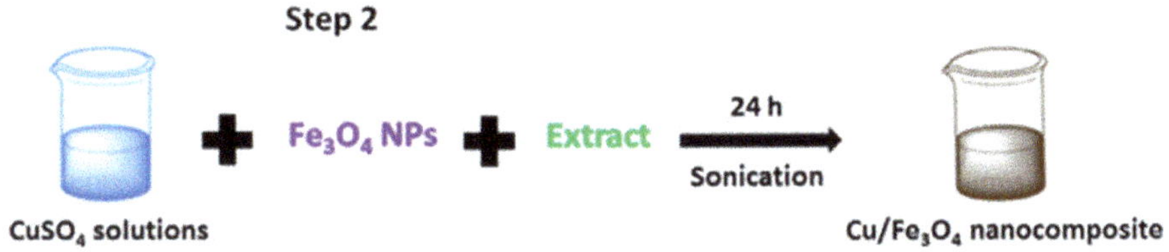

Fig. 1 The overall process for GS of Cu/Fe_3O_4NC (Joharian et al., 2022)

Cu/Fe_3O_4 NCs were synthesized utilizing the seed extract of Eriobotrya japonica (Fig. 1). Various parameters, including the amount of extract, temperature, concentration of copper sulfate salt, quantity of Fe_3O_4, and sonication time, were optimized to achieve the required particle size of the Cu/Fe_3O_4 nanocomposite for antibacterial applications. Subsequently, the researchers assessed the antibacterial efficacy of smaller and more biocompatible Cu/Fe_3O_4 NCs against Pseudomonas aeruginosa, E. coli, and S. aureus, employing disk diffusion and MIC testing techniques (Joharian et al., 2022).

Using *Hibiscus sabdariffa* flower extract which is an environmentally friendly reducing agent of the immobilized Cu^{2+}, Wang and co-workers successfully prepared a novel biocompatible, phytochemically improved, and magnetically separable Cu-containing nanocatalyst at the surface of Fe_3O_4NPs. They investigate the high catalytic activity of the Cu-*Hibiscus*@Fe_3O_4 nanocomposite, which was found to be extremely genuine in more components at a one-pot synthesis of various pyrano-[3,2-c] chromenes. By using the inhibitory activity against the colon cancer cells, they also investigated the biological activities of Cu-Hibiscus@Fe_3O_4 nanocomposite, which has an amazing dose-dependent reduction in cell sustainability in contrast to the cell lines, excellent antioxidant capabilities, and antioxidant characteristics (Wang et al., 2022). A novel catalyst, Cu(II) poly-5-vinyltetrazolate nanocomposite, was created by Zuraev and co-workers for the production of 1,4- and 1,4,5-substituted 1H-triazoles. The synthesized Cu-poly-5-vinyltetrazolate nanocomposite catalyst was more effective, recyclable, and affordable. It is resistant to oxidation in air, employed at low catalytic loading, has a fast reaction time, and permits the reaction to occur in H_2O at 25 °C. The unique aspect of the synthesis technique is that it uses stabilizers, extra ligands, and supports while preventing the formation of potentially dangerous chemicals (Zuraev et al., 2018). Nasrollahzadeh and co-workers synthesized Cu/MgO nanocomposite quickly and easily via the GS of CUNPs using an extract from *Cassytha filiformis* fruits. The flavonoids found in *C. filiformis* fruits extract are used for stabilizing, reducing, and capping agents. By using $NaBH_4$ in an aqueous medium at 25 °C, the Cu/MgONC demonstrates significant catalytic action for the reduction of 4-NP, 2,4-DNPH, MB, and CR. The absence of hazardous ingredients, mild reaction environments, quick reaction times, and straightforward work-up procedures are some of this methodology's key benefits, and multiple recycling attempts of the Cu/MgONC catalyst were made without any significant loss of catalytic action (Nasrollahzadeh et al., 2018). In general, the most widely known applications of copper

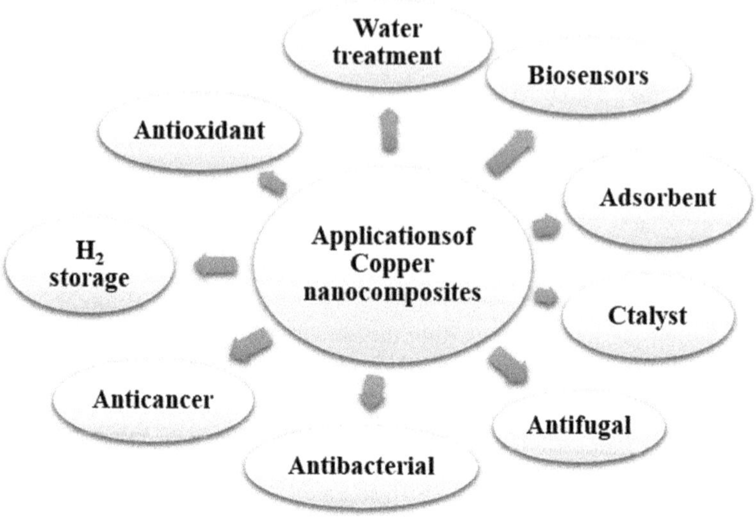

Fig. 2 Applications of plant-based green synthesized copper nanocomposites in biomedical and environmental fields

nanocomposite applications are summarized as shown in Fig. 2. Further, the green synthesized copper-based nanocomposites and copper nanoparticles from different plant materials, the role of plant materials in synthesis, and their applications are presented in Table 1.

3 Characterization Techniques for Copper Nanocomposites

To ensure the validity of different functions and to better understand how the physical and chemical characteristics of nanocomposites affect biological impacts, it is crucial to characterize a nanocomposite's physical and chemical properties. Additionally, the characterization of materials that are made from nanocomposites enables researchers to pinpoint a material's structure, how this structure relates to its macroscopic characteristics, and how it will behave in specific technological applications (Wu et al., 2020). Therefore, the characterization techniques of copper nanocomposites depend on different properties, which are characterized by different techniques. The common properties, which are characterized by various advanced techniques, are size, surface charge, shape, optical properties, and magnetic properties of nanocomposites, which are presented hereinafter.

3.1 Size

The size of nanocomposites is characterized by XRD, DLS, TEM, and UV–Vis spectroscopy. DLS is used to measure the size of NPs in colloidal suspension and is a potent method when used for actual time observation of the combination progression. The measurements can be more accurate when DLS is joined with DCS (differential centrifugal sedimentation). Even if the weakness of DLS is lacking the necessary resolution during functioning with fewer amount aggregates, it has the great capability of measuring dissimilar sizes of nanoparticles at once. It can also use the XRD characterization technique for the determination of nanocomposites because the XRD characterization can offer evidence about the grain size, structure of crystallinity, and phase nature of nanocomposites and also lattice parameters. But XRD has some limitations such as crystal developing trouble, enlarged energy to get data on single conformation as well as less strength (Mourdikoudis et al., 2018). The biologically synthesized Cu/GO/MnO_2NC TEM image shows that the majority of the CuNPs have around 9 nm size and have spherical morphology. GO/MnO_2NP surface is also extremely dispersed on it by CuNPs. Although TEM has trouble quantifying large quantities of particles, it can provide accurate estimates of the homogeneity of nanoparticles. Another drawback is that TEM creates unreliable images because of orientation issues. Nowadays this problem is solved by using the HRTEM method due to its high-resolution ability to characterize the internal structure of nanoparticles (Crisan et al., 2021; Naghdi et al., 2018). Additionally, the surface area of the Cu/GO/MnO_2NC was calculated using Brunauer–Emmett–Teller, with promising outcomes. The average pore diameter, BET surface area, and total pore volume of the Cu/GO/MnO_2NC are 9.3749 nm, 37.21 m^2/g, and 87.209×10^{-3} cm^3/g, respectively (Naghdi et al., 2018).

3.2 Shape

Similar to particle size, shape also affects biodistribution, cellular membrane deformability, internalization kinetics, and cargo loading effectiveness in NP-based medication delivery. It has been demonstrated that smaller NPs are better able to evade the body's normal clearance processes and stay in the bloodstream for a longer period. In studies showing that cylindrical nanoparticles interact with cells quite differently from spherical ones, leading to dramatic variations in bioavailability, form has therefore played a significant role in addition to size. Especially the use of nanoparticles for medicinal purposes can be made more varied and flexible by the large range of shapes that they can take (Robertson et al., 2016). HRTEM, AFM, and TEM are used to describe the shape/morphology of nanocomposites. AFM can be used to study the interaction of NPs in real time. It is inexpensive, space-saving, and mainly used to identify the shape/morphology of NPs (Crisan et al., 2021).

3.3 Surface Charge

Surface charge is a key factor in regulating the interaction of NPs with biological components. Few studies have examined the influence of surface charge on the bioavailability and absorption of NPs after oral administration, although significant studies have shown that surface charge affects the fate of NPs after intravenous administration. To study the influence of surface charge on the oral absorption of polymer NPs, researchers XJ DU and co-workers prepared polymer NPs with the same particle size and surface density as polyethylene glycol (PEG), but with different surface charges (neutral, negative, and positive). Polyethylene glycol-block-polylactic acid (PEG-PLA) was used to make the nanoparticles, and then lipid components with various charges were added. In this study, the small intestine revealed a favorable dispersion of the positively charged nanoparticles, which dramatically increased oral bioavailability (Du et al., 2018). High-charged particles tend to resist one another and create stable colloidal solutions. To assess the stability of the produced colloidal solutions, zeta potential is used. Therefore, electrophoretic mobility (EPM) and zeta potential can be used to characterize the surface charge of CuNPs or nanocomposites. NP aggregations are related to low levels of EPM, whereas stable nanoparticles for extended periods are associated with high values of EPM (Mourdikoudis et al., 2018).

3.4 Optical Properties

Another property of copper nanoparticles or nanocomposites is optical properties which can be also characterized by photoluminescence (PL) and UV–Vis (Crisan et al., 2021). Due to their nanoscale size and surface plasmon resonance nature, the optical characteristics of nanomaterials are highly motivating to research. The form, doping, surface functionalization, size, interactions with other materials, and other factors all have a high influence on these properties. Changes in the optical energy bandgap cause them to exhibit size-dependent optical properties, which in turn affect the nanomaterials' surface plasmon resonance. In the case of semiconductor nanomaterials, in particular, the optical bandgap increases with decreasing particle size. Here it is important to note that surface plasmon resonance is the result of coherent excitation of the free electrons present in the conduction band of nanomaterials and their in-phase resonant vibrations with the applied light energy. With this in mind, surface plasmon resonance can only be observed when the particle size of the material is smaller than the wavelength of the incident radiation. Surface plasmon resonance is thus only possible in nanomaterials and not in bulk materials. According to the quantity of these excited electrons and the medium's dielectric constant, the intensity of such surface plasmon resonance is directly proportional to both of these variables (Huang, 2019; Wei & Liu, 2017).

3.5 Magnetic Properties

Both vibrational sample magnetometry (VSM) and superconducting quantum interference device (SQUID) magnetometry can be used to analyze the magnetic properties of NPs and/or nanocomposites. SQUID is used to measure the magnetic properties of NPs, such as magnetization saturation, resonance, and blocking temperature, whereas VSM is employed as a function of time, temperature, and magnetic field (Crisan et al., 2021). To prepare Cu-ferrite NPs, Subha and co-workers varied the ratio of metal nitrate to citric acid. The NPs were then annealed at different temperatures. According to the XRD analyzes of this study, the samples prepared with metal nitrate to citric acid (M: C) ratios of 1:1, 1:2, and 2:1 are predominantly cubic samples as prepared, and the annealing leads to a transition to the tetragonal phase with an increase in the citric acid ratio. Cubic Cu-ferrite is associated with the citric acid ratio and displayed stronger magnetism and less coercivity than the tetragonal phase. The cubic Cu-ferrite sample generated M:C = 2:1 had a maximum magnetization value of 45.6 emu/g and the lowest coercivity of 190 Oe at 300.0 K. The tetragonal Cu-ferrite sample constructed M:C = 1:3 and annealed at TA = 800.0 °C was found to have the maximum coercivity 1530 Oe at 300.0 K. These Cu-ferrite nanoparticles' magnetic behavior is explained by phase change, rise in tetragonal distortion, and grain development with higher annealing temperatures (Subha et al., 2018).

4 Potential Applications of Plant Material-Mediated Green Synthesized Copper Nanoparticles

4.1 Removal of Water Pollutants

Green synthesized nanoparticles that are mediated by plant extracts are particularly significant because they don't need specialized machinery or hazardous chemicals to produce them. It, therefore, has no negative environmental effects. Recently, several techniques for the plant-based production of CuNPs and their applications for the removal of pollutants from water have been reported (Ali et al., 2019). Chandra and Khan's study revealed that it uses nanoscale zero-valent copper for this, which is produced by the green synthesis method from Anacardium occidentale testa extract in a completely eco-friendly and cost-effective process without the use of hazardous chemicals to remove uranium from water samples. The purpose of removing uranium from water because of its carcinogenic and toxic properties, as, uranium ion [U(VI)] is extremely harmful to all living things when present in amounts that are higher than the permissible limit (Chandra & Khan, 2020).

CuNPs and Ag–CuNPs were created by Afolayan and co-workers using a biological process that was environmentally benign. The study compared and characterized these NPs' antibacterial activity against bacteria obtained from pharmaceutical effluent. Additionally, Wistar rats' liver and kidney function indices were examined for any toxicological effects associated with wastewater treatment using NPs. The created CuNPs and Ag–CuNPs were found to be unagglomerated spheres with hydrodynamic diameters ranging from 70.0 to 90.0 nm and an overall surface zeta potential of −28.0 to −31.0 mV. According to the inhibition zone, the minimum inhibitory concentrations, and also bactericidal concentration values found in this research, it can be concluded that the NPs significantly enhanced the antibacterial effects of bacteria, with Ag–CuNPs having the strongest effect in comparison to CuNPs at 50.0 g/ml. For assessment of subacute toxicity, the NPs at 200.0 g/kg showed signals of progressive onset of toxicity, indicated by a decrease in aspartate transaminase, alkaline phosphatase, and gamma-glutamyl transaminase activities in the kidney and liver of the NPs-treated groups compared to the control group. Because of this finding, it is advised that safety precautions be taken while using Ag–Cu NPs, which may not be harmless for direct application at 200.0 g/kg (Afolayan et al., 2022).

4.2 Catalytic Activities

Using *Euphorbia esula* L. leaf extract as a reducing and stabilizing agent, Nasrollahzadeh and co-workers demonstrated a straightforward and environmentally friendly synthetic approach to produce stable CuNPs. In this procedure, no capping agent, template, or surfactant was employed. They also have shown that the prepared CuNPs have the potential for a range of catalytic uses and the study offers a green synthesis method for producing ethers and 4-aminophenol. The prepared CuNPs are used to catalyze processes, which is of significant importance for green chemistry science. CuNPs can be used at incredibly low loadings and can be separated for recycling without contaminating the product because they are in a distinct phase from the product (Nasrollahzadeh et al., 2014). According to Nguyen and coworkers' study, the CuNPs are synthesized from cocoa pod extract, which is used as a reducing and stabilizing agent for catalytic activities. Later synthesized they impregnated it on different supports such as CeO_2, TiO_2, and Al_2O_3 to prove an effective and fast method for deep oxidation of aromatic hydrocarbons. The catalytic activity of CuNPs in benzene, toluene, ethylbenzene, and xylene (BETX) deep oxidation was increased as a result of the ability to generate them on the support surface, which prevented CuNPs from clumping together and reduced their size. Their study shows that it was highly intriguing to learn that the optimum catalytic performance for BTEX deep oxidation was demonstrated by the 7.50 weight percent of CuNPs supported on CeO_2 and the development of catalytic systems for the oxidation of aromatic hydrocarbons using *cocoa* pod extracts as reducing agents can be a promising strategy (Nguyen et al., 2020).

4.3 Antimicrobial Activity

Shende and co-workers used lemon juice for the synthesis of CuNPs, which represents a new phase in the biogenic synthesis of CuNPs. This synthesis method is environmentally friendly, rapid, and non-toxic. Additionally, they disclosed the biogenic CuNPs used against various human and plant pathogens and their potency against a variety of microbes. So the exploration of CuNPs from lemon (citron) juice is employed in a variety of formulations, including nanofertilizers, nanoantimicrobials, and nanofungicides, which could serve a dual purpose by defending crop plants from infections and giving the plants nutrients. The most crucial point is that it is an economical strategy because the raw materials used in the synthesis are quite inexpensive (Shende et al., 2015). CuNPs with a size of 63.3 nm were synthesized using *Tinospora cordifolia* leaf extract. The synthesized nanoparticles were subsequently incorporated into fabrics, and their antimicrobial activity was investigated. In the antimicrobial study conducted on both CuNPs and fabrics coated with CuNPs, it was observed that both the nanoparticles and the nano-coated fabrics exhibited antimicrobial activity. However, this activity was found to be more potent against gram-positive bacteria than against gram-negative bacteria. At the maximum concentration, the copper nanoparticle-coated fabric demonstrated 101% efficacy against gram-positive bacteria, while its efficacy against gram-negative bacteria was 74% (Sharma et al., 2019).

Using *Prunus mahaleb* L. stem, leaves, and fruit extracts, Dashtizadeh and co-workers created CuNPs, which they then examined using XRD, UV–vis, EDX, and FE-SEM techniques. According to the analysis, the CuNPs were spherical and ranged in size from 20 to 30 nm. After that, the anticancer, antioxidant, antibacterial, and antimutagenicity properties of CuNPs were evaluated. CuNPs displayed only little antioxidant activity (15.9% inhibition) in the DPPH assay. The brine shrimp lethality experiment was employed to assess the anticancer potential of CuNPs, revealing a robust cytotoxic effect with an LC_{50} value of 3.60 g/mL. Additionally, results from the Ames test indicated that CuNPs exhibited significant anti-mutagenic properties, displaying over 40% inhibition of mutagenicity and confirming their non-mutagenic nature. Notably, CuNPs had a more pronounced impact on gram-positive bacteria. Furthermore, the antimicrobial activity of Ca alginate beads in CuNCs was found to be lower compared to that of CuNPs (Dashtizadeh et al., 2021).

4.4 Biosensor

Biosensors are autonomous analytical tools that can transform a biological response into a quantifiable and interpretable analytical output. A biological recognition component, like enzymes, tissues, or nucleic acids, specifically reacts with the analyte. A physical transducer like electrochemical, optical, or thermal converts the analyte-based receptors into the output signals and a signal output display makes up the biosensor. Due to their affordability, simplicity, speed, selectivity, and high sensitivity, biosensors have found a comprehensive variety of uses in the biomedical industry for medical diagnostics. NPs have been employed as transducer materials in the development of biosensors because of their extraordinary qualities. This has improved analytical signals and the performance of biosensors, including their selectivities and sensitivities. The creation of ecologically friendly biosensors depends heavily on green nanomaterials (Noah & Ndangili, 2022). CuNPs produced through biological means have been applied to the creation of biosensors. According to Dayakr and colleagues, *Ocimum tenuiflorum* leaf extract was used to create CuNPs, which were then coated on a glass carbon electrode to create a non-enzymatic electrochemical-based glucose biosensor. The electrochemical response was then measured using amperometry and cyclic voltammetry. A broad linear array of 1–7 mM and a detection array of 0.0380 μM were all displayed by the biosensor, along with impressive sensitivity, selectivity, and fast response time of fewer than 3 s (Dayakar et al., 2017). Wang and colleagues' study illustrates the utilization of terminal protection of small molecules linked to DNA and polythymine-templated CuNPs in constructing a fluorescent biosensor for protein detection. The small chemical, biotin, and the protein streptavidin (SA), known for its binding affinity to biotin, were utilized as examples to elucidate the functionality of this technique. In this investigation, the target protein SA exhibited selective and high-affinity binding to the biotin-linked poly-T probe (biotin-T30). It was verified that the BiotinT30 probe is resistant to hydrolysis by exonuclease I (Exo I) due to the protective binding events, which efficiently guide the synthesis of fluorescent CuNPs. The study's results indicated that the developed technique had a detection limit for SA ranging from 0.5 to 1000 nM, showcasing high sensitivity. Furthermore, the repeatability of the approach was deemed adequate, displaying a 3.6% relative standard deviation across five independent assays of 50 nM SA. The prepared biosensor exhibited desirable sensitivity, coupled with excellent selectivity, low cost, and straightforward operations. Consequently, there is substantial potential for the development of a simple, targeted, and sensitive fluorescence platform for detecting small molecule–protein interactions in molecular diagnostics and genomic research (Wang et al., 2015).

4.5 Drug Delivery

Because of their small sizes (between 1 and 100 nm), NPs have a high surface area-to-volume ratio as a distinguishing feature. They can access and permeate cell membranes thanks to this characteristic, making it possible to administer medications to specific bodily parts. Because of this, several kinds of NPs have been created, altered, and used as medication transporters throughout the body. Because of this, it is now possible to promptly begin treatment after diagnosing new diseases. These days, NPs are used in medicine as agents against disease-causing bacteria or as nanocarriers of drugs to specific infection sites (Alao et al., 2022). In their study, Gopalakrishnan and Muniraj used the neem flowers' aqueous extracts as the stabilizing, capping, and reducing agent in a straightforward, affordable, and repeatable method for the synthesis of stable CuNps. XRD, SEM, FT-IR, and UV–VIS spectroscopy were used to characterize the biosynthesized NPs. The extract's phytoconstituents caused the decrease and stabilization of CuNps. The copper nanoparticles show potential use in medicinal applications due to their positive antibacterial activity. The drug delivery technique makes excellent use of these biosynthesized copper nanoparticles. The green synthesized CuNps could replace several antibiotic medications used to treat human pathogenic microorganisms and be more affordable for the pharmaceutical sector because of their excellent antibacterial action (Gopalakrishnan & Muniraj, 2021).

5 Conclusion and Future Directions

The increasing demand for green chemistry and nanotechnology has driven the usage of green synthetic pathways for the synthesis of nanomaterials employing bioactive agents like plant materials, microbes, diverse biowastes, and others over the past few decades. In recent years, scientists have focused their attention on developing environmentally friendly nanocomposites. Plant extract-based nanocomposites have drawn a lot of attention in research due to their affordability, availability, eco-friendliness, and non-toxic technique, as well as their prospective uses in a range of sectors. Copper nanocomposites are used in a variety of biomedical sectors including water treatment, catalysis, dye degradation, optics, sensors, electronics, and so forth. The use of plant-extracted materials in the green synthesis of nanocomposites is a fascinating and speedily growing area of nanotechnology. Green synthesis of nanocomposites has a significant impact on the environment in the direction of sustainability and more advancement in the area of nanoscience. In the future, copper nanocomposite synthesis is expected to take a greener path and this will likely lead to exponential growth in applications. However, there is cause for concern regarding the long-term effects of these on human beings and animals, as well as the environment's accumulation of these materials, both of which must be addressed in the future. These biogenic copper nanocomposites can be utilized to make nanoweapons that fight an organism parasitic on a plant host, as well as other water purification techniques for environmental cleanup. These nanocomposites could be the biomedical industry's next big thing in medication delivery systems. The effects of different particle properties on the toxicity and delivery of copper ions can potentially be examined in further research.

References

Afolayan, J. S., Ajani, E., Saheed, S., Folorunsho, R. D., & Abdullateef, M. A. (2022). Green synthesis of copper and silver nanoparticles and their comparative toxicity and antibacterial evaluation in pharmaceutical wastewater treatment. *Nanotechnology for Environmental Engineering*, 8, 333–346.

Alao, I. I., Oyekunle, I. P., Iwuozor, K. O., & Emenike, E. C. (2022). Green synthesis of copper nanoparticles and investigation of its anti-microbial properties. *Advanced Journal of Chemistry-Section B, 4*(1), 39–52.

Ali, J. A., Kolo, K., Sajadi, S. M., Khaksar Manshad, A., & Stephen, K. D. (2019). Green synthesis of ZnO/SiO_2 nanocomposite from *pomegranate* seed extract: coating by natural xanthan polymer and its characterizations. *Micro and Nano Letters, 14*(6), 638–641.

Amer, M., & Awwad, A. (2020). Green synthesis of copper nanoparticles by *Citrus* limon fruits extract, characterization and antibacterial activity, *7*(1), 1–8.

Ameta, R., Rai, A. K., Vyas, S., Bhatt, J. P., & Ameta, S. C. (2021). *Green synthesis of nanocomposites using plant extracts and their applications Handbook of Greener Synthesis of Nanomaterials and Compounds* (pp. 663–682). Elsevier.

Amjad, R., Mubeen, B., Ali, S. S., Imam, S. S., Alshehri, S., Ghoneim, M. M., Alzarea, S. I., Rasool, R., Ullah, I., Nadeem, M. S., & Kazmi, I. (2021). Green synthesis and characterization of copper nanoparticles using *Fortunella margarita* leaves. *Polymers, 13*(24), 4364.

Bachheti, R. K., & Bachheti, A. (Eds.). (2023). *Secondary metabolites from medicinal plants: nanoparticles synthesis and their applications*. CRC Press. https://doi.org/10.1201/9781003213727

Bhat, A., Budholiya, S., Aravind Raj, S., Sultan, M. T. H., Hui, D., Md Shah, A. U., & Safri, S. N. A. (2021). Review on nanocomposites based on aerospace applications. *Nanotechnology Reviews, 10*(1), 237–253.

Bhavyasree, P. G., & Xavier, T. S. (2021). Adsorption studies of methylene blue, coomassie brilliant blue, and congo red dye onto CuO/C nanocomposites synthesized via *Vitex negundo* Linn leaf extract. *Current Research in Green and Sustainable Chemistry, 4*, 100161.

Bhavyasree, P. G., & Xavier, T. S. (2020). Green synthesis of Copper Oxide/Carbon nanocomposites using the leaf extract of *Adhatoda vasica* Nees, their characterization, and antimicrobial activity. *Heliyon, 6*(2), e03323.

Biresaw, S. S., & Taneja, P. (2022). Copper nanoparticles green synthesis and characterization as anticancer potential in breast cancer cells (MCF7) derived from *Prunus nepalensis* phytochemicals. *Materials Today: Proceedings, 49*, 3501–3509.

Bordbar, M., Negahdar, N., & Nasrollahzadeh, M. (2018). *Melissa Officinalis* L. leaf extract assisted green synthesis of CuO/ZnO nanocomposite for the reduction of 4-nitrophenol and Rhodamine B. *Separation and Purification Technology, 191*, 295–300.

Chandra, C., & Khan, F. (2020). Nano-scale zerovalent copper: green synthesis, characterization and efficient removal of uranium. *Journal of Radioanalytical and Nuclear Chemistry, 324*, 589–597.

Chung, I. M., Abdul Rahuman, A., Marimuthu, S., Vishnu Kirthi, A., Anbarasan, K., Padmini, P., & Rajakumar, G. (2017). Green synthesis of copper nanoparticles using *Eclipta prostrata* leaves extract and their antioxidant and cytotoxic activities. *Experimental and Therapeutic Medicine, 14*(1), 18–24.

Crisan, M. C., Teodora, M., & Lucian, M. (2021). Copper nanoparticles: Synthesis and characterization, physiology, toxicity, and antimicrobial applications. *Applied Sciences, 12*(1), 141.

Dashtizadeh, Z., Kashi, F. J., & Ashrafi, M. (2021). Photosynthesis of copper nanoparticles using *Prunus mahaleb* L. and its biological activity. *Materials Today Communications, 27*, 102456.

Dayakar, T., Rao, K. V., Bikshalu, K., Rajendar, V., & Park, S. H. (2017). Novel synthesis and characterization of pristine Cu nanoparticles for the non-enzymatic glucose biosensor. *Journal of Materials Science: Materials in Medicine, 28*, 1–8.

Du, X. J., Wang, J. L., Iqbal, S., Li, H. J., Cao, Z. T., Wang, Y. C., Du, J. Z., & Wang, J. (2018). The effect of surface charge on oral absorption of polymeric nanoparticles. *Biomaterials Science, 6*(3), 642–650.

Garibo, D., Borbón-Nuñez, H. A., de León, J. N., García Mendoza, E., Estrada, I., Toledano-Magaña, Y., Tiznado, H., Ovalle-Marroquin, M., Soto-Ramos, A. G., Blanco, A., & Rodríguez, J. A. (2020). Green synthesis of silver nanoparticles using *Lysiloma acapulcensis* exhibits high-antimicrobial activity. *Scientific Reports, 10*(1), 1–11.

Ghosh, M. K., Sahu, S., Gupta, I., & Ghorai, T. K. (2020). Green synthesis of copper nanoparticles from an extract of *Jatropha curcas* leaves Characterization, optical properties, CT-DNA binding and photocatalytic activity. *RSC Advances, 10*(37), 22027–22035.

Gopalakrishnan, V., & Muniraj, S. (2021). Neem flower extract assisted green synthesis of copper nanoparticles–optimization, characterization, and anti-bacterial study. *Materials Today: Proceedings, 36*, 832–836.

Huang, M. H. (2019). Facet-dependent optical properties of semiconductor nanocrystals. *Small, 15*(7), 1804726.

Husen, A. (2022). *Engineered nanomaterials for sustainable agricultural production, soil improvement and stress management*. Elsevier Inc. ISBN 978-0-323-91933-3. https://doi.org/10.1016/C2021-0-00054-7

Husen, A. (2023a). *Nanomaterials from agricultural and horticultural products*. Springer Nature Singapore Pte Ltd. ISBN: 978–981–99–3437–9

Husen, A. (2023b). *Nanomaterials and Nanocomposites exposures to plants (Response, Interaction, Phytotoxicity and Defense Mechanisms)*. Springer Nature Singapore Pte Ltd. https://doi.org/10.1007/978-981-99-2419-6

Husen, A., Bachheti, R. K., Bachheti, A. (2023a). *Nanomaterials for environmental and agricultural sectors*. Springer Nature Singapore Pte Ltd. https://doi.org/10.1007/978-981-99-2874-3

Husen, A., Bachheti, R. K., Bachheti, A. (2023b). *Current trends in green nano-emulsions (Food, Agriculture and Biomedical Sectors)*. Springer Nature Singapore Pte Ltd.

Husen, A., & Iqbal, M. (2019) *Nanomaterials and plant potential*. Springer International Publishing AG. https://doi.org/10.1007/978-3-030-05569-1

Husen, A., & Siddiqi, K. S. (2023) *Advances in smart nanomaterials and their applications*. Elsevier Inc. https://doi.org/10.1016/C2021-0-02202-1

Jain, A., & Banik, B. K. (2020). *Green synthesis of nanoparticles and nanocomposites: Medicinal aspects green approaches in medicinal chemistry for sustainable drug design* (pp. 231–242). Elsevier.

Joharian, M., Mojarab, M., Darvishi, E., Khosravi, H., Nazari, V., Varnamkhasti, B. S., & Mirsadeghi, S. (2022). Green synthesis of biogenic Cu/Fe_3O_4 nanocomposite using the *Eriobotrya japonica* seed extract against pathogenic bacteria. *Journal of Molecular Structure, 1264*, 133229.

Kalantari, E., Khalilzadeh, M. A., Zareyee, D., & Shokouhimehr, M. (2020). Catalytic degradation of organic dyes using green synthesized Fe_3O_4-cellulose-copper nanocomposites. *Journal of Molecular Structure, 1218*, 128488.

Khalilzadeh, M. A., Tajik, S., Beitollahi, H., & Venditti, R. A. (2020). Green synthesis of magnetic nanocomposite with iron oxide deposited on cellulose nanocrystals with copper (Fe_3O_4@ CNC/Cu): investigation of catalytic activity for the development of a venlafaxine electrochemical sensor. *Industrial and Engineering Chemistry Research, 59*(10), 4219–4228.

Khani, R., Roostaei, B., Bagherzade, G., & Moudi, M. (2018). Green synthesis of copper nanoparticles by fruit extract of *Ziziphus spina-christi* (L.) Willd.: Application for adsorption of triphenylmethane dye and antibacterial assay. *Journal of Molecular Liquids, 255*, 541–549.

Liu, Q., Ma, P., Liu, P., Li, H., Han, X., Liu, L., & Zou, W. (2020). Green synthesis of stable Fe, Cu oxide nanocomposites from *Loquat* leaf extracts for removal of *Norfloxacin* and *Ciprofloxacin*. *Water Science and Technology, 81*(4), 694–708.

Mali, S. C., Dhaka, A., Githala, C. K., & Trivedi, R. (2020). Green synthesis of copper nanoparticles using *Celastrus paniculatus* Willd. leaf extract and their photocatalytic and antifungal properties. *Biotechnology Reports, 27*, e00518.
Mourdikoudis, S., Pallares, R. M., & Thanh, N. T. (2018). Characterization techniques for nanoparticles: comparison and complementarity upon studying nanoparticle properties. *Nanoscale, 10*(27), 12871–12934.
Naeimi, A., Honarmand, M., & Sedri, A. (2019). Ultrasonic assisted fabrication of first MoO_3/copper complex bio-nanocomposite based on *Sesbania sesban* plant for green oxidation of alcohols. *Ultrasonics Sonochemistry, 50*, 331–338.
Naghash-Hamed, S., Arsalani, N., & Mousavi, S. B. (2023). The catalytic performance of $CuFe_2O_4$@ CQD nanocomposite as a high-perform heterogeneous nanocatalyst in nitroaniline group reduction. *Scientific Reports, 13*(1), 3329.
Naghdi, S., Sajjadi, M., Nasrollahzadeh, M., Rhee, K. Y., Sajadi, S. M., & Jaleh, B. (2018). *Cuscuta reflexa* leaf extract mediated green synthesis of the Cu nanoparticles on graphene oxide/manganese dioxide nanocomposite and its catalytic activity toward the reduction of nitroarenes and organic dyes. *Journal of the Taiwan Institute of Chemical Engineers, 86*, 158–173.
Nasrollahzadeh, M., Issaabadi, Z., & Sajadi, S. M. (2018). Green synthesis of a Cu/MgO nanocomposite by *Cassytha filiformis* L. extract and investigation of its catalytic activity in the reduction of methylene blue, congo red, and nitro compounds in aqueous media. *RSC Advances, 8*(7), 3723–3735.
Nasrollahzadeh, M., Momeni, S. S., & Sajadi, S. M. (2017). Green synthesis of copper nanoparticles using *Plantago asiatica* leaf extract and their application for the cyanation of aldehydes using K_4Fe (CN) 6. *Journal of Colloid and Interface Science, 506*, 471–477.
Nasrollahzadeh, M., Sajadi, S. M., & Khalaj, M. (2014). Green synthesis of copper nanoparticles using aqueous extract of the leaves of *Euphorbia esula* L and their catalytic activity for ligand-free Ullmann-coupling reaction and reduction of 4-nitrophenol. *RSC Advances, 4*(88), 47313–47318.
Nasrollahzadeh, M., Sajjadi, M., & Sajadi, S. M. (2018). Biosynthesis of copper nanoparticles supported on manganese dioxide nanoparticles using *Centella asiatica* L. leaf extract for the efficient catalytic reduction of organic dyes and nitroarenes. *Chinese Journal of Catalysis, 39*(1), 109–117.
Nguyen, P. A., Nguyen, A. V., Dang-Bao, T., Phan, H. P., Nguyen, T. T., Tran, B. A., Huynh, T. L., Hoang, T. C., Huynh, V. T., Nguyen, T. (2020). Green synthesis of copper nanoparticles using *Cocoa pod* extract and its catalytic activity in deep oxidation of aromatic hydrocarbons. *SN Applied Sciences, 2*, 1–13.
Noah, N. M., & Ndangili, P. M. (2022). Green synthesis of nanomaterials from sustainable materials for biosensors and drug delivery. *Sensors International, 3*, 100166.
Padma, P. N., Banu, S. T., & Kumari, S. C. (2018). Studies on green synthesis of copper nanoparticles using Punica granatum. *Annual Research and Review in Biology, 23*(1), 1–10.
Parveen, K., Banse, V., & Ledwani, L. (2016). Green synthesis of nanoparticles: Their advantages and disadvantages. In *Paper Presented at the AIP Conference Proceedings*.
Pozdnyakov, A. S., Emel'yanov, A. I., Korzhova, S. A., Kuznetsova, N. P., Bolgova, Y. I., Trofimova, O. M., Semenova, T. A., & Prozorova, G. F. (2021). Green synthesis of stable nanocomposites containing copper nanoparticles incorporated in poly-N-vinylimidazole. *Polymers, 13*(19), 3212.
Rajeshkumar, S., Menon, S., Kumar, S. V., Tambuwala, M. M., Bakshi, H. A., Mehta, M., Satija, S., Gupta, G., Chellappan, D. K., Thangavelu, L., & Dua, K. (2019). Antibacterial and antioxidant potential of biosynthesized copper nanoparticles mediated through *Cissus arnotiana* plant extract. *Journal of Photochemistry and Photobiology B: Biology, 197*, 111531.
Reddy, B. C., Manjunatha, S., Manjunatha, H. C., Vidya, Y. S., Sridhar, K. N., Seenappa, L., Munirathnam, R., Rajesh, K., Thirunavukkarasu, V., Gupta, P. D., & Aluri, S. (2022). *Neem* leaves mediated green synthesis of copper ferrite decorated reduced graphene oxide nanocomposite for photoluminescence, gamma/X-ray radiation shielding, antimicrobial and anticancer properties. *Solid State Sciences, 134*, 107029.

Revathi, K., Palantavida, S., & Vijayan, B. K. (2019). Effective reduction of p-Nitroaniline to p-Phenylenediamine using Cu-CuO nanocomposite. *Materials Today: Proceedings, 9*, 633–638.
Robertson, J. D., Rizzello, L., Avila-Olias, M., Gaitzsch, J., Contini, C., Magoń, M. S., Renshaw, S. A., Battaglia, G. (2016). Purification of nanoparticles by size and shape. *Scientific Reports, 6*(1), 1–9.
Sarwar, N., Humayoun, U. B., Kumar, M., Zaidi, S. F., Yoo, J. H., Ali, N., Jeong, D. I., Lee, J. H., & Yoon, D. H. (2021). Citric acid mediated green synthesis of copper nanoparticles using *cinnamon* bark extract and its multifaceted applications. *Journal of Cleaner Production, 292*, 125974.
Sathiyavimal, S., Vasantharaj, S., Kaliannan, T., Garalleh, H. A., Garaleh, M., Brindhadevi, K., Chi, N. T., Sharma, A., & Pugazhendhi, A. (2023). Bio-functionalized copper oxide/chitosan nanocomposite using *Sida cordifolia* and their efficient properties of antibacterial, and anticancer activity against breast and lung cancer cell lines. *Environmental Research, 218*, 114986.
Sebeia, N., Jabli, M., & Ghith, A. (2019). Biological synthesis of copper nanoparticles, using *Nerium oleander* leaves extract characterization and study of their interaction with organic dyes. *Inorganic Chemistry Communications, 105*, 36–46.
Sharma, P., Pant, S., Dave, V., Tak, K., Sadhu, V., & Reddy, K. R. (2019). Green synthesis and characterization of copper nanoparticles by *Tinospora cardifolia* to produce nature-friendly copper nano-coated fabric and their antimicrobial evaluation. *Journal of Microbiological Methods, 160*, 107–116.
Shende, S., Ingle, A. P., Gade, A., & Rai, M. (2015). Green synthesis of copper nanoparticles by *Citrus medica* Linn.(Idilimbu) juice and its antimicrobial activity. *World Journal of Microbiology and Biotechnology, 31*, 865–873.
Sirisha, N. G. D., & Asthana, S. (2018). Microwave-mediated green synthesis of copper nanoparticles using aqueous extract of *Piper nigrum* seeds and particle characterization. *IAETSD Journal of Advanced Research in Applied Sciences, 5*(2), 859–870.
Subha, A., Shalini, M. G., Sahu, B., & Sahoo, S. C. (2018). Structural transformation and magnetic properties of copper ferrite nanoparticles prepared by sol–gel method. *Journal of Materials Science: Materials in Electronics, 29*, 20790–20799.
Tamayo, L., Azócar, M., Kogan, M., Riveros, A., & Páez, M. (2016). Copper-polymer nanocomposites: An excellent and cost-effective biocide for use on antibacterial surfaces. *Materials Science and Engineering: C, 69*, 1391–1409.
Vinyas, M., Athul, S. J., Harursampath, D., Loja, M., & Thoi, T. N. (2019). A comprehensive review on analysis of nanocomposites: from manufacturing to properties characterization. *Materials Research Express, 6*(9), 092002.
Wang, C., Karmakar, B., Awwad, N. S., Ibrahium, H. A., El-Kott, A. F., Abdel-Daim, M. M., Oyouni, A. A., Al-Amer, O., & Batiha, G. E. S. (2022). Bio-supported of Cu nanoparticles on the surface of Fe_3O_4 magnetic nanoparticles mediated by *Hibiscus sabdariffa* extract: Evaluation of its catalytic activity for the synthesis of pyrano [3, 2-c] chromenes and study of its anti-colon cancer properties. *Arabian Journal of Chemistry, 15*(6), 103809.
Wang, H. B., Zhang, H. D., Chen, Y., & Liu, Y. M. (2015). A fluorescent biosensor for protein detection based on poly (thymine)-templated copper nanoparticles and terminal protection of small molecule-linked DNA. *Biosensors and Bioelectronics, 74*, 581–586.
Wei, C., & Liu, Q. (2017). Shape-, size-, and density-tunable synthesis and optical properties of copper nanoparticles. *CrystEngComm, 19*(24), 3254–3262.
Wu, S., Rajeshkumar, S., Madasamy, M., & Mahendran, V. (2020). Green synthesis of copper nanoparticles using *Cissus vitiginea* and its antioxidant and antibacterial activity against urinary tract infection pathogens. *Artificial Cells, Nanomedicine, and Biotechnology, 48*(1), 1153–1158.
Xu, L., Zhang, L., Ren, D., Peng, Y., Liu, Z., Meng, Y., Deng, W., & Zhang, Y. (2022). Green synthesis of Cu/Fe_3O_4 nanoparticles using green tea extract: Evaluation of its catalytic activity, antioxidant, and anti-colon cancer effects. *Inorganic Chemistry Communications, 144*, 109927.

Zuraev, A. V., Grigoriev, Y. V., Budevich, V. A., & Ivashkevich, O. A. (2018). Copper-polymer nanocomposite: an efficient catalyst for green Huisgen click synthesis. *Tetrahedron Letters, 59*(16), 1583–1586.

Antimicrobial Applications of Zinc Oxide Nanoparticles in Food Packaging Industry

Nisha Sharma, Prakash Chandra Gupta, Swarnakshi Upadhyay, Sweta Rai, and Priyanka Mishra

Abstract Foodborne diseases affect the world. Foodborne illnesses and fatalities occur worldwide. Thus, foodborne pathogen control technology development is urgent. Food packaging ensures food safety and quality. Active packaging meets the needs of consumers for safe, fresh-like goods with extended shelf life, excellent cost-to-benefit ratios, and ease. Including antimicrobials in packaging components permits the progressive diffusion of specific bactericidal or bacteriostatic chemicals into a food matrix, eliminating the requirement for elevated preservative concentrations in food formulations. Antimicrobial packaging prevents food contamination during nonthermal methods. This ingenious packaging reduces disease contamination and extends food shelf life. Zinc oxide (ZnO), one of the metal oxides, is one of the most frequently used substances in a variety of industries due to its exceptional antibacterial and photocatalytic qualities. In addition to being excellent semiconductors with outstanding biocompatibility, material stability, and biocidal activities both in vitro and in vivo, zinc oxide nanoparticles (ZnO NPs) are also reasonably priced. They are antibacterial and effective against microscopic organisms and foodborne diseases. ZnO NPs have been employed in a variety of food-packaging coatings to maintain nutrient colors, stop food from spoiling, and enhance the mechanical quality, impedance qualities, and stability of packaging materials. A greater focus has been given to applying potent antimicrobial substances built into packaging materials as bacteria-controlling agents in food packing.

Keywords Antimicrobial · Packaging · Microbes · Diseases · Nanofabrication · Factors · Toxicity response · Uses

---

N. Sharma · P. C. Gupta (✉) · S. Upadhyay · S. Rai · P. Mishra
School of Pharmaceutical Sciences, Chhatrapati Shahu Ji Maharaj University, Kalyanpur, Kanpur 208024, India
e-mail: prakashchandragupta@csjmu.ac.in

R. K. Bachheti et al. (eds.), *Metal and Metal-Oxide Based Nanomaterials*, Smart Nanomaterials Technology, https://doi.org/10.1007/978-981-99-7673-7_4

1 Introduction

The biggest problems in the food sector in recent years have been food waste and contamination. Worldwide, foodborne illness outbreaks caused by bacteria have occurred often. People's life and health are inextricably connected to the food supply chain. However, the public's trust in the food supply's reliability has been eroded due to the reported outbreaks. In addition to promoting the growth of dangerous germs and rendering food unusable, food rotting causes a loss of color, texture, and nutritional value. Exposure to the environment during slaughtering, packaging, and processing might contaminate food (Appendini & Hotchkiss, 2002). Heating, freezing, and salting are common methods used to preserve food for longer periods of time, but they do not prevent recontamination, which could put consumers at risk. In recent innovations in food packaging, antimicrobial-active packaging solutions combine antimicrobial chemicals into a polymer film to inhibit the growth of specific pathogens. Antimicrobial packaging refers to packaging that interacts with the product to prevent the growth of bacteria, mold, and other microbes that might otherwise contaminate the food.

Agents with antimicrobial properties can be added to packaging films to give them antimicrobial properties. This can make the lag phase of microbes that have been contaminated last longer and lower the number of microbes that are still alive, which stops their growth or slows it down a lot (Cho et al., 2011). Antimicrobial packaging materials are utilized to inhibit and control the proliferation of microorganisms, retain moisture, ensure safety, prevent the penetration of liquids or gases, and regulate the shelf life (Hameed et al., 2016). Antimicrobials are categorized as organic chemicals, like quaternary ammonium compounds, halogenated compounds, phenols, chitosan, and chitin, or inorganic materials like metals and metal oxides. Chitosan & chitin are also included in this category (Hosseinnejad & Jafari, 2016).

Metals and metal oxides are essential because of the exceptional antibacterial properties that they possess, as well as the high level of atmospheric stability that their compounds have. It is important to note that metal oxide particles do not necessarily enter the cells to generate toxicity; they react as they contact the cell walls of bacteria or the abdomen of crustaceans (Prabhu et al., 2015). Since its introduction by Richard Feynman in 1959, nanotechnology has been put to use in many different scientific disciplines, and the realm of nanoscience has grown to encompass many different sectors and industries, including those dealing with electronics, optics, medicine, energy, defense, pharmaceuticals, food, and agriculture (Bachheti et al., 2023; Chawla et al., 2023; Husen & Iqbal, 2019; Husen & Jawaid, 2020; Husen, 2022, 2023a, 2023b, 2023c; Husen et al., 2023, 2023a, 2023b, 2023c; Husen & Siddiqi, 2023; Jin-Chul et al., 2021; Mandal et al., 2023; Sharma, 2017; Taghiyari et al., 2023). In one of the dimensions, materials of a nanometer scale, between about 1 and 100 nm, are often referred to as nanomaterials (Bratovcic et al., 2015; Godeto et al., 2023; Joshi et al., 2019). Because of the growing problem of antibiotic-resistant microorganisms, there is much interest in many materials with antimicrobial properties, such as inorganic materials and zinc oxide (ZnO). The use of zinc oxide confers a number of benefits, one of which is that it maintains its antibacterial activity over a broad pH range from 3 to 10, and this is a characteristic that is extremely useful for the packaging sector to a wide variety of food products (Saral et al., 2019). Zinc oxide nanoparticles (ZnO NPs) have found widespread use as an antibacterial and antimicrobial agent in modern food packaging. Introducing these particles into packaging reduces food microbial growth (Kazemi et al., 2020). ZnO NPs also help prevent food spoilage by reducing the prevalence of harmful microorganisms, and they can lengthen the life of perishable goods. It is believed that antibacterial action is caused by a mechanism involving hydrogen peroxide (H_2O_2) formation and that this mechanism is mediated by zinc oxide (Nikolic et al., 2021).

1.1 Zinc Oxide Nanoparticles

Since zinc is a vital micronutrient and plays crucial functions in developmental processes and wellness in humans and animals, food industries use ZnO as a dietary source of zinc element. ZnO is one of five zinc compounds stated to be widely accepted as safe (GRAS) by the US Food and Drug Administration (FDA) (Robertson, 2012). ZnO NPs are expressed as a multipurpose inorganic substance with various uses. Nanoscale ZnO particles have excellent optical transparency, high electrical conductivity, and piezoelectricity (Mihindukulasuriya & Lim, 2014). Research on the antibacterial properties of ZnO was initiated in the early 1950s. In 1995, the powders containing magnesium oxide (MgO), calcium oxide (CaO), and zinc oxide (ZnO) possessed antibacterial properties against some types of bacteria (Sawai et al., 1998). This discovery marked the real beginning of using ZnO as an antimicrobial agent.

2 Methods of Synthesis

ZnO NPs have drawn much interest because of their distinct characteristics and wide range of uses (Husen & Iqbal, 2019; Siddiqi et al., 2018). These nanoparticles are created by carefully forming tiny ZnO particles with the necessary properties.

2.1 Sol–Gel Method

One approach that is frequently used to create ZnO NPs is the sol-gel method. It entails the hydrolysis and condensation of metal precursors to transform a sol (colloidal solution) into a solid gel. For instance, by hydrolyzing zinc acetate in a solvent and then aging and calcining the solution, Li et al. (2009a, 2009b) created ZnO NPs using the sonochemical sol-gel process.

2.2 Precipitation Method

Zinc ions are precipitated using the precipitation method by adding a precipitating agent to a solution. The precipitant is typically sodium hydroxide or ammonium hydroxide. To create zinc oxide nanoparticles, the precipitate is gathered, cleaned, and calcined. By hydrolyzing zinc acetate in an aqueous solution, Xu et al. (2006) created ZnO NPs using a precipitation technique.

2.3 Hydrothermal Method

When producing zinc oxide nanoparticles hydrothermally, a hydrothermal solution and zinc salts are combined in an environment of high pressure and temperature. This approach allows particle size and morphology to be controlled by altering reaction variables such as reaction time, pressure, and temperature. In order to create nanostructured zinc oxide particles utilizing a thermal decomposition strategy, Sankar et al. (2015) used a hydrothermal process.

2.4 Microemulsion Method

Using a microemulsion system as a template, the microemulsion approach creates zinc oxide nanoparticles. This technique combines zinc precursors with water, oil, and surfactants. The reaction is aided by the condensed nanoscale droplets in the microemulsion, which lead to the production of nanoparticles. Using a microemulsion method, Zhang and Zhang (2001) effectively created almost monodisperse zinc oxide nanoparticles.

2.5 Vapor Deposition Methods

Zinc oxide nanoparticles are created using vapor deposition techniques, including chemical vapor deposition (CVD) and physical vapor deposition (PVD). In CVD, zinc oxide is deposited on a substrate as a result of a precursor gas's breakdown in a high-temperature reactor. PVD involves the physical vaporization of a zinc oxide target and subsequent deposition on a substrate. Hu and Chen (2008) examined the use of CVD to create nanoparticles made of zinc oxide (Fortunato et al., 2018).

2.6 Green Methods of Synthesis of Zinc Oxide Nanoparticles

Recent years have seen a significant increase in interest in "green synthesis," which refers to the production of ZnO NPs utilizing sustainable and ecologically friendly processes. Green synthesis techniques strive to reduce the energy use, waste production, and use of dangerous chemicals that come with traditional synthesis techniques.

Plant extracts are used as reducing and capping agents in plant-mediated synthesis to create ZnO NPs. *Aloe vera* extract was used in the manufacture of ZnO NPs that (Mofid et al., 2020) claimed strong antibacterial activity. As reducing agents and stabilizers for the manufacture of zinc oxide nanoparticles, plant extracts, mainly aqueous extracts taken from various plant parts (leaves, stems, or seeds), are utilized in this process (Sathishkumar et al., 2009).

In biogenic synthesis using the plant biomass technique, plant biomass, like leaves, stems, or roots, is a starting material to create zinc oxide nanoparticles. For the creation of nanoparticles, the components obtained from plants serve as both a precursor and a reducing agent (Gopi et al., 2021). In phytochemical-assisted synthesis, certain phytochemicals from plant extracts are isolated and used as stabilizers and reducing agents to create zinc oxide nanoparticles. The process to create zinc oxide nanoparticles is then started under controlled conditions by combining the separated phytochemicals with a zinc precursor solution (Bindhu & Umadevi, 2013). Microorganisms like bacteria or fungi are used in microbial-mediated synthesis to create ZnO NPs. For instance, ZnO NPs with substantial antibacterial effectiveness were synthesized sustainably using *Bacillus cereus* bacteria (Verma et al., 2019).

3 Structure and Morphology of ZnO

Due to their small size and high surface-to-volume ratio, ZnO NPs display various intriguing characteristics. Depending on the synthesis techniques used, ZnO NPs can have a variety of shapes and morphologies (Look, 2001).

3.1 Crystal Structure

The most stable phase of ZnO at room temperature, hexagonal wurtzite, is frequently adopted by ZnO NPs. Layers of zinc and oxygen atoms alternate in the wurtzite structure (Zhang & Wong, 2017).

3.2 Particle Size and Shape

The size of ZnO NPs can range from a few to several hundred nanometers. The chosen synthesis methodology, such as sol-gel, precipitation, hydrothermal, or vapor phase procedures, determines the size and form of ZnO NPs. Spherical, rod-like, and hexagonal forms are typical (Kumar & Venkateswarlu, 2016).

3.3 Surface Modifications

The size of ZnO NPs can range from a few to several hundred nanometers. The chosen synthesis methodology, such as sol-gel, precipitation, hydrothermal, or vapor phase procedures, determines the size and form of ZnO NPs. Spherical, rod-like, and hexagonal forms are typical (Azizian-Kalandaragh et al., 2018; Verma & Kumar, 2018)

4 Factors Affecting the Microbial Activity of ZnO Nanoparticles

Because of their antibacterial qualities, ZnO NPs are attractive candidates for use in various fields, including medicine, environmental cleanup, and food preservation. However, a number of factors can affect the microbial activity of ZnO NPs. To maximize these elements' antimicrobial effectiveness while assuring safety and minimizing any potential adverse effects, it is essential to understand them.

4.1 Surface Area and Particle Size

Their size dramatically influences the antibacterial activity of ZnO NPs. Since smaller nanoparticles have a higher surface area-to-volume ratio, they often have better antibacterial effectiveness. The improved antibacterial and antifungal activities result from the increased surface area's enhancement of the interaction between the nanoparticles and microbial cells (Brayner et al., 2006). For instance, research has shown that smaller ZnO NPs can successfully stop *Escherichia coli* germs from growing (Brayner et al., 2006). ZnO NPs' antibacterial capabilities can be optimized by tailoring their particle size utilizing various production techniques.

4.2 Surface Charge and Chemistry

ZnO NPs' surface charge and chemicals make a big difference in how they interact with microbial cells. Surface alterations and functionalizations can change the surface charge, which then influences how well nanoparticles adhere to surfaces and enter microbial cells. ZnO NPs' antibacterial activity may change depending on surface charge. By increasing their attraction to negatively charged bacterial surfaces, positively charged nanoparticles, for example, have demonstrated enhanced bactericidal effects (Borkow & Gabbay, 2005). Additionally, by changing their mode of action and interactions with microbial targets, ZnO NPs' chemical makeup and the surface functional groups on them might affect their antibacterial activity (Zhang & Burn, 2006).

4.3 Concentration and Dose

Their concentration or dose significantly influences ZnO NP's antibacterial effectiveness. Due to increased interaction with microbial cells, higher nanoparticle concentrations typically provide better antibacterial effects. The concentration must be carefully optimized to prevent cytotoxicity to both host cells and microorganisms. Unintended harmful effects could result from too high doses (Ma et al., 2013). Therefore, it is important to compromise between maximizing antimicrobial effectiveness and lowering possible cytotoxicity.

4.4 Surface Coating and Functionalization

ZnO NP's interaction with microbes can be greatly influenced by the surface coating or functionalization applied to them. The physicochemical properties of the nanoparticles can be changed, and their antibacterial activity can be improved by coating them with organic or inorganic compounds. As an illustration, it has been demonstrated that surface functionalizing ZnO NPs with amino groups improves their antibacterial activities by enhancing adherence to bacterial cell surfaces (Raghupathi et al., 2011).

4.5 Environmental Factors

The microbial activity of ZnO NPs can be considerably influenced by environmental factors such as pH, temperature, and the presence of organic matter. The stability and reactivity of nanoparticles can be modified by alteration in pH and temperature, which can change how effective they are at fighting microbes. Additionally, organic matter may affect how nanoparticles and bacteria interact, thereby affecting the antimicrobial effectiveness of such interactions (Gunawan et al., 2011). To forecast and optimize antimicrobial activity in various application contexts, it is essential to comprehend these environmental parameters and how they affect nanoparticle performance.

5 Antimicrobial Activity and Mechanism

Due to their potential use in various industries, including medicine, food safety, and environmental cleanliness, ZnO NPs have drawn much attention. Broad-spectrum antibacterial activity of ZnO NPs has been shown against various species, including bacteria, viruses, fungi, and protozoa. ZnO NPs have been shown in numerous studies to have potent antibacterial effects on both Gram-positive and Gram-negative bacteria. ZnO NPs have strong antibacterial action against *Escherichia coli* and *Staphylococcus aureus* (Sharma et al., 2017a, 2017b).

The generation of reactive oxygen species (ROS), destruction of cell membranes, and interference with microbial enzymes and DNA are only a few of the activities that go into the antibacterial mechanisms of ZnO NPs.

5.1 Generation of Reactive Oxygen Species (ROS)

When in contact with moisture or biological fluids, ZnO NPs can produce ROS such as superoxide radicals ($O_2^{\bullet}$) and hydrogen peroxide (H_2O_2), among others. ROS cause oxidative stress in microbial cells, which harms DNA, proteins, and lipids. ROS buildup ultimately impairs cellular processes and may even cause cell death. ZnO NPs produced ROS, which was crucial to their antibacterial action against *Escherichia coli* (Li et al., 2013).

5.2 Disruption of Cell Membranes

ZnO NPs can interact with the membranes of microorganisms and harm their structural and functional integrity. The nanoparticles can cause membrane permeabilization, which allows ions and intracellular substances to flow out. As a result of this disruption, cell viability, food absorption, and energy production are compromised as a result of this disruption, ultimately leading to cell death. *Staphylococcus aureus* cell membrane was harmed by ZnO NPs, which killed the germs (Gurunathan et al., 2014).

5.3 *Interference with Microbial Enzymes and DNA*

ZnO NPs can inhibit or degrade DNA and microbial enzymes through interactions. The nanoparticles can obstruct crucial enzyme functions and DNA replication and repair processes, ultimately resulting in cell death. ZnO NPs damaged *Escherichia coli's* DNA and prevented the action of bacterial enzymes (Li et al., 2011) (Fig. 1).

6 Antimicrobial Food Packaging

The most apparent problems in the food sector in recent years have been food deterioration and safety. Worldwide, foodborne illness outbreaks caused by bacteria have occurred often. The health and safety of customers are intimately linked to the safety of the food supply chain. However, because of the reported outbreaks, consumers' worries have escalated, and public trust in the safety of the food supply has eroded (Natrajan et al., 2000). Food spoiling fosters the spread of harmful germs and causes the food to lose its color, texture, and nutritional value, rendering it inedible. Exposure to the environment during processes including slaughtering, packing, and processing may introduce harmful bacteria into food (Malhotra et al., 2015). Common food preservation methods include heating, freezing, and salting; however, these processes do not prevent recontamination, which might put consumers in danger. A recent innovation in food packaging, antimicrobial-active packaging systems (Sung et al., 2013), embeds antimicrobial chemicals into a polymer film to inhibit the growth of specific pathogens. By including compounds with antimicrobial activity in packing films, it is possible to prevent the development of contaminated microorganisms and

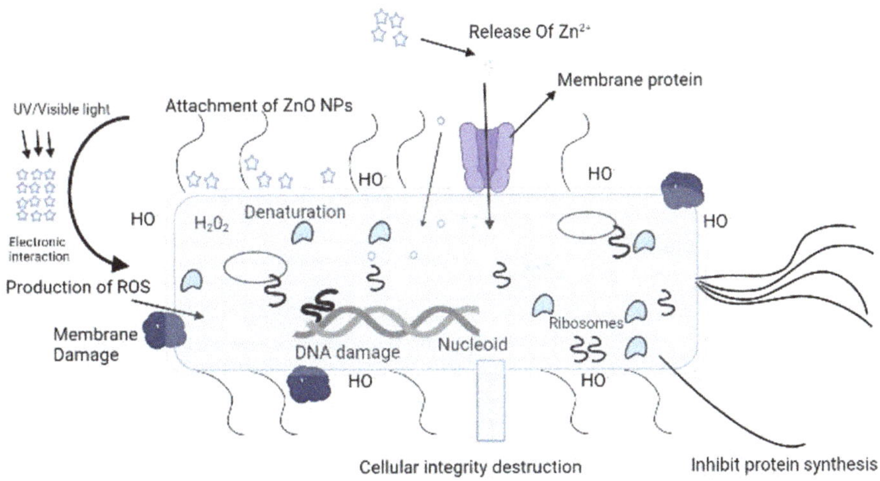

Fig. 1 Antimicrobial mechanism of ZnO NPs

significantly slow their growth rate (Han, 2000) by prolonging the lag phase of the bacteria and decreasing their live counts. The FDA strictly regulates the process of formulating food with these antimicrobial agents to establish acceptable quantities of added antibacterial compounds. Another option is to include these compounds in packaging sheets, which would immediately suppress target contaminated bacteria. However, once these substances are no longer present, some survived bacteria may resume their growth. This depletion might be the result of the natural breakdown of these chemicals over time, which would reduce their storage life (Kester et al., 1986). Polyvinyl chloride's (PVC) excellent qualities, including durability at high temperatures, cheap cost, and resistance to acids, have piqued the interest of scientists (Chi-Zhang et al., 2004). This polymer's hard thermoformed foil characteristic makes it an excellent choice for packaging various foods. Adding nanoparticles to PVC may provide the polymer with improved antibacterial characteristics. The increasing problem of bacteria resistant to medicines has stoked interest in several materials with antibacterial characteristics, including inorganic minerals like ZnO. Zinc oxide's addition to food packaging has various benefits, including its ability to maintain its antibacterial action throughout a broad pH range (from 3 to 10), which is crucial for many foods. It is believed that zinc oxide's antibacterial action may be attributed to its role in the formation of H_2O_2 (Li et al., 2011). This study aims to develop an improved method of inhibiting the development of *E. coli* bacteria by treating polyvinyl chloride (PVC) sheets with increasing concentrations of ZnO NPs as an antibacterial addition. Antimicrobial zinc oxide-coated PVC sheets might be used for food packaging.

Pathogens and spoilage bacteria may be prevented from contaminating food using antimicrobial packaging technology (Han et al., 2005). The purpose of antimicrobial food packaging is to prevent the development of microorganisms to preserve the food's quality and extend its shelf life. To accomplish the antimicrobial function, antimicrobial agents may be introduced into polymers for food-contact goods or added to the packaging system for indirect food contact (Appendini & Hotchkiss, 2002).

6.1 Methods to Prepare Antimicrobial Food Packaging

6.1.1 Solvent Casting

Casting in a solvent allows a polymer composite to be produced with greater mixing efficiency and at a lower temperature than when using the melt-casting process (Siemann et al., 2005). Metal particles have a greater propensity to disperse when mixed correctly. The solvent loss and solvent recovery constraints of the solution casting technique result from the off-gassing process. If the particle dispersion in this system is optimized, we can attain well-dispersion. This technique benefits from optical films, medical film sheets (electronic applications), and engineering plastics.

6.1.2 Coating Method

Polymer coating, applied by printing or painting methods, may provide materials with enhanced performance. To make it, just cover an organic substrate with inorganic chemicals. Coatings are used to increase surface qualities, including conductivity, magnetism, adsorption, and optical perfection, to stabilize the substrate materials and minimize unwanted interactions with external influences. Coatings arise by either the adsorption of monomers on the surfaces, leading to a direct polymerization, or the interaction of preformed polymers with inorganic cores. In all approaches, the chemical compatibility of the coated materials is essential. High-quality materials also exploit the strong bond between the coating and the substrate. Antimicrobial agents on polymers are only one example of how often the coating process is employed in packaging materials.

Usually, packing materials with an antimicrobial coating may inhibit microbial development via direct contact with the product. Antimicrobial packaging coatings are often made from low-density polyethylene (LDPE), chitosan, or methylcellulose. Liquid food in packaging may also benefit from coating techniques. This led to several applications of ZnO NPs to create a nanocomposite antimicrobial covering for packaging (Saekow et al., 2019).

6.1.3 Extrusion Method

The first lead pipes were made using the melt extrusion process around the end of the eighteenth century. As a result, this technique has seen extensive usage in fabricating materials, including bags, sheets, and pipes in the food processing, plastics, and rubber industries. By forcing a mixture of raw materials through a heated barrel at high pressure and temperature, extrusion may produce products with consistent thickness and shape. Using a heat-melting process, powders, and granules may be transformed into a more consistent form. Due to its solvent-less, time-efficient, and non-ambient procedure, extrusion has widespread usage in pharmaceutical, food, and plastic manufacturing. Much research has used the extrusion technique to create antimicrobial packaging using ZnO NPs (Polat et al., 2018).

6.1.4 Injection Molding

Plastics and metals, both thermoplastic and thermosetting, may be molded using the injection technique, which involves heating the material to a fluid condition before injecting it into molds. Injection techniques require three distinct phases of production:

(i) plasticizing or injecting units are heated to flow under pressure,
(ii) the melted material is allowed to cool in the mold, and
(iii) the finished goods are ejected.

To solidify and reshape components under pressure, molten metals or polymers may be pumped into molds having interior voids. Accordingly, molds may have a single cavity, many cavities, or even radically different cavities, all of which can be linked to flow channels or runners transporting molten materials to their respective interiors. This technique has many potential applications, from the automotive industry to manufacturing plastic toys, mobile phone covers, containers, and water bottles (Altan & Yildirim, 2014).

7 Application of ZnO NPs in Food Packaging

Advanced forms of packaging materials are becoming more significant as the demands placed on packaging technology rise. This new development has brought up concerns about pollution and garbage disposal. This is why eco-friendly, biodegradable packaging manufactured from renewable resources is a relevant topic in the scientific community. Edible packaging seems to be a viable replacement for plastic materials and has significant application potential in various settings (Zhang et al., 2021). The convergence of expertise in biodegradable materials and food packaging is crucial to the progress of edible packaging. Several studies have shown that edible packaging has a significant impact on preserving quality and extending shelf life, especially for perishable items like fresh or sliced food (Darmajana et al., 2017). Research also suggests that edible packaging might serve as a barrier on the surface of food, preventing moisture loss and altering the interior environment to delay spoilage (Maringgal et al., 2020). One other possible advantage of edible packaging is that it may be used to transport active compounds that boost the food's natural defenses, such as antioxidants or bacteria-fighting chemicals (La et al., 2021). The use of antibacterial agents in food packaging for foodborne pathogens may be a key and in-demand field in the food industry (Sirelkhatim et al., 2015).

Consequently, a novel kind of secure, dynamic, nutritious, and smart food packaging is emerging due to developments in nanofood packaging technology. ZnO NPs provide a safer and more cost-effective solution for pathogen-free food packaging and processed components than other metal oxides (Sahani et al., 2021). ZnO NPs are used in processed food because they are a rich source of metallic elements (Zn) essential to human and animal development, advancement, and well-being. The antibacterial, antifungal, and antiviral properties of ZnO NPs make them a valuable addition to food packaging against foodborne infections. Meat is a great way to get your daily protein, fat, vitamins, and minerals. However, it may also stimulate the development of spoilage and pathogenic bacteria due to its high protein content and low carbohydrate content. Fresh meat is home to various microorganisms, but some may become more prevalent depending on pH, composition, texture, storage, temperature, and even transportation (Omerovic et al., 2021). Inadequate handling and containment of these pathogens may lead to foodborne diseases. Raw meat is a potential breeding ground for harmful bacteria, including Salmonella spp., *E. coli*, and *S. aureus*. Inactivating germs via different possible methods, zinc oxide nanoparticles (ZnO-NPs) are considered a safe and stable antibacterial. Our goal was to use ZnO NPs in food packaging to reduce the prevalence of Campylobacter in uncooked chicken. When it comes to high-risk foods like raw meat and fresh fruit, active antimicrobial packaging is an efficient means of management. Controlling freshness and preventing spoiling organisms are two of its primary uses (Panea et al., 2014) in quality preservation, shelf-life extension, and food safety.

Despite extensive in vitro research on the antibacterial activity of ZnO nanocomposites, only a few studies have shown their efficacy when used in direct contact with food. Antimicrobial Low Density Polyethylene (LDPE) films containing silver (Ag) and zinc oxide nanoparticles have been evaluated in an orange juice food matrix (Emamifar et al., 2010). In contrast to making the active packaging contains silver nanoparticles, which had adverse effects on the sensory ability of both Gram-positive and Gram-negative bacteria to zinc oxide nanoparticles, the advantages of active packaging containing ZnO NPs prolonged the shelf life of fresh orange juice up to 28 days without causing any negative effects on sensory parameters. Antibacterial activity was also shown against S. aureus in chitosan and polyvinyl alcohol (PVA) films that had zinc oxide nanoparticles (25–30 nm) inserted (Vicentini et al., 2010). Antifungal activity against Candida albicans was shown using Zinc oxide nanorods (30 nm in diameter and 500 nm in length) placed on a glass surface (Eskandari et al., 2011). Researchers also found that the nanocomposite material they created remained stable after being stored for 2 months without losing its antifungal properties. The antifungal action of ZnO was proposed to be due to the generation of reactive oxygen species (ROS), namely hydrogen peroxide, on its surface (Eskandari et al., 2011). Recent research, however, suggests that ZnO's mode of action in fungi is distinct from that described for bacteria (He et al., 2011). However, although several researchers have examined ZnO nanocomposites' antibacterial activity in vitro, only a few have shown the materials' efficacy when used in direct contact with food. Antimicrobial LDPE films containing silver and zinc oxide nanoparticles have been evaluated in an orange juice food matrix (Emamifar et al., 2010). Researchers found that fresh orange juice could be stored for up to 28 days when active packaging containing ZnO NPs was used, with no negative effects on sensory parameters. However, active packaging developed with silver nanoparticles negatively affected the sensory quality of fresh juice.

Cheese is one of the most sought-after components in the food business, particularly in the fast-food industry and among producers of convenient, on-the-go meals. When shipping or storing cheese, it is crucial to use packing materials explicitly made for that purpose. As a result of its many applications, the packaging industry is expected to expand rapidly in the coming years. By 2029's end, analysts predict that the worldwide cheese packaging market will have grown at a Compound Annual Growth Rate (CAGR) of 5.6% (Market foresight: Cheese, 2022–2029).

8 Toxicity of ZnO NPs

8.1 Toxicity to Cancer Cells

The most significant and likely mechanism causing NP toxicity, according to in vitro data, is the production of oxidative stress. A decline in the activity of internal enzymes and transcription variables relying on Zn^{2+} results from ZnO NPs dissolving in the environment outside the cell, raising the Zn^{2+} content in the extracellular segment. This pathway has been proven with the help of A549 and BEAS-2B cells (cell line). In contrast, ZnO NPs penetrate the cell after dissolving in the lysosomal compartment. The modification of Ca^{2+} intracellular flow, production of reactive oxygen species, membrane damage, and mitochondrial dysfunction are some of the harmful effects of ZnO NPs (Vandebriel & Jong, 2012).

Human lung cancer H1355 cells and human leukemia Jurkat cells were used in in vitro experiments to show that ZnO NPs treatment increased the content of free Zn^{2+} in the cytosol and mitochondria. ZnO NPs activated caspase-3 and caused the release of LDH in H1355 cells and depression of the mitochondrial membrane potential (Kao et al., 2012).

Compared to ZnO, which is 0.44 micrometers in size, ZnO NPs (8–10 nm) show a more significant harmful effect on human colon cancer cells (RKO). In cell culture medium, both particle types have been reported to aggregate into micrometer-sized particles and cause toxicity by activating apoptotic pathways (Moos et al., 2010).

ZnO NPs (11.5 g/ml) injected into LoVo cells (human colon cancer) for 24 h caused a significant decline in the survival of cells, growth of H_2O_2/OH, reduction in $O_2(-)$, depolarization of the internal mitochondrial membrane, apoptosis, and release of IL-8. Over 24 h of therapy, elevated dosages resulted in an elevated level of cytotoxicity. According to experimental evidence, cellular oxidative stress might be a major mechanism in colon cancer cells' production of cytotoxicity by ZnO NPs. Additionally, research using analytical electron microscopy to examine the relationship between ZnO NPs' toxicological effects and physical-chemical makeup reveals that their cytotoxicity is not primarily caused by their surface (Berardis, 2010).

8.2 Toxicity of ZnO NPs on Dermal Cells

Skincare items like sunscreen contain ZnO NPs. In vitro, researchers have looked into how skin cells react to exposure because of this. In human dermal fibroblasts, Meyer et al. have shown that ZnO NPs induce apoptosis through the cell cycle regulator protein p53 and mitogen-triggered protein kinase p38 pathways (Meyer et al., 2011).

8.3 Toxicity to Human Immune System Cells

Human immune cell categories responded to ZnO NPs differently in terms of cytotoxicity, with lymphocytes constituting the most resilient and monocytes being the most vulnerable. Additionally, there were notable variations between memory cells that had previously been activated and novice lymphocytes, suggesting a connection between NPs vulnerability and cell-cycle capability. The production of ROS, which is harmful in high concentrations in monocytes, and connections to cellular membranes are two of the causes of toxicity. Both NPs size and generation of ROS showed a negative correlation with each other, as well as with cytotoxicity. IFN-γ, TNF-α, and IL-12 were produced in response to ZnO NPs, albeit at levels below those that would normally result in significant cell death (Hanley et al., 2009; Huang et al., 2010).

8.4 Toxicity of ZnO NPs on Experimental Animal Models

Invertebrates like *D. melanogaster* and vertebrates (rodents and zebrafish) have shown the detrimental impacts of in vivo exposure. There is evidence that ZnO NPs may be genetically toxic in other research. As oxidative stress has been proposed to be the primary mechanism creating the toxicity of ZnO NPs, it still needs to be apparent what function the produced ROS play in the toxicity. Studies have demonstrated that rodents' subchronic cytotoxicity of ZnO NPs is relatively low. Furthermore, a prior investigation employing *D. melanogaster* found no harm following ZnO nanoparticle consumption. In a different investigation, ZnO NPs were found to be poorly genotoxic and to cause oxidative stress in *D. melanogaster* (Ng et al., 2017).

8.5 Important Factors and Toxicity Mechanisms

The size, surface properties, dissolution, and exposure pathways of ZnO NPs are the main contributing factors to their toxicity. The relationship between the size of NPs and many qualities, such as surface characteristics, solubility, chemical reactivity, and interactions between nanomaterials and biomolecules, influence the nanotoxicological actions of NPs in vivo (Zhao et al., 2007)

The surface properties of NPs, such as surface charge and rugosity, are important additional elements that affect ZnO NPs' toxicity and size (Hu et al., 2009). The accidental release of ions of metal is a significant contributor to toxicity. Investigations indicate that plant cell walls in the roots and leaves received solute metallic ions emitted from NPs but rejected big NPs (Proseus & Boyer, 2005). According to Brunner et al., soluble metal/metal oxide nanoparticles had a more harmful six-day effect than insoluble ones when used at the same dose. The temperature and pH of the solution have an impact on ZnO solubility (Brunner et al., 2006; Chang et al., 2012).

9 Characterization Techniques of ZnO NP Packaging Material

X-ray diffraction (XRD), scanning electron microscopy (SEM), transmission electron microscopy (TEM), atomic force microscopy (AFM), Fourier transform infrared spectroscopy (FTIR), mechanical and thermal resistance, as well as permeability assessments against water vapor and oxygen are some of the techniques used to characterize antimicrobial packaging materials containing ZnO NPs.

9.1 X-Ray Diffraction (XRD)

The nanocomposite's crystallographic structure, chemical makeup, and physical characteristics are all revealed by XRD. This method is widely used in material characterization because it is non-invasive and does not necessitate sophisticated sample preparation. XRD is an extensively utilized technique to assess the distribution of nanosized fillers inside a polymeric matrix (Koo, 2006). ZnO nanorods deposited on the glass surface were examined using XRD to determine their crystal structure. It only revealed one diffraction peak, assigned the index (002), and associated with the wurtzite structure (Eskandari et al., 2011). ZnO nanocrystals were sonochemically produced on a glass slide as a result, and a very strong, sharp diffraction peak was found at about 34° that was attributed to the (002) plane of wurtzite (Applerot et al., 2009).

9.2 Scanning Electron Microscope (SEM), Transmission Electron Microscopy (TEM), and Atomic Force Microscopy (AFM)

Although XRD can disclose some structural characteristics of ZnO, image acquisition of individual nanoparticles can only be accomplished via microscopy methods such as SEM, TEM, and AFM microscopies. SEM showed ZnO NPs to be spherical in shape, and multilayers were developed in a pyramidal structure. The ZnO NPs' spherical form and the degree of roughness levels obtained by AFM also supported the development of a pyramid-like structure over multiple layers (Applerot et al., 2009). Similarly, SEM and TEM were employed to assess the shape and dispersion quality of BOPP films with ZnO addition. Spherical nanoparticles and one-dimensional nanorods were the two ZnO morphologies that were seen in this investigation. Additionally, the authors discovered that spherical nanoparticles of ZnO had a stronger tendency to aggregate than nanorods composed of ZnO of identical diameter due to their larger surface area (Lepot et al., 2011).

9.3 Ultraviolet–Visible Spectroscopy

Through the use of ultraviolet-visible spectroscopy, the optical characteristics of ZnO-containing composite glass were investigated, and it was discovered that the glass, in its pure form, had the highest transmittance. Due to ZnO's capacity to absorb ultraviolet light, greater ZnO levels resulted in lower transmittance in both the UV and visible wavelength ranges (Applerot et al., 2009).

9.4 Fourier Transform Infrared Spectroscopy (FTIR)

The photographic stability of ZnO-containing PE nanotechnology and the interaction of bacteria with chitosan-capped ZnO NPs have been investigated using FTIR. Since the FTIR method relies on vibrational spectroscopy, it is possible to recognize and quantify the functional groups detected in the sample (Bhadra et al., 2011; Yang et al., 2010).

9.5 Mechanical Properties

Their mechanical qualities greatly influence polymeric nanocomposites' ability to be employed as food packaging. Uniaxial tensile tests are frequently used to assess mechanical properties such as Young's modulus, strength of tensile, and elongation at break. Studies demonstrated that adding ZnO NPs (27 nm) strengthened the mechanical properties of PU-based coatings and films. Particularly, the tensile strength and Young's modulus were raised to an ideal level of 2.0% (wt). Elongation at break was, as predicted, negatively associated with Young's modulus and strength of tensile, showing that ZnO NPs may boost the durability yet decrease the malleability of nanocomposite films (Li et al., 2009a, 2009b). Additionally, the mechanical characterization of BOPP nanocomposites, including nanofiller ZnO NPs (50–70 nm), revealed that nanofillers increased stiffness as Young's modulus steadily rose (Lepot et al., 2011).

9.6 Barrier Properties

Barrier qualities, especially those relating to moisture and oxygen, are another criterion for polymeric materials designed for food packaging applications. In this context, investigations found that the presence of ZnO NPs improved the oxygen barrier qualities of Biaxially Oriented Polypropylene (BOPP) films. Studies linked the conclusion to a reduction in the area accessible for oxygen transport caused by impermeable nanoparticles replacing the polymer's permeable gaps (Lepot et al., 2011). In experiments, ZnO nanoparticle coating has been shown to reduce the thin water vapor transfer of thin PE films (Xing et al., 2012).

9.7 Thermal Properties

The thermal characteristics of ZnO-filled nanocomposites have been researched using thermoanalytical methods, including differential scanning calorimetry (DSC) and thermogravimetric analysis (TGA). TGA, which tracks weight changes as an indicator of temperature and time, is frequently used in the field of polymer science because it makes it possible to identify the temperature range at which the sample acquires a chemical composition that is fixed, defined, and constant as well as the temperature at which the sample begins to decompose (Matos & Machado, 2003). TGA experiments revealed that chitosan/PVA films with ZnO added were more thermally stable than control films (as much as 150 °C) (Vicentini et al., 2010).

Through DSC tests, research demonstrates how ZnO NPs affect BOPP's melting point and crystallinity. Because the crystal lamellae in the nanocomposites are thinner than those in pure BOPP, they exhibit lower melting temperatures, which is understandable. Furthermore, it has been demonstrated that crystallinity decreases as ZnO content rises (Lepot et al., 2011).

10 Safety Issues

Migration: Packing nanocomposites with zinc nanoparticles can improve their mechanical, barrier, and antibacterial attributes. However, based on the nanomaterial attributes (concentration, size, shape, and dispersion), environmental attributes (mechanical stress, temperature), condition of food (pH and composition), characteristics of polymer (viscosity and structure), and extent of contact, nanoparticles may migrate through packaging and into/into food items (Bradley et al., 2011). Due to the sensitivity of migration to these and other variables, there is no universally accepted method or standard for researching it in wrapping nanocomposites.

Therefore, packaging materials should be evaluated on a case-by-case basis to decrease expulsion and its effects on the human body and food standards. Migration is defined as the unintended and unintentional conversion of inorganic or organic compounds into consumable ones via contact. Primarily, there are three types of dietary migration.

1. maximum permitted quantity (QM) is an evaluation of the standard of the remaining elements in food contact materials;
2. specific migration limit (SML) is the amount of distinct matter in food; and
3. the overall migration limit (OML) is the maximum quantity of all non-volatile matter that may enter into food.

Substances' inertness is frequently measured using the OML, which is 60 mg/kg of material food. Regulation (EU) 2016/1416 of the European Commission and Regulation (EU) No. 10/2011 of the European Parliament and of the Council of the European Union establishing maximum zinc levels in food contact materials of 25 mg/kg food and 5 mg/kg food, respectively, in accordance with the SML).

11 Future Trends and Challenges

As previously indicated, nanomaterials can improve packaging performance regarding the expiry of fresh food goods owing to the large surface-to-volume ratio. Due to the exceptional qualities of ZnO NPs, efforts have been made to develop ZnO NP-based antimicrobial packaging. In the coming years, antimicrobial packaging nanocomposites will increasingly be used in the food business, illustrating the importance of such systems.

The next packaging phase can be based on polymeric nanocomposite containing active agents, and due to their potent antibacterial properties, ZnO NPs may play a significant part in this. In light of this, adding a sufficient number of ZnO NPs to the packaging can offer adequate antibacterial action. ZnO NPs can also enhance the mechanical, barrier, and thermal qualities of packaging. The packaging sector may pay more attention to ZnO NPs because they are less expensive than specific nanoparticles, such as titanium oxide and silver (Polat et al., 2018). Therefore, the use of packaging nanocomposite incorporating ZnO nanofillers has high potential as an application for ephemeral or short-shelf-life products, including fresh juice, vegetable or fruit products, and meat-based products (Saekow et al., 2019; Suo et al., 2017). Active packaging bio-nanocomposites can extend the storage period of fresh food and decrease the amount of food packaging disposed of since package disposal, as the primary waste in landfills, is still a significant concern in waste management. Accordingly, synthetic biopolymers such as poly(lactic acid), polycaprolactone, polyglycolic acid, polyvinyl alcohol, polybutylene succinate, and polybutylene adipate terephthalate, as well as antimicrobial bio-packaging containing ZnO NPs, may be accurately examined. In addition, natural biopolymers (such as cellulose, gelatin, chitosan, and starch) are eco-friendly materials that may be routinely researched to develop good packaging with antimicrobial properties based on ZnO NPs (Shankar & Rhim, 2018).

ZnO NPs tend to improve the biocidal efficiency of the packaging materials; however, there are still worries about migration and toxicity. The knowledge that is currently known about the ZnO traces in packaging for food and their effects on human health still needs to be improved. The human body requires Zn^{+2} trace, and the ZnO NPs are characterized as a low-toxic and biocompatible substance (Zhang et al., 2013). It has been strongly advised to take into account the six key aspects of nanomaterial emigration from food contact materials, including migration rate, mechanism of migration, how to analyze the trace amount, how to predict migration simulation, suitable food simulant, and the danger of migration (Jokar et al., 2017).

Concerning the use of nanoparticles in food, there are no worldwide safety standards or regulations. It is imperative to consider novel regulations for antimicrobial nano-agents, notably ZnO NPs as a commonly utilized biocidal material since the physiochemical characteristics of macro and nano-scale materials differ. The following are the key factors that must be taken into account when creating complete regulations for nanomaterial applications:

i. It has been suggested that analytical techniques such as Single particle inductively coupled plasma—mass spectrometry (SP-ICP-MS) and Transmission Electron Microscope- Energy Dispersive X-ray (TEM–EDX) characterization be used to produce a reliable migration detection strategy. These techniques can identify nanomaterial traces in low amounts due to high-accurate detection.
ii. It is vital to utilize chemicals and techniques comparable to food content materials as possible because the substances used in food replicas are not chemically identical to food ingredients.
iii. The adverse effects of nanomaterials on the human body in food-contact products have yet to be given much thought. Human body entry points such as inhalation, penetration via the skin, and ingestion must be carefully considered. As a result, thorough regulation can help employees, manufacturers, and consumers limit risk (Kim et al., 2022).

The worldwide demand for packaging with antimicrobial properties is expected to grow from 6.51 million tons in 2015 reach 10.09 million tons in 2024 (11 billion USD) at a constant rate. A lot of interest has also been paid to nano-ZnO, and it is anticipated that by 2022, sales will rise from over 2,099 million USD in 2015 to 7,677 million USD. Antimicrobial packaging is becoming more and more popular, and ZnO NPs, an affordable and biocompatible biocidal, can play a significant role in this trend. In order to meet the demand for fresh food, antimicrobial packaging can extend the lifespan of fresh products. Food goods' shelf lives and quality may be preserved by using antimicrobial packaging throughout and packaging substances containing antimicrobial compounds in particular. Powerful antibacterial properties make ZnO NPs useful in various applications; however, their antimicrobial effectiveness is diminished once they are incorporated into the packaging matrix. It could be explained by the fact that polymer segments cover the bulk ZnO NPs, whereas ZnO NPs require immediate contact with food to increase shelf life. As a result, it might prevent ZnO NPs from being used in a wider variety of meals.

Furthermore, the diffusion of nanofiller in its matrix is a crucial component in creating an innovative packaging nanocomposite. Drop-in polymer performance is susceptible to ZnO NPs' imperfect dispersion, which results in subpar packaging material qualities. As a result, the industry still struggles to strike a compromise between the qualities needed for packaging, including ZnO NPs. As was already indicated, nanoparticle concentrations of 5% can produce an acceptable level of biocidal activity, but applications of ZnO NPs are only effective at concentrations of 5% or less due to poor interaction and clumping of ZnO NPs at greater concentrations (Honarvar et al., 2016). Prior to the feasibility and usability of packaging materials, consumer acceptance is the final consideration. In order to discover potential safety concerns, customers are likely to take ZnO NP migration and side effects into account. There are several reports on the compatibility and safety of ZnO NPs; however, caution must be made to limit the transition below the hazardous threshold and allay customer concerns about safety (Wahab et al., 2010). Nanoparticles are frequently used in packaging materials to attain sophisticated characteristics, including antibacterial activity. As a result, the expense of antimicrobial packaging technologies, particularly those based on Ag, TiO_2, and ZnO, remains a barrier. Modern methods are needed to produce metal-based nanoparticles and scale up packaging nanocomposites, which could increase final costs and lower market acceptability (Zare, 2016)

12 Conclusion

Zinc oxide nanoparticles have strong antibacterial qualities. Nanoparticles' physical and morphological attributes alter when made differently, changed with chemicals, or used with other nanomaterials. This, in consequence, changes their antibacterial abilities. It is feasible to utilize packaging that has zinc oxide nanoparticles in it. This will make it feasible to stop germs and food from worsening. To keep the adverse effects of ZnO NPs on packed food and the human body to a minimum, attention must be paid to how they move and how safe they are. ZnO NPs can be used to keep up with the growing trend of antimicrobial packaging and meet the market's high demand for fresh goods. Even though ZnO NPs tend to make the packing materials more biocidal, there remain problems with toxicity and migration. The knowledge available about ZnO traces in food packaging and how they affect the human body must be improved. Nanomaterials must be regulated as soon as possible before being used in packing.

References

Ahvenainen, R. (Ed.). (2003). *Novel food packaging techniques*. Woodhead Publishing Limited.

Altan, M., & Yildirim, H. (2014). Comparison of antibacterial properties of nano TiO_2 and ZnO particle filled polymers. *Acta Physsica. Polonica A, 125*, 645–647.

Appendini, P., & Hotchkiss, J. H. (2002). Review of antimicrobial food packaging. *Innovative Food Science Emerging Technology, 3*, 113–126.

Applerot, G., Perkas, N., Amirian, G., Girshevitz, O., & Gedanken, A. (2009). Coating of glass with ZnO via ultrasonic irradiation and a study of its antibacterial properties. *Applied Surface Science, 256*, S3–S8.

Azizian-Kalandaragh, Y., Ebrahimzadeh, M. A., & Soofivand, F. (2018). Recent advances in the synthesis and surface modification of ZnO nanoparticles for biomedical applications: A review. *Ceramics International, 44*, 1249–1267.

Bachheti, R. K., & Bachheti, A. (Eds.). (2023). *Secondary metabolites from medicinal plants: Nanoparticles synthesis and their applications*. CRC Press. https://doi.org/10.1201/9781003213727

De Berardis, B., Civitelli, G., Condello, M., Lista, P., Pozzi, R., Arancia, G., & Meschini, S. (2010). Exposure to ZnO nanoparticles induces oxidative stress and cytotoxicity in human colon carcinoma cells. *Toxicology and Applied Pharmacology, 246*, 116–127.

Bhadra, P., Mitra, M. K., Das, G. C., Dey, R., & Mukherjee, S. (2011). Interaction of chitosan-capped ZnO nanorods with *Escherichia coli*. *Materials Science and Engineering C, 31*, 929–937.

Bindhu, M. R., & Umadevi, M. (2013). Antibacterial studies of zinc oxide nanoparticles synthesized from *Sargassum longifolium*. *Arabian Journal of Chemistry, 6*, 113–119.

Borkow, G., & GabbayJ. (2005). Putting copper into action: Copper-impregnated products with potent biocidal activities. *The FASEB J, 19*, 1–11.

Bradley, E. L., Castle, L., & Chaudhry, Q. (2011). Applications of nanomaterials in food packaging with a consideration of opportunities for developing countries. *Trends in Food Science and Technology, 22*, 604–610.

Bratovcic, A, Odobasi, C. A., Cati, C. S., & Sestan, I. (2015). Application of polymer nanocomposite materials in food packaging. *Croatian Journal of Food Science and Technology, 7*, 86–94.

Brayner, R., Ferrari-Iliou, R., Brivois, N., Djediat, S., Benedetti, M. F., & Fievet, F. (2006). Toxicological impact studies based on Escherichia coli bacteria in ultrafine ZnO nanoparticles colloidal medium. *Nano Letters, 6*, 866–870.

Brunner, T. J., Wick, P., Manser, P., Spohn, P., Grass, R. N., Limbach, L. K., Bruinink, A., & Stark, W. J. (2006). In vitro cytotoxicity of oxide nanoparticles: Comparison to asbestos, silica, and the effect of particle solubility. *Environmental Science and Technology, 40*, 4374–4438.

Chang, Y. N., Zhang, M., Xia, L., Zhang, J., & Xing, G. (2012). The toxic effects and mechanisms of Cuo and ZnO nanoparticles. *Materials (basel), 5*, 2850–2871.

Chawla, S., Singh, S., & Husen, A. (2023). *Smart nanomaterials targeting pathological hypoxia*. Springer Nature Singapore Pte Ltd. ISBN: 978-981-99-1717-4. https://doi.org/10.1007/978-981-99-1718-1

Chi-Zhang, Y. D., Yam, K. L., & Chikindas, M. L. (2004). Effective control of Listeria monocytogenes by combination of nisin formulated and slowly released into a broth system. *International Journal of Food Microbiology, 90*, 15–22.

Cho, I. J., Lee, H. S., Lim, S. J., Koh, J. Y., Kwak, S. H., & Hwang, G. I. (2011). Detection and distribution of food-borne bacteria in ready-to-eat foods in Korea. *Food Science and Biotechnology, 20*(2), 525e529.

Darmajana, D. A., Afifah, N., Solihah, E., & Indriyanti, N. (2017). Effects of carrageenan edible coating on fresh cut melon quality in cold storage. *Agritech, 37*, 280–287.

Deng, X., Luan, Q., Chen, W., Wang, Y., Wu, M., Zhang, H., & Jiao, Z. (2009). Nanosized zinc oxide particles induce neural stem cell apoptosis. *Nanotechnol, 20*, 115101.

Emamifar, A., Kadivar, M., Shahedi, M., & Soleimanian-Zad, S. (2010). Evaluation of nanocomposite packaging containing Ag and ZnO on shelf life of fresh orange juice. *Innovative Food Science and Emerging Technologies, 11*, 742–748.

Eskandari, M., Haghighi, N., Ahmadi, V., Haghighi, F., & Mohammadi, S. R. (2011). Growth and investigation of antifungal properties of ZnO nanorod arrays on the glass. *Physica b: Condensed Matter, 406*, 112–114.

Fortunato, E., Barquinha, P., & Martins, R. (2018). *Zinc oxide materials for electronic and optoelectronic device applications*. Elsevier.

Godeto, Y. G., Ayele, A., Ahmed, I. N., Husen, A., & Bachheti, R. K. (2023). *Medicinal plant-based metabolites in nanoparticles synthesis and their cutting-edge applications: An overview. secondary metabolites from medicinal plants* (pp. 1–34). CRC Press.

Gopi, D., Kanimozhi, K., Kavitha, L., Venkatesh, K. S., & Karthikeyan, B. (2021). Green synthesis of zinc oxide nanoparticles using Tridax procumbens and their antibacterial activity. *Materials Today: Proceedings, 42*, 398–402.

Gunawan, C., Marquis, C. P., & Amal, R. (2011). Influence of the aggregation behavior on the antibacterial efficacy of ZnO nanoparticles. *Langmuir, 27*, 1–9.

Gurunathan, S., Han, J. W., Kwon, D. N., & Kim, J. H. (2014). Enhanced antibacterial and anti-biofilm activities of silver nanoparticles against Gram-negative and Gram-positive bacteria. *Nanoscale Research Letters, 9*(1), 1–11.

Han, J. H. (2000). Antimicrobial food packaging. *Food Technology, 54*, 56–65.

Han, J. H. (2005). Antimicrobial packaging systems. In Han, J. H (Ed.), *Innovations in food packaging* (pp. 80–107). New York: Elsevier Academic Press.

Hameed, A. S. H., Karthikeyan, C., Ahamed, A. P., Thajuddin, N., Alharbi, N. S., Alharbi, S. A., & Ravi, G. (2016). In vitro antibacterial activity of zno and nd doped zno nanoparticles against esbl-producing *Escherichia coli* and *Klebsiella pneumoniae. Scientific Reports, 6*, 24312.

Hanley, C., Thurber, A., Hanna, C., Punnoose, A., Zhang, J., & Wingett, D. G. (2009). The influences of cell type and Zno nanoparticle size on immune cell cytotoxicity and cytokine induction. *Nanoscale Research Letters, 4*(12), 1409–1420.

He, L., Liu, Y., Mustapha, A., & Lin, M. (2011). Antifungal activity of zinc oxide nanoparticles against Botrytis cinerea and Penicillium expansum. *Microbiological Research, 166*, 207–215.

Honarvar, Z., Hadian, Z., & Mashayekh, M. (2016). Nanocomposites in food packaging applications and their risk assessment for health. *Electronic Physician, 8*(6), 2531.

Hosseinnejad, M., & Jafari, S. M. (2016). Evaluation of different factors affecting antimicrobial properties of chitosan. *International Journal of Biological Macromolecules, 85*, 467–475.

Hu, Z., & Chen, Y. (2008). Titanium dioxide-mediated heterogeneous photocatalysis: Recent progress and issues. *Chemical Engineering Journal, 144*, 386–395.

Hu, X., Cook, S., Wang, P., & Hwang, H. M. (2009). In vitro evaluation of cytotoxicity of engineered metal oxide nanoparticles. *Science of the Total Environment, 407*, 3070–3072.

Huang, C. C., Aronstam, R. S., Chen, D. R., & Huang, Y. W. (2010). Oxidative stress, calcium homeostasis, and altered gene expression in human lung epithelial cells exposed to ZnO nanoparticles. *Toxicology in Vitro, 24*(1), 45–55.

Husen, A. (2022). *Engineered nanomaterials for sustainable agricultural production, soil improvement and stress management*. Elsevier Inc. https://doi.org/10.1016/C2021-0-00054-7

Husen, A. (2023a). *Secondary metabolites based green synthesis of nanomaterials and their applications*. Springer Nature Singapore Pte Ltd. https://doi.org/10.1007/978-981-99-0927-8

Husen, A. (2023b). *Nanomaterials from agricultural and horticultural products*. Springer Nature Singapore Pte Ltd. https://link.springer.com/book/9789819934348

Husen, A. (2023c) *Nanomaterials and nanocomposites exposures to plants (Response, Interaction, Phytotoxicity and Defense Mechanisms)*. Springer Nature Singapore Pte Ltd. https://doi.org/10.1007/978-981-99-2419-6

Husen, A., Bachheti, R. K., & Bachheti, A. (2023). *Current trends in green nano-emulsions (Food, Agriculture and Biomedical Sectors)*. Springer Nature Singapore Pte Ltd.

Husen, A., & Iqbal, M. (2019). *Nanomaterials and plant potential*. Springer International Publishing AG. https://doi.org/10.1007/978-3-030-05569-1

Husen, A., & Jawaid, M. (2020). *Nanomaterials for agriculture and forestry applications*. Elsevier Inc. https://doi.org/10.1016/C2018-0-02349-X

Husen, A., & Siddiqi, K. S. (2023). *Advances in smart nanomaterials and their applications*. Elsevier Inc. https://doi.org/10.1016/C2021-0-02202-1

Jin-Chul, K., Madhusudhan, A., & Husen, A. (2021). *Smart nanomaterials in biomedical applications*. Springer Nature Switzerland AG. https://doi.org/10.1007/978-3-030-84262-8

Jokar, M., Pedersen, G. A., & Loeschner, K. (2017). Six open questions about the migration of engineered nano-objects from polymer-based food-contact materials: A review. *Food Additives and Contaminants: Part A, 34*, 434–450.

Joshi, A., Sharma, A., Bachheti, R. K., Husen, A. & Mishra, V. K. (2019). Plant-Mediated Synthesis of Copper Oxide Nanoparticles and Their Biological Applications. In Nanomaterials and Plant Potential, Springer, Cham, pp. 221–237. https://doi.org/10.1007/978-3-030-05569-1_8

Kao, Y. Y., Chen, Y. C., Cheng, T. J., Chiung, Y. M., & Liu, P. S. (2012). Zinc oxide nanoparticles interfere with zinc ion homeostasis to cause cytotoxicity. *Toxicological Sciences, 125*(2), 462–472.

Kazemi, M. M., Hashemi-Moghaddam, H., Mohammadi Nafchi, A., & Ajodnifar, H. (2020). Application of modified packaging and nano ZnO for extending the shelf life of fresh pistachio. *Journal of Food Process Engineering, 43*(12), e13548.

Kester, J. J., & Fennema, O. R. (1986). Edible films and coatings: A Review. Food Technology. *Chicago*, *40*, 47– 59.

Kim, I., Viswanathan, K., Kasi, G., Thanakkasaranee, S., Sadeghi, K., & Seo, J. (2022). ZnO nanostructures in active antibacterial food packaging: Preparation methods, antimicrobial mechanisms, safety issues, future prospects, and challenges. *Food Review International, 38*, 537–565.

Koo, J. (2006). *Polymer nanocomposites: Processing, characterization, and application*. McGraw-Hill Professional.

Kumar, S. V., & Venkateswarlu, P. (2016). Synthesis and characterization of ZnO nanoparticles for antibacterial and photocatalytic applications. *Advances in Natural Sciences: Nanoscience and Nanotechnology, 7*(3), 035011.

La, D. D., Nguyen-Tri, P., Le, K. H., Nguyen, P. T. M., Nguyen, M. D. B., Vo, A. T. K., Nguyen, M. T. H., Chang, S. W., Tran, L. D., & Chung, W. J. (2021). Effects of antibacterial zno nanoparticles on the performance of a chitosan/gum arabic edible coating for post-harvest banana preservation. *Progress in Organic Coatings, 151*, 106057.

Lepot, N., Van Bael, M. K., Van den Rul, H., D'Haen, J., Peeters, R., Franco, D., & Mullens, J. (2011). Influence of incorporation of ZnO nanoparticles and biaxial orientation on mechanical and oxygen barrier properties of polypropylene films for food packaging applications. *Journal of Applied Polymer Science, 120*, 1616–1623.

Li, M., Zhu, L., Lin, D., & Zhu, X. (2013). Toxicity of ZnO nanoparticles to Escherichia coli: Mechanism and the influence of medium components. *Environmental Science and Technology, 47*, 1745–1752.

Li, J. H., Hong, R. Y., Li, M. Y., Li, H. Z., Zheng, Y., & Ding, J. (2009a). Effects of ZnO nanoparticles on the mechanical and antibacterial properties of polyurethane coatings. *Progress in Organic Coatings, 64*, 504–509.

Li, W., Xu, C., Zhao, X., Cong, D., Sun, Y., Li, Y. (2009b). Synthesis and characterization of zinc oxide nanoparticles via a novel sonochemical method. *Ultrasonsonochem, 16*, 469–474.

Li, W., Li, X., Zhang, P., & Xing, Y. (2011). Development of nano-ZnO coated food packaging film and its inhibitory effect on Escherichia coli in vitro and in actual tests. *Advanced Materials Research*, 152–153, 489–492.

Look, D. C. (2001). Recent advances in ZnO materials and devices. *Materials Science and Engineering B, 80*, 383–387.

Ma, Y., Zhang, P., Zhang, T., & Zhou, J. (2013). Enhanced photocatalytic performance of ZnO microspheres sensitized by CdSe quantum dots. *Journal of Nanoparticle Research, 15*, 1–9.

Malhotra, B., Keshwani, A., & Kharkwal, H. (2015). Antimicrobial food packaging: Potential and pitfalls. *Frontiers in Microbiology, 6*, 1–9.

Mandal, A. K., Ghorai, S., & Husen, A. (2023). *Functionalized smart nanomaterials for point-of-care testing*. Springer Nature Singapore Pte Ltd.

Maringgal, B., Hashim, N., Tawakkal, I. S. M. A., & Mohamed, M. T. M. (2020). Recent advances in edible coating and its effect on fresh/fresh-cut fruit quality. *Trends in Food Science and Technology, 96*, 253–267.

Matos, J. R., Machado, L. D. B. (2003). Analise termica-termogravimetria. In: S. V. Canevarolo Jr. (Ed.), *Tecnicas de caracterizacao de polimeros* (pp. 209–228). Artliber.

Meyer, K., Rajanahalli, P., Ahamed, M., Rowe, J. J., & Hong, Y. (2011). ZnO nanoparticles induce apoptosis in human dermal fibroblasts via p53 and p38 pathways. *Toxicology in Vitro, 25*, 1721–1726.

Mihindukulasuriya, S. D. F., & Lim, L. T. (2014). Nanotechnology development in food packaging: A review. *Trends in Food Science and Technology, 40*, 149–167.

Mofid, H., Sadjadi, M. S., Sadr, M. H., Banaei, A., & Farhadyar, N. (2020). Green synthesis of zinc oxide nanoparticles using *Aloe vera* plant for investigation of antibacterial properties. *Advances in Nanochemistry, 1*, 32–35.

Moos, P. J., Chung, K., Woessner, D., Honeggar, M., Cutler, N. S., Veranth, J. M. (2010). ZnO particulate matter requires cell contact for toxicity in human colon cancer cells. *Chemical Research in Toxicology, 23*, 733–739.

Natrajan, N., & Sheldon, B. W. (2000). Efficacy of nisin-coated polymer films to inactivate salmonella typhimurium on fresh broiler skin. *Journal of Food Protection, 63*, 1189–1196.

Ng, C. T., Yong, L. Q., Hande, M. P., Ong, C. N., Yu, L. E., Bay, B. H., & Baeg, G. H. (2017). Zinc oxide nanoparticles exhibit cytotoxicity and genotoxicity through oxidative stress responses in human lung fibroblasts and Drosophila melanogaster. *International Journal of Nanomedicine, 12*, 1621–1637.

Nikolic, M. V., Vasiljevic, Z. Z., Auger, S., & Vidic, J. (2021). Metal oxide nanoparticles for safe active, and intelligent food packaging. *Trends in Food Science, 116*, 655–668.

Omerovic, N., Djisalov, M., Zivojevic, K., Mladenovic, M., Vunduk, J., Milenkovic, I., Knezevic, N., Gadjanski, I., & Vidic, J. (2021). Antimicrobial nanoparticles and biodegradable polymer composites for active food packaging applications. *Comprehensive Reviews in Food Science and Food Safety, 20*, 2428–2454.

Panea, B., Ripoll, G., Gonzalez, J., Fernandez-Cuello, A., & Albertí, P. (2014). Effect of nanocomposite packaging containing different proportions of ZnO and Ag on chicken breast meat quality. *Journal of Food Engineering, 123*, 104–112.

Polat, S., Fenercioglu, H., UnalTurhan, E., & Guclu, M. (2018). Effects of nanoparticle ratio on structural migration properties of polypropylene films and preservation quality of lemon juice. *Journal of Food Processing and Preservation, 42*, 13541.

Prabhu, Y. T., Rao, K. V., Kumari, B. S., & Pavani, T. (2015). Decoration of magnesium oxide nanoparticles on o-mwcnts and its antibacterial studies. *Rend Lincei, 26*, 263–270.

Premanathan, M., Karthikeyan, K., Jeyasubramanian, K., & Manivannan, G. (2011). The selective toxicity of ZnO nanoparticles toward Gram-positive bacteria and cancer cells by apoptosis through lipid peroxidation. *Nanomed, 7*, 184–192.

Proseus, T. E., & Boyer, J. S. (2005). Turgor pressure moves polysaccharides into the growing cell walls of characorallina. *Annals of Botany, 95*, 967–979.

Raghupathi, K. R., Koodali, R. T., & Manna, A. C. (2011). Size-dependent bacterial growth inhibition and mechanism of antibacterial activity of zinc oxide nanoparticles. *Langmuir, 27*, 4020–4028.

Robertson, G. L. (2012). Food Packaging: Principles and Practice (3rd ed.). CRC Press.

Saekow, M., Naradisorn, M., Tongdeesoontorn, W., & Hamauzu, Y. (2019). Effect of carboxy methyl cellulose coating containing ZnO-nanoparticles for prolonging the shelf life of persimmon and tomato fruit. *Journal of the Science of Food and Agriculture, 5*, 41–48.

Sahani, S., & SharmaYC. (2021). Advancements in applications of nanotechnology in the global food industry. *Food Chemistry, 342*, 128318.

Sankar, R., Manikandan, A., Malarvizhi, V., & Fathima, N. N. (2015). Synthesis and characterization of nanostructured zinc oxide particles by a simple thermal decomposition method. *Materials Research Bulletin, 62*, 21–26.

Saral, S. K., Indumathi, M. P., & Rajarajeswari, G. R. (2019). Mahua oil-based polyurethane/chitosan/nano ZnO composite films for biodegradable food packaging applications. *International Journal of Biological Macromolecules, 124*, 163–174.

Sathishkumar, M., Sneha, K., Kwak, I. S., Mao, J., Tripathy, S. J., & Yun, Y. S. (2009). Phytocrystallization of palladium through reduction process using *Cinnamomum zeylanicum* bark extract. *Journal of Hazardous Materials, 171*, 400–404.

Sawai, J., Shoji, S., Igarashi, H., Hashimoto, A., Kokugan, T., Shimizu, M., & Kojima, H. (1998). Hydrogen peroxide as an antibacterial factor in zinc oxide powder slurry. *Journal of Fermentation and Bioengineering, 86*(5), 521–522.

Shankar, S., & Rhim, J. W. (2018). Bionanocomposite films for food packaging applications. *Reference Module in Food Science,* 1–10.

Sharma, C., Dhiman, R., Rokana, N., & Panwar, H. (2017). Nanotechnology: An untapped resource for food packaging. *Frontiers in Microbiology, 8*, 1735.

Sharma, D., Rajput, J., Kaith, B. S., & Kaur, M. (2017). Antibacterial activity of zinc oxide nanoparticles synthesized by aqueous route. *Journal of Nanoparticle Research*, 1–6.

Siddiqi, K. S., Rahman, A., & Tajuddin, H. A. (2018). Properties of zinc oxide nanoparticles and their activity against microbes. *Nanoscale Research Letters, 13*(141), 1–13. https://doi.org/10.1186/s11671-018-2532-3

Siemann, U. (2005). Solvent cast technology– a versatile tool for thin film production. In *Scattering methods and the properties of polymer materials* (1–14), Berlin, Heidelberg, Germany: Springer.

Sirelkhatim, A., Mahmud, S., Seeni, A., Kaus, N. H. M., Ann, L. C., Bakhori, S. K. M., Hasan, H., & Mohamad, D. (2015). Review on zinc oxide nanoparticles: Antibacterial activity and toxicity mechanism. *Nano-Micro Letters, 7*, 219–242.

Sung, S. Y., Sin, L. T., Tee, T. T., Bee, S. T., Rahmat, A. R., Rahman, W. A ., Tan, A. C., & Vikhraman, M. (2013). Antimicrobial agents for food packaging applications. *Trends in Food Science & Technology, 33*, 110–123.

Suo, B., Li, H., Wang, Y., Li, Z., Pan, Z., & Ai, Z. (2017). Effects of ZnO nanoparticle-coated packaging film on pork meat quality during cold storage. *Journal of the Science of Food and Agriculture, 97*, 2023–2029.

Taghiyari, H. R., Morrell, J. F., & Husen, A. (2023). *Emerging nanomaterials* (*Opportunities and Challenges in Forestry Sectors*). Springer Nature. https://doi.org/10.1007/978-3-031-17378-3

Vandebriel, R. J., & De Jong, W. H. (2012). A review of mammalian toxicity of ZnO nanoparticles. *Nanotechnology, Science and Applications, 5*, 61–71.

Verma, A., Mehata, M. S., & Pathak, M. (2019). Synthesis of ZnO nanoparticles using *Bacillus cereus* and evaluation of their antimicrobial efficacy. *Der Pharma Chemica, 11*, 16–22.

Verma, A., & Kumar, S. (2018). Surface modification of ZnO nanoparticles for biomedical applications. In *Advanced nanostructured materials for environmental remediation* (pp. 263–282). Elsevier.

Vicentini, D. S., Smania, A., & Laranjeira, M. C. M. (2010). Chitosan/poly (vinyl alcohol) films containing ZnO nanoparticles and plasticizers. *Materials Science and Engineering C, 30*, 503–508.

Wahab, R., Mishra, A., Yun, S. I., Kim, Y. S., & Shin, H. S. (2010). Antibacterial activity of ZnO nanoparticles prepared via non-hydrolytic solution route. *Applied Microbiology and Biotechnology, 87*(5), 1917–1925.

Xing, Y., Li, X., Zhang, L., Xu, Q., Che, Z., Li, W., Bai, Y., & Li, K. (2012). Effect of TiO_2 nanoparticles on the antibacterial and physical properties of polyethylene-based film. *Progress in Organic Coatings, 73*, 219–224.

Xu, C., Sun, Y., & Li, Y. (2006). Sonochemical synthesis of zinc oxide nanoparticles from a zinc acetate aqueous solution. *Ultrasonics Sonochemistry, 13*(4), 313–316.

XuS, Z. X. (2015). Hydrothermal synthesis of ZnO nanoparticles and their applications. *Materials Science in Semiconductor Processing, 31*, 428–434.

Yang, R., Christensen, P. A., Egerton, T. A., & White, J. R. (2010). Degradation products formed during UV exposure of polyethylene-ZnO nano-composites. *Polymer Degradation and Stability, 95*, 1533–1541.

ZareY. (2016). Study of nanoparticles aggregation/agglomeration in polymer particulate nanocomposites by mechanical properties. *Composites. Part a, Applied Science and Manufacturing, 84*, 158–164.

Zhang, H., & Burn, P. L. (2006). Luminescent and phosphorescent iridium (III) complexes in OLEDs: Materials and strategies for performance optimization. *Advanced Materials, 18*, 2325–2338.

Zhang, Y. R., Nayak, T., Hong, H., & Cai, W. (2013). Biomedical applications of zinc oxide nanomaterials. *Current Molecular Medicine, 13*, 1633–1645.

Zhang, H., & Wong, K. S. (2017). Synthesis of ZnO nanomaterials: From nanostructures to nanodevices. *Journal of Materials Chemistry C, 5*(40), 10357–10378.

Zhang, Z., & Zhang, Y. (2001). Formation of nearly monodisperse inorganic nanocrystals in microemulsions. *Journal of the American Chemical Society, 123*, 7451–7452.

Zhang, W., Zhang, Y., Cao, J., & Jiang, W. (2021). Improving the performance of edible food packaging films by using nanocellulose as an additive. *International Journal of Biological Macromolecules, 166*, 288–296.

Zhang, Z., Zhang, Q., & Li, Y. (2012). Synthesis of zinc oxide nanoparticles by microemulsion method and their characterization. *Materials Letters.*

Zhao, Y., Meng, H., Chen, Z., Feng, Z., & Chai, Z. (2007). Dependence of nanotoxicity on nanoscale characteristics and strategies for reducing and eliminating nanotoxicity. In *Nanotoxicol* (pp. 265–280).

Role of Zinc Nanoparticles for the Management of Post-harvest Diseases

Priyanshi Sharma, K. Aggarwal, Anish Kumar Pal, Kundan Kumar Chaubey, Shweta Sharma, Deepak Kumar Verma, Rajneesh Tripathi, and Archana Bachheti

Abstract Food is an important part of life, and we cannot survive without it. Every year, a large part of the crop is spoiled by post-harvest diseases. Sometimes it happens because of climate change, long shipping, bad handling, and lack of packaging service. Post-harvest disease control is essential to reduce food and crop losses, and there are many ways. However, the role of zinc nanoparticles, such as zinc

Equal Contribution to Priyanshi Sharma1 and Kalash Aggarwal.

P. Sharma
School of Basic & Applied Sciences, Shri Guru Ram Rai University, West Patel Nagar, Patel Nagar, Dehradun 248001, Uttarakhand, India

K. Aggarwal
School of Biotechnology, Gautam Buddha University, Greater Noida 201312, Uttar Pradesh, India

Department of Biotechnology, School of Basic and Applied Sciences, Sanskriti University, Mathura 281401, Uttar Pradesh, India

A. K. Pal · K. K. Chaubey (✉)
Division of Research and Innovation, School of Applied and Life Sciences, Uttaranchal University, Arcadia Grant, P.O. Chandanwari, Premnagar, Dehradun 248007, Uttarakhand, India
e-mail: kundan2006chaubey@gmail.com

K. K. Chaubey
Department of Biotechnology, School of Basic and Applied Sciences, Sanskriti University, Mathura 281401, Uttar Pradesh, India

S. Sharma
College of Biotechnology, DUVASU, Mathura, Uttar Pradesh, India

D. K. Verma
Centre for Health Research and Innovation, Apeejay Business Centre, 6th Floor, Arunachal Building, Barakhamba Road, New Delhi 110001, India

R. Tripathi
Health System Strengthening, PATH, Regus Business Center, 10th Floor, Levana Cyber Heights, Vibhuti Khand, Gomti Nagar, Lucknow, Uttar Pradesh, India

A. Bachheti
Department of Environmental Science, Graphic Era University, Dehradun 248002, Uttarakhand, India

R. K. Bachheti et al. (eds.), *Metal and Metal-Oxide Based Nanomaterials*, Smart Nanomaterials Technology, https://doi.org/10.1007/978-981-99-7673-7_5

oxide, zinc ferrite, and zinc sulfide, is important, but generally, zinc oxide is used for this purpose. There are two methods to prepare ZnO nanoparticles, electrolysis and physical vapor synthesis (PVS). In the mechanochemical process, zinc chloride reacts with sodium carbonate and sodium chloride to form zinc oxide nanoparticles. In the PVS method, plasma arc energy is used with solid precursors to produce zinc nanoparticles. They have various sensors to detect urea, uric acid, DNA, protein, phenol, glucose, and lactic acid levels. It can be used in biosensing, is important in bioimaging, and has antibacterial and antiviral properties. Zinc nanoparticles should be used correctly as they are important resources and not be wasted. It can spray zinc nanoparticles to create a vegetation layer and reduce loss. Zinc nanoparticles are an emerging field in controlling post-harvest diseases in plants.

Keywords Post-harvest disease management · Zinc nanoparticles · Zinc oxide · Long shelf life · Various uses

1 Introduction

Post-harvest diseases are a major problem for the global agriculture industry, causing significant economic losses and are caused by various diseases, such as fungi, bacteria, and viruses (Abdullaeva, 2017). Post-harvest diseases can significantly impact the quality and value of agricultural products. Post-harvest diseases can cause spoilage, reduce shelf life in fruits, vegetables, and grains that is economic losses for growers and producers. But with effective management strategies, the risk of post-harvest disease can be reduced, and good quality can be maintained (Manjunatha et al., 2016). Many factors are involved in the occurrence of post-harvest diseases. These include misuse, improper storage, and the presence of insects or other pests. In addition, some crops may suffer from certain diseases, such as fruit rot or mold caused by fungi or bacteria (Vidal, 2012). Growers and distributors should take adequate measures to control post-harvest diseases. This includes taking steps to reduce the risk of contamination during collection and processing, such as using clean equipment and thoroughly cleaning products. Many post-harvest diseases thrive in hot and humid conditions so; there is a need of proper storage. Refrigeration or controlled storage can prevent fungal and bacterial growth. In addition to take preventive measures, many treatments are available for post-harvest disease control, which includes the use of antibiotics and physical treatments such as heat or radiation. But it is very note that overuse may lead to the development of resistant bacteria (Sasson et al., 2007). Finally, effective post-harvest disease management requires an integrated approach with all the factors contributing to disease development. By implementing preventive measures, using good storage and handling strategies, and monitoring disease symptoms, growers and suppliers can control post-harvest diseases and provide their best products. Post-harvest disease management includes strategies to reduce the incidence and severity of these diseases. One of the best strategies is to implement Good Agricultural Practices (GAP) at the production stage including carefully harvesting, processing, and storage of crops and minimize damage and exposure to disease. It also includes integrated management methods (IPM), such as health management and cultural practices which reduce the risk of pest infestations. Another strategy for post-harvest disease control is using physical and chemical treatments to control the disease. Overall, post-harvest disease management is a very complex and multifaceted process that requires proper planning and the implementation to ensure that crops are harvested, processed, and stored and minimize the risk of contamination. The incidence and severity of post-harvest diseases can be reduced, ensuring a safe, healthy, and sustainable food supply with proper management for the coming year.

Over the years, scientists have explained various planning, including the use of fungicides to control post-harvest diseases but fungicides associated with environmental pollution and health hazards. Therefore, there is a requirement for another method, which should be eco-friendly. In this connection, nanotechnology has shown potential applications (Chawla et al., 2023; Husen & Iqbal, 2019; Husen & Jawaid, 2020; Husen & Siddiqi, 2023; Husen et al., 2023, 2023a, 2023b, 2023c; Husen, 2022, 2023a, 2023b, 2023c; Jin-Chul et al., 2021; Mandal et al., 2023; Taghiyari et al., 2023). Like other nanomaterials, zinc nanoparticles can act as a promising tool to solve this problem. Zinc nanoparticles have antibacterial properties that can inhibit the growth of various post-harvest diseases. This is because of unique properties of nanoparticles that allow them to penetrate into the cell walls of bacteria and influence their metabolic processes (Naderi & Abedi, 2012). A zinc nanoparticle improves the defense mechanisms of plants that means when zinc nanoparticles are applied to plants, they can resist and protect plants against diseases caused by bacteria. Using zinc nanoparticles to control post-harvest diseases is safe and sustainable, which is better than medicine. These nanoparticles also can be used for coating fruits and vegetables that prevent the growth of pathogenic bacteria (Naderi & Abedi, 2012). Nanoparticles reduce the risk of post-harvest disease by a physical barrier that prevents the entry of pathogenic bacteria into crops. Zinc nanoparticles extend the shelf life of fruits and vegetables and reducing food waste by reducing respiration and ethylene production in the product, which is the leading cause of spoilage (Sozer & Kokini, 2009). This feature makes them especially useful for fruit and vegetable storage and shipping. Zinc nanoparticles in post-harvest disease control have many advantages over fungicides. They penetrate the walls of fungal cells and interfere with their metabolic processes and inhibiting fungal growth and development. This functionality makes them an ideal for crop post-harvest disease control. Many studies have shown that zinc nanoparticles can control post-harvest diseases in various crops.

The role of zinc nanoparticles in post-harvest disease management is an exciting and promising area of research. For example, zinc nanoparticles in tomatoes reduce the incidence of *Botrytis cinerea* caused by the fungus *Botrytis cinerea*. Similarly, zinc nanoparticles in strawberries are effective against the anthracnose-causing fungus *Colletotrichum acutatum* (Agrawal & Rathore, 2014). The role of zinc nanoparticles will more likely to become a necessary cause of the need of safe and sustainable agriculture in crop protection. With further research, these nanoparticles will likely become essential to sustainable agriculture worldwide (Fig. 1).

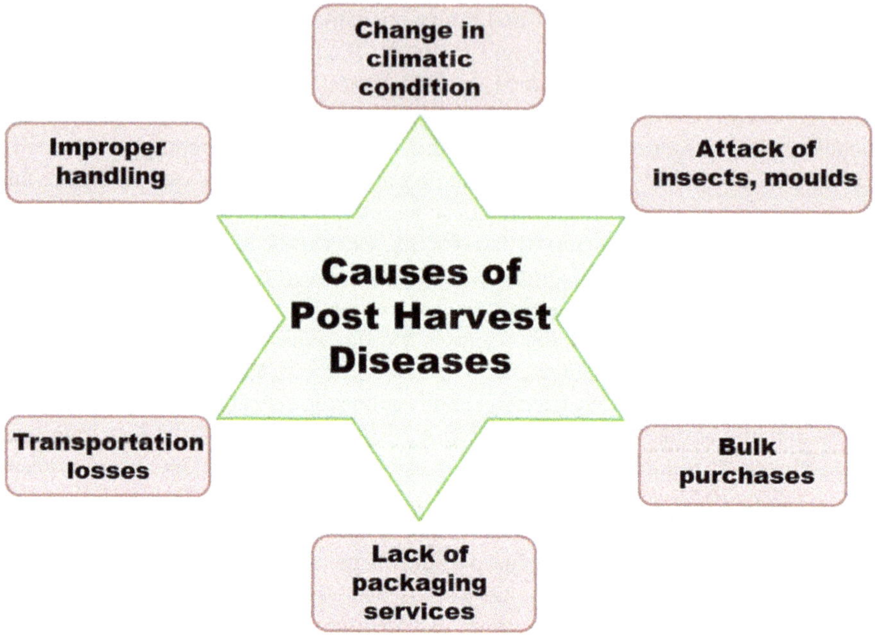

Fig. 1 Causes of post-harvest diseases

2 Synthesis of Zinc Nanoparticles

The synthesis of zinc nanoparticles is an exciting research area in the world of nanotechnology. These small particles composed of zinc molecules, which have unique properties that allow to be used in various fields (Agarwal et al., 2023; Dhewa, 2015; Husen, 2019; Siddiqi et al., 2018). Synthesis of zinc nanoparticles steps should be done accurately and carefully. An important method used to make zinc nanoparticles is the sol–gel method, which includes converting a liquid sol into a solid gel and done by adding precursors to the solvent, hydrolysis, and condensation to form a gel. Then, gel dried and calcined at high temperatures and formed zinc oxide nanoparticles. After that, particles can be reduced to form zinc nanoparticles. Chemical reduction is another method, which includes the reduction of zinc ions by using a reducing agent to form nanoparticles (Bachheti & Bachheti, 2023; Gao et al., 2006). This method is commonly used because of its simplicity, low cost, and large producibility. The reduction can be done by using various reducing agents, like sodium borohydride, hydrazine, and sodium citrate. They also have excellent catalytic properties, which make them useful in catalysis operations. The synthesis method also depends on the desired nanoparticle size, shape, and properties. The size of nanoparticles can be done by adjusting the concentration of reactants, reaction temperature, or reaction time and shape can be done by using different surfactants or stabilizers. When zinc nanoparticles are combined, they can be characterized using

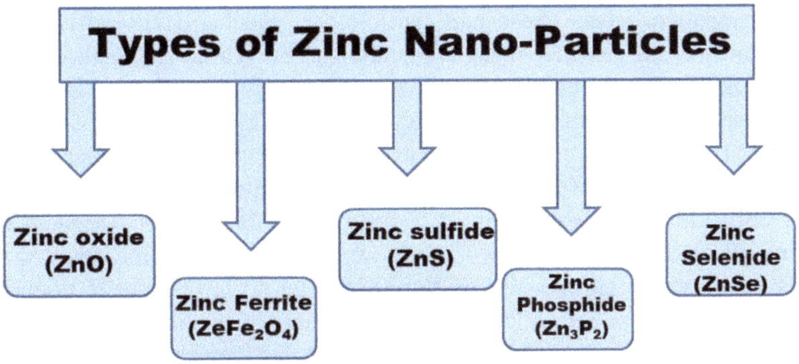

Fig. 2 Types of zinc nanoparticles

various techniques such as X-ray diffraction, transmission electron microscopy, and dynamic light scattering (Casey, 2006). These techniques help determine nanoparticle size, crystal structure, and morphology.

Other methods are thermal separation, sol–gel method, and electrochemical methods. Each method has its own advantages and disadvantages, and the using of each method depends on the particular application of the nanoparticles. Whatever the method used for synthesizing zinc nanoparticles requires attention and good understanding of basic chemistry. The resulting nanoparticles should be uniform in size and shape, with their high purity and stability. This requires precise reaction control, such as temperature, pH, and reaction time. Despite the difficulties of synthesizing zinc nanoparticles benefits of these small particles are enormous (Swihart, 2003). Zinc nanoparticles have antibacterial properties, making them more useful in medical applications such as wound healing and drug delivery. They can also be used in electronic products, including electronic products and solar panels. With the continuous research and development, the synthesis of zinc nanoparticles will play an important role in many industries and applications. Zinc oxide is a multifunctional inorganic material. It is a semiconductor and is divided into two groups. Since ZnO has many applications, we need to prepare it commercially. There are two methods for commercially preparing ZnO nanoparticles. The first method is mechanochemical processing (MCP), and the second method is physical vapor phase synthesis (PVS) (Casey, 2006). There are many ways to study chemical processes, such as thermal decomposition, precipitation, and hydrothermal synthesis (Fig. 2).

2.1 Mechanochemical Processing (MCP)

Mechanochemical processing is the best way to prepare nanomaterials. This method is used for ZnS, ZnO, CeO_2, CdS, and SiO_2 (Aghababazadeh et al., 2006). In this model, the process simultaneously reduces the size of physical and chemical reactions to produce nanoparticles. These nanoparticles are mechanically activated during nanoscale milling. MCPs have many applications. This is a simple method to prepare crystalline ZnO nanoparticles.

In this process, sodium carbonate (Na_2CO_3) and zinc chloride ($ZnCl_2$) act as precursors. Two pioneers ground into one stone at the same time. After collapsing, they produce sodium chloride (NaCl) and zinc carbonate ($ZnCO_3$) through the ball-powder collision mechanism. This is a drug swap. Stone is a chemical reactor involving chemical reactions. It works at low temperatures, so the reaction occurs in areas of hot and high pressure on the nanometer scale that are in contact with the surface. NaCl was also added at the beginning of the reaction. It acts as an inert diluent necessary to complete the reaction. After the reaction is complete, the product is considered a nanocomposite with NaCl produced as the matrix phase (Casey, 2006).

The reactions that occur in this process are:

$$ZnCl_2 + Na_2CO_3 + 8NaCl \rightarrow ZnCO_3 + 10NaCl$$

The product we get after the reaction is a mixture of nanostructured products that are then processed into ZnO nanoparticles. The mixture is subjected to heat treatment at 170–380 °C. With this heat treatment, $ZnCO_3$ decomposes into ZnO. Now wash to separate the NaCl from the ZnO. Now it is dry for further processing. The nanoparticles are ready after drying. This method produces an average particle size of 20–30 nm (Aghababazadeh et al., 2006). The sizes of nanoparticles can vary depending on heat treatment, time, and grinding time.

According to the experiment, if collision time is inversely proportional to the size of the nanoparticles. If the grinding time is increased, the size of the nanoparticles decreases (Ao et al., 2006). Optimum grinding time is in middle for the production of medium-sized nanoparticles. If the milling time is increased from 5 to 40 min, this time change will reduce the size of nanoparticles from 40 to 24 nm (Shen et al., 2006). If the milling time is increased to 70 min above the temperature, the size of the nanoparticles increases by 27 nm.

On the other hand, if the temperature of the heat treatment process changes, this will also cause a change in the size of the nanoparticles. If the temperature rises to 200 °C, the size of the nanoparticles will increase by 3 nm. At 400 °C, the size of the nanoparticles is 18 nm, and if the temperature rises to 600 °C, the size of the nanoparticles will be 21 nm. If the temperature is increased to 800 °C, the size of the nanoparticles reaches 36 nm (Ao et al., 2006). This is a simple and low-cost method. From an environmental point of view, this process is acceptable as it does not contain organic solvents that can harm the environment (Lu et al., 2008). This method also has some disadvantages. Because the product forms groups or clumps during grinding. However, it can be reduced with the help of the salt matrix. After grinding, the matrix must be removed by a simple cleaning method (Fig. 3).

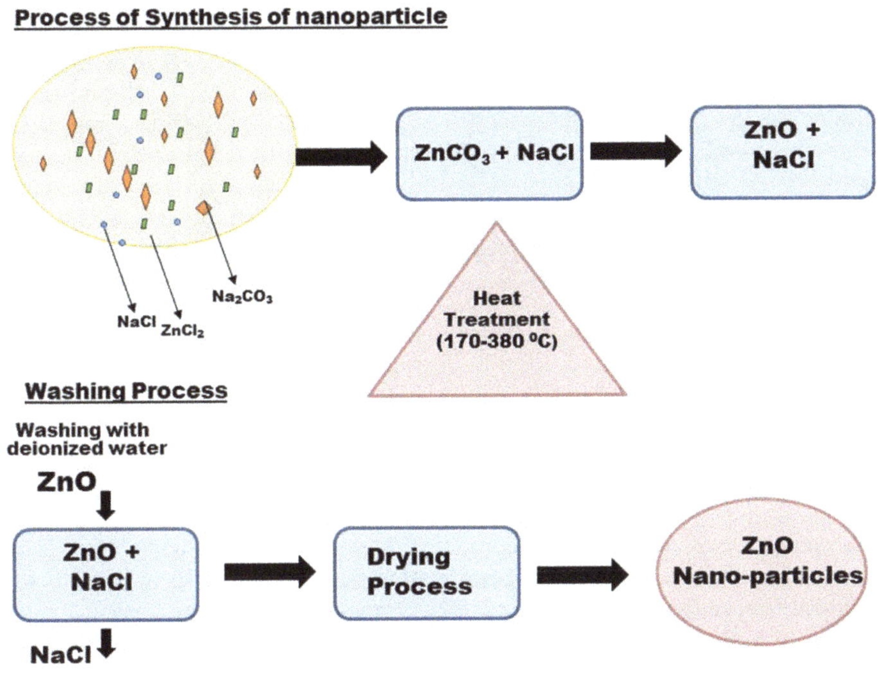

Fig. 3 Process of synthesis of nanoparticles by mechanochemical method

2.2 *Physical Vapor Synthesis (PVS)*

Physical vapor phase synthesis (PVS) is a cost-effective method for nanoparticle synthesis. In this way, solid precursors are used to synthesize nanoparticles. At high temperatures, plasma arc is energy that preheats the product to evaporate. The energy of plasma arc provides used to initiate the reaction, when injected into the precursor blood, which causes supersaturation and particle nucleation. This energy of plasma arc is used to divide the precursors and form atoms. Later, the atoms react and condense and form particles. This product can be cooled with the help of an air conditioner or with an expanding nozzle (Swihart, 2003) (Fig. 4).

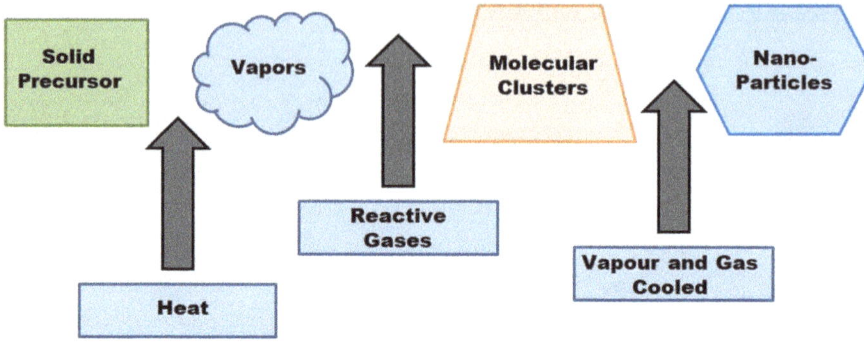

Fig. 4 Process of synthesis of zinc nanoparticles by physical vapor synthesis

3 Physical and Chemical Properties of Zinc Nanoparticles

Zinc nanoparticles are an interesting topic of research in the nanotechnology. It has different physical and chemical properties that make them useful in many fields. Different size, shape, and surface area of zinc nanoparticles and their physical properties help in determining their behavior in various environments. Zinc nanoparticles are generally smaller than 100 nm and may be different shapes like spherical, rod-shaped, or geometric. Their small size makes their surface area larger, making them more strong and prone to oxidation (Espitia et al., 2017). They exhibit quantum confinement effects that can alter their optical and electrical properties make them useful in optoelectronics, photovoltaics, and biological imaging applications (Manjunatha et al., 2016; Naderi & Abedi, 2012).

The chemistry of zinc nanoparticles is equally interesting. Another most important physical property is color of zinc nanoparticles. It has a unique silver-gray color that distinguishes them from other nanoparticles. Zinc can form stable compounds with many other elements because it has +2 transition metal. It is also an essential micronutrient for living organisms and vital in many biological processes. Zinc nanoparticles have unique chemical properties that are why it acts as antibacterial, antiviral, and antifungal agent. They can be used in food packaging as well (Wyser et al., 2016). Zinc nanoparticles act as essential catalysts increasing their activities with high surface area. It increases the rate of chemical reactions by providing more surface area for reactant interactions during reaction. Their stability in different conditions makes them ideal for many industrial and medical purposes and their biocompatibility, which makes them unique and ideal for medical applications. It is non-toxic and non-carcinogenic, making it suitable for drug delivery, cancer treatment, and other medical applications (Kumar & Krishnamoorti, 2010). Zinc nanoparticles also combine with other molecules to target cells or tissues, work good in delivering drugs to specific areas. Zinc nanoparticles easily interact with other molecules, so they can be used in many drugs. They are stable with other materials, which making them helpful in making many composites and they are less likely to degrade or break

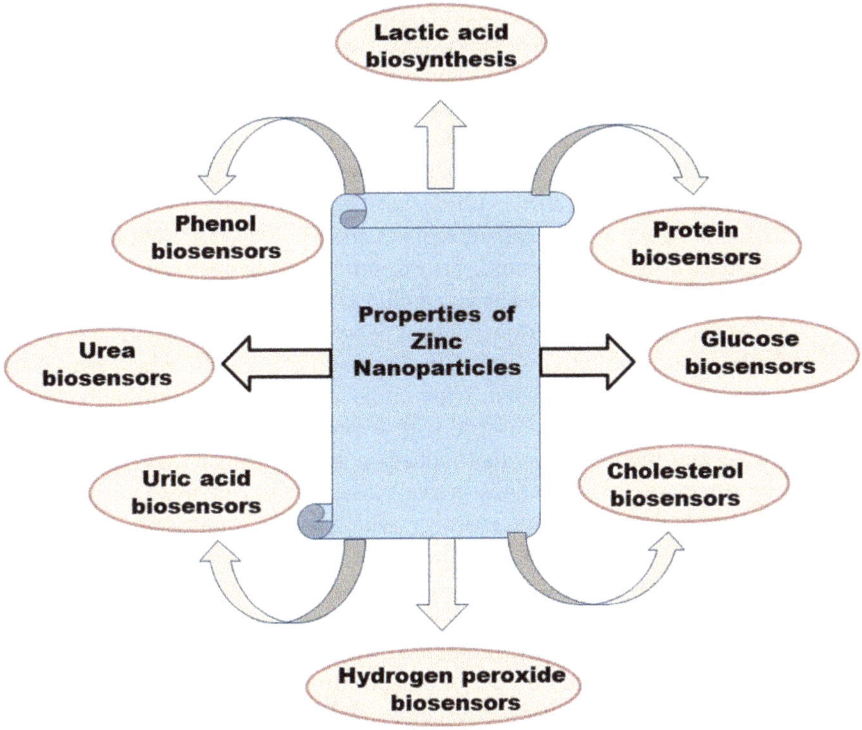

Fig. 5 Properties of zinc nanoparticles

with time (Espitia et al., 2013; Wyser et al., 2016). Zinc is an essential element for the body; it does not contain any chemicals that cause it to decompose or lose its properties. Overall, zinc nanoparticle has physical and chemical properties, which makes them incredibly versatile for many applications. From their unique colors to their high performance and durability, these miniatures have become popular among researchers and manufacturers in many industries (Siddiqui et al., 2015). Whether used in electronics, medicine, or other applications, zinc nanoparticles are ready to set and play an important role in shaping the future of technology and innovation (Fig. 5).

4 How to Use Zinc Nanoparticles in Post-Harvest Disease Management?

Zinc nanoparticles are a promising tool and these tiny particles have unique properties that effectively control various fungal and bacterial diseases that damage crops during storage and transportation for controlling post-harvest disease (Prasanna,

2007). First, zinc nanoparticles should be used correctly otherwise it can be toxic. Therefore, it is important to follow all the practices for effective and proper working (Cortes-Lobos, 2013). Nanoparticles can used as a spray or coating on products by mixing them with water and surfactants to improve the dispersion of the drug. Second, the pH value and temperature of the drug, and the temperature of the nanoparticles are important factors affecting the treatment. Therefore, it is important to optimize all instructions for the best results (Liu & Lal, 2015). Finally, it is very important to note that zinc nanoparticles cannot replace good post-harvest practices. Proper processing, storage, and transportation are essential to prevent post-harvest diseases. Zinc nanoparticles should be used as an additional treatment for these industries.

Improving the natural defense mechanisms of crops with zinc nanoparticles allows them to be less susceptible to disease. Zinc nanoparticles can be mixed with water and sprayed on crops before harvest. After crops are harvested, they can be treated with a zinc nanoparticle solution mixed with water. This helps to protect the crops during storage and transportation that reduces post-harvest diseases. Another way the use zinc nanoparticles against post-harvest disease is to combine with packaging materials such as plastic films or coatings, which act as a barrier preventing the growth of pathogens in crops helps to transportation. This helps to extend the life shelf of the crop and helps to control of post-harvest loss. With the proper and good post-harvest practices and application, zinc nanoparticles can help reduce post-harvest diseases and improve agricultural product quality. Although the applications of zinc nanoparticles in disease control are promising, more needed in research. It is important to remember that these nanoparticles do not negatively impact the environment or human health (Aslani et al., 2014; Liu & Lal, 2015). With the help of research and development, zinc nanoparticles have potentially revolutionized the way of managing the post-harvest diseases and improving food safety for people around the world.

5 Applications of Zinc Nanoparticles in Post-Harvest Disease Management

A recent advancement in agricultural science that has multiple potential benefits is zinc nanoparticles. Zinc nanoparticles demonstrate promising results in the control of post-harvest diseases. Post-harvest diseases may seriously damage crops, result in economic losses for farmers, and damage the quality of consumer goods (DeRosa et al., 2010). Zinc nanoparticles can help reduce these losses and improve crop overall quality and health. These tiny particles can potentially revolutionize how we protect crops from diseases and pests and reduce the number of pesticides and fungicides used in agriculture (Sozer & Kokini, 2009). Fungal diseases can spread rapidly to stored produce, causing rot and decay. An important benefit of zinc nanoparticles is their ability to act as anti-inflammatory agents.

Zinc nanoparticles have the ability to inhibit the growth of fungal infections, thereby reducing the incidence of these infections. There is a synergistic effect when zinc nanoparticles are combined with other antibiotics, which increase their effects and reduce the need for heavy drugs (Kumar & Krishnamoorti, 2010; Naderi & Abedi, 2012). Another benefit of zinc nanoparticles acts as a preservative by extending the crop's shelf life especially important for shipping and storage. Research has shown that treating crops with zinc nanoparticles helps preserve their quality and nutrients for longer, reduce waste, and increase access to new products. In addition to its antibacterial and antifungal properties, zinc nanoparticles have shown beneficial effects on plant growth and development (Akbar et al., 2020; Akintelu & Folorunso, 2020).

Zinc is an essential nutrient, and its nanoparticles can improve the absorption and utilization of this nutrient. This can result in stronger, healthier, disease-resistant, and more productive plants. Zinc nanoparticles in post-harvest disease control have ability to target specific pathogens. Zinc nanoparticles only affect bacteria unlike insecticides and fungicides, which can harm insects, fungi, and microbes too. This makes them safer and environmentally friendly crop protection. Another benefit of using zinc nanoparticles is their ability to prolong the life of fruits and vegetables (Espitia et al., 2013; Wyser et al., 2016). Studies have shown that the treatment of agricultural products with zinc nanoparticles can reduce the occurrence of diseases such as post-harvest rot and mold and extend the shelf life of products. This not only benefits farmers and suppliers by reducing waste but also benefits consumers by providing fresh food and health. Zinc nanoparticles have many applications in post-harvest disease control. From reducing the incidence of fungal diseases to extending crop shelf life and improving plant growth and development, zinc nanoparticles hold promise for both farmers and customers. As research in this area continues to evolve, we are likely to see new uses for this new technology (Fig. 6).

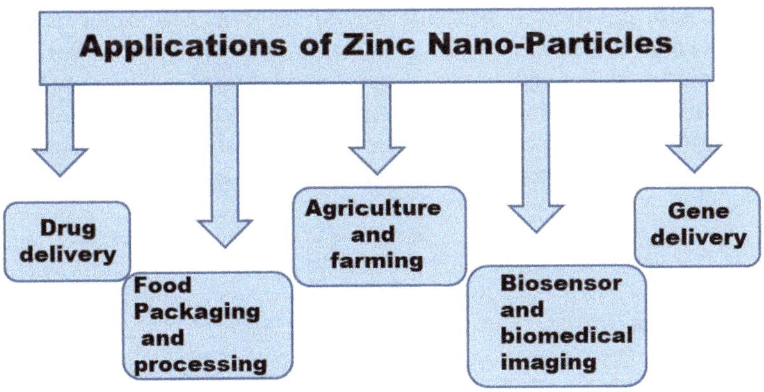

Fig. 6 Applications of zinc nanoparticles

6 Future of Zinc Nanoparticles in Post-Harvest Disease Management

The introduction of zinc nanoparticles makes the future bright for post-harvest disease control. This small crop shows the ability to reduce the spread of disease after the crop is harvested. They inhibit the growth of bacteria and viruses, which are an essential element in post-harvest diseases. Zinc nanoparticles can be used in other methods rather than chemical treatments which used for post-harvest disease control (Bhagat et al., 2015; Sharifan et al., 2021). The use of zinc nanoparticles is practical and environmentally friendly. Unlike Pharmaceuticals, Nano Zinc is not toxic and can be easily disposed of without affecting the environment. They also do not release harmful substances into the crops so that they can be safely eaten. Zinc nanoparticles have also extended the shelf life of crops, which is good for farmers and consumers (Sharifan et al., 2021). The longer shelf life means less waste and more profit for farmers, while consumers can enjoy fresh produce for longer.

One of the most exciting facts about zinc nanoparticles is their ability to combine with other natural substances to create better treatments. One of the main advantages of using zinc nanoparticles in post-harvest disease control is their ability to provide long-term protection (Sorrentino et al., 2007). Unlike other conventional pesticides that need to be recycled, zinc nanoparticles remain active for a long peroid, providing continuous protection against diseases after harvesting. For example, researchers have found that combining zinc nanoparticles with essential oils can improve their antibacterial and antifungal properties. This increases the variety of approaches for preventing post-harvest disease.

The future of post-harvest disease control may be depends on zinc nanoparticles. Their efficiency, safety, and potential for further development benefit farmers and consumers (Bhagat et al., 2015). As research continues, expect more ways to use zinc nanoparticles in agriculture. With sustainable agriculture continues to increase, the need for safe and effective pesticide use will increase. Zinc nanoparticles show great potential in this area, and further research and development will definitely lead to innovative and effective insect control and post-harvest disease (Valadkhan et al., 2005). As our understanding of its properties and applications continues to evolve, researchers can expect to see nanoparticles widely used in agriculture. This will increase crop yields, reduce waste, and ultimately lead to excellent health and productivity.

7 Conclusion

Zinc nanoparticles have shown up as a promising tool for post-harvest disease control; a zinc nanoparticle has the ability to combat fungal diseases that cause decline after harvest. These nanoparticles break down the cell membranes of the bacteria and preventing them from growing and spreading. These nanoparticles have shown to antibacterial and antifungal properties, which make them a good choice for controlling the growth of post-harvest pathogens. They can also trigger resistance in plants, which help plants protect themselves against diseases. Zinc nanoparticles increase the quality and yield of fruits and vegetables by preventing the disease. Zinc nanoparticles can use for coating or as spray on fruits and vegetables to control post-harvest diseases. It penetrates in bacterial cell wall and disrupts metabolic processes, inhibits growth and reproduction. In addition, the use of zinc nanoparticles is safe and environmentally friendly alternative to drug therapy. They do not leave any residue on the product as they are safe for human consumption. They also do not harm the environment while contributing to the development of the immune system. The role of zinc nanoparticles in post-harvest disease control is revolutionary for agriculture. It provides a safe and effective solution for controlling post-harvest diseases and extending the shelf life of agricultural products. As more research is done, we can expect to see more benefits and applications of zinc nanoparticles in agriculture.

References

Abdullaeva, Z. (2017). Nanomaterials in food industry and packaging. *Nanomaterials in daily life* (pp. 23–46). Springer.

Agarwal, S., Chauhan, A., Anand, J., Abate, L., & Rai, N. (2023). *Medicinal plant-based alkaloids: A suitable precursor for nanoparticle synthesis and their various applications* (pp. 111–132). CRC Press.

Aghababazadeh, R., Mazinani, B., Mirhabibi, A., & Tamizifar, M. (2006). ZnO nanoparticles synthesized by mechanochemical processing. *Journal of Physics Conference Series, 26*(1), 312.

Agrawal, S., & Rathore, P. (2014). Nanotechnology pros and cons to agriculture: A review. *International Journal of Current Microbiology and Applied Sciences, 3*(3), 43–55.

Akbar, S., Tauseef, I., Subhan, F., Sultana, N., Khan, I., Ahmed, U., et al. (2020). An overview of the plant-mediated synthesis of zinc oxide nanoparticles and their antimicrobial potential. *Inorganic and Nano-Metal Chemistry, 50*(4), 257–271. https://doi.org/10.1080/24701556.2019.1711121

Akintelu, S. A., & Folorunso, A. S. (2020). A Review on green synthesis of zinc oxide nanoparticles using plant extracts and its biomedical applications. *BioNanoScience, 10*(4), 848–863. https://doi.org/10.1007/s12668-020-00774-6

Ao, W., Li, J., Yang, H., Zeng, X., & Ma, X. (2006). Mechanochemical synthesis of zinc oxide nanocrystalline. *Powder Technology, 168*(3), 148–151.

Aslani, F., Bagheri, S., Muhd Julkapli, N., Juraimi, A.S., Hashemi, F.S.G., & Baghdadi, A. (2014). Effects of engineered nanomaterials on plants growth: an overview. *The Scientific World Journal, 28*. https://doi.org/10.1155/2014/641759

Bachheti, R. K., & Bachheti, A. (Eds.). (2023). *Secondary metabolites from medicinal plants: Nanoparticles synthesis and their applications*. CRC Press. https://doi.org/10.1201/9781003213727

Bhagat, Y., Gangadhara, K., Rabinal, C., Chaudhari, G., & Ugale, P. (2015). Nanotechnology in agriculture: A review. *Journal of Pure and Applied Microbiology, 9*, 737–747.

Casey, P. (2006). Nanoparticle technologies and applications. In R. H. J. Hannink & A. J. Hill (Eds.), *Nanostructure control of materials* (pp. 1–27). Woodhead Publishing Limited.

Chawla, S., Singh, S., & Husen, A. (2023). *Smart nanomaterials targeting pathological hypoxia*. Springer Nature Singapore Pte Ltd. ISBN: 978-981-99-1717-4. https://doi.org/10.1007/978-981-99-1718-1

Cortes-Lobos, R. (2013). Nanotechnology research in the US agri-food sectoral system of innovation: toward sustainable development.
DeRosa, M. C., Monreal, C., Schnitzer, M., Walsh, R., & Sultan, Y. (2010). Nanotechnology in fertilizers. *Nature Nanotechnology, 5*(2), 91.
Dhewa, T. (2015). Nanotechnology applications in agriculture: An update. *Octa Journal of Environmental Research, 3*(2), 204–211.
Espitia, P. J. P., Soares, Nd. F. F., Teóflo, R. F., dos Reis Coimbra, J. S., Vitor, D. M., Batista, R. A., Ferreira, S. O., de Andrade, N. J., & Medeiros, E. A. A. (2013). Physical–mechanical and antimicrobial properties of nanocomposite films with pediocin and ZnO nanoparticles. *Carbohydrate Polymers, 94*(1), 199–208.
Gao, F., Hong, F., Liu, C., Zheng, L., Su, M., Wu, X., Yang, F., Wu, C., & Yang, P. (2006). Mechanism of nano-anatase TiO_2 on promoting photosynthetic carbon reaction of spinach. *Biological Trace Element Research, 111*(1–3), 239–253.
Husen, A. (2019). Natural product-based fabrication of zinc oxide nanoparticles and their application. In A. Husen & M. Iqbal (Eds.), N*anomaterials* and plant potential (pp. 193–291). Springer International Publishing AG. https://doi.org/10.1007/978-3-030-05569-1_7
Husen, A. (2022). *Engineered nanomaterials for sustainable agricultural production, soil improvement and stress management*. Elsevier Inc. https://doi.org/10.1016/C2021-0-00054-7
Husen, A. (2023a). *Secondary metabolites based green synthesis of nanomaterials and their applications.* Springer Nature Singapore Pte Ltd. https://doi.org/10.1007/978-981-99-0927-8
Husen, A. (2023b). *Nanomaterials from agricultural and horticultural products.* Springer Nature Singapore Pte Ltd. https://link.springer.com/book/9789819934348
Husen, A. (2023c). *Nanomaterials and nanocomposites exposures to plants (Response, Interaction, Phytotoxicity and Defense Mechanisms).* Springer Nature Singapore Pte Ltd. https://doi.org/10.1007/978-981-99-2419-6
Husen, A., Bachheti, R. K., & Bachheti, A. (2023). *Current trends in green nano-emulsions (Food, Agriculture and Biomedical Sectors).* Springer Nature Singapore Pte Ltd.
Husen, A., & Iqbal, M. (2019). *Nanomaterials and plant potential.* Springer International Publishing AG. https://doi.org/10.1007/978-3-030-05569-1
Husen, A., & Jawaid, M. (2020). *Nanomaterials for agriculture and forestry applications.* Elsevier Inc. https://doi.org/10.1016/C2018-0-02349-X
Husen, A., & Siddiqi, K. S. (2023). *Advances in smart nanomaterials and their applications.* Elsevier Inc. https://doi.org/10.1016/C2021-0-02202-1
Jin-Chul, K., Madhusudhan, A., & Husen, A. (2021). *Smart nanomaterials in biomedical applications.* Springer Nature Switzerland AG. https://doi.org/10.1007/978-3-030-84262-8
Kumar, S. K., & Krishnamoorti, R. (2010). Nanocomposites: Structure, phase behavior, and properties. *Annual Review of Chemical and Biomolecular Engineering, 1*(1), 37–58. https://doi.org/10.1146/annurev-chembioeng-073009-100856
Liu, R., & Lal, R. (2015). Potentials of engineered nanoparticles as fertilizers for increasing agronomic productions. *Science of the Total Environment, 514*(Supplement C), 131–139. https://doi.org/10.1016/j.scito tenv.2015.01.104
Lu, J., Ng, K. M., & Yang, S. (2008). Efficient, one-step mechanochemical process for the synthesis of ZnO nanoparticles. *Industrial and Engineering Chemistry Research, 47*(4), 1095–1101.
Mandal, A. K., Ghorai, S., & Husen, A. (2023). *Functionalized smart nanomaterials for point-of-care testing.* Springer Nature Singapore Pte Ltd.
Manjunatha, S. B., Biradar, D. P., & Aladakatti, Y. R. (2016). Nanotech nology and its applications in agriculture: A review. *Journal of Farm Sciences, 29*(1), 1–13.
Naderi, M. R., & Abedi, A. (2012). Application of nanotechnology in agriculture and refinement of environmental pollutants. *Journal of Nanotechnology, 11*(1), 18–26.
Prasanna, B. M. (2007). *Nanotechnology in agriculture.* ICAR National Fellow, division of Genetics, IARI

Sasson, Y., Levy-Ruso, G., Toledano, O., & Ishaaya, I. (2007). Nanosuspensions: Emerging novel agrochemical formulations. *Insecticides design using advanced technologies* (pp. 1–39). Springer.
Shen, L., Bao, N., Yanagisawa, K., Domen, K., Gupta, A., & Grimes, C. A. (2006). Direct synthesis of ZnO nanoparticles by a solutionfreemechanochemical reaction. *Nanotechnology, 17*(20), 5117.
Siddiqi, K. S., Rahman, A., & Tajuddin, H. A. (2018). Properties of zinc oxide nanoparticles and their activity against microbes. *Nanoscale Research Letters, 13*(141), 1–13. https://doi.org/10.1186/s11671-018-2532-3
Siddiqui, M. H., Al-Whaibi, M. H., Firoz, M., & Al-Khaishany, M. Y. (2015). Role of nanoparticles in plants. In: M. H. Siddiqui, M. H. AlWhaibi, F. Mohammad (Eds.), *Nanotechnology and plant sciences: nanoparticles and their impact on plants* (pp. 19–35). Springer. https://doi.org/10.1007/978-3-319-14502-0_2
Sharifan, H., Noori, A., Bagheri, M., & Moore, J. M. (2021). Postharvest spraying of zinc oxide nanoparticles enhances shelf life qualities and zinc concentration of tomato fruits. *Crop and Pasture Science*. https://doi.org/10.1071/CP21191
Sorrentino, A., Gorrasi, G., & Vittoria, V. (2007). Potential perspectives of bio-nanocomposites for food packaging applications. *Trends in Food Science and Technology, 18*(2), 84–95.
Sozer, N., & Kokini, J. L. (2009). Nanotechnology and its applications in the food sector. *Trends in Biotechnology, 27*(2), 82–89.
Swihart, M. T. (2003). Vapor-phase synthesis of nanoparticles. *Current Opinion in Colloid and Interface Science, 8*(1), 127–133.
Taghiyari, H. R., Morrell, J. F., & Husen, A. (2023). *Emerging nanomaterials* (*Opportunities and Challenges in Forestry Sectors*). Springer Nature Switzerland AG. https://doi.org/10.1007/978-3-031-17378-3
Valadkhan, M., Mohammadi, K., & Nezhad, M. T. K. (2005). Effect of priming and foliar application of nanoparticles on agronomic traits of chickpea. *Biological Forum—an International Journal, 7*(2), 599–602.
Vidal, J. (2012). Food scarcity: the time bomb setting nation against nation. The Guardian. https://www.theguardian.com/global-development/2012/Oct/14/food-climate-change-population-water
Wyser, Y., Adams, M., Avella, M., Carlander, D., Garcia, L., Pieper, G., Rennen, M., Schuermans, J., & Weiss, J. (2016). Outlook and challenges of nanotechnologies for food packaging. *Packaging Technology and Science, 29*(12), 615–648.

Role of Zinc and Zinc Oxide Nanofertilizer in Enhancing Crop Production

Sathiyanarayanan Anusuya and Kilimas Rajan

Abstract Food demand is rising as the world's population expands, yet agricultural yield declines due to climate change, low soil fertility, and nutrient deficiency are serious concerns in crop production. Nanomaterials, especially metal and metal oxide nanoparticles used as nanopesticides, herbicides, fungicides, and fertilizers, have become a new-age material in the last decade, transforming modern agriculture. Among other metal nanoparticles, zinc and zinc oxide nanoparticles (ZnONPs) have attracted undoubted attention for sustainable plant performance as it stimulates germination of seed, early flowering, enzyme activity, and higher yield. Although physiological indices determine ZnONPs, they affect structural modifications, including stomatal and trichome morphology, induced vacuole in root cortex cells, protoplast shrinkage, and thylakoid degradation, indicating the toxicity of ZnONPs in the photosynthetic apparatus. In addition, applying these nanoparticles to plants arrests the cell cycle and induces apoptosis, which causes severe damage to DNA. The beneficial or adverse response usually depends on the species' type, size, concentration, treatment methods, stage of development, and the genotype of the species or environmental conditions. However, increased use certainly leads to the accumulation of ZnONPs in the ecosystem. In order to properly use and regulate the release of nanopreparations, it is necessary to understand how they change and behave in complex systems.

Keywords Zinc oxide nanoparticles · Plant growth · Productivity · Photosynthesis · Toxicity · Seed germination

1 Introduction

Nanotechnology holds great promise to significantly impact the agricultural and food sectors by improving food security and productivity, which are required by the projected growth of the world's population (Husen, 2022, 2023; Husen & Siddiqi,

S. Anusuya (✉) · K. Rajan
Department of Botany, St. Joseph's College, Tiruchirappalli, Tamilnadu, India
e-mail: anusathsar.rajesh@gmail.com

R. K. Bachheti et al. (eds.), *Metal and Metal-Oxide Based Nanomaterials*, Smart Nanomaterials Technology, https://doi.org/10.1007/978-981-99-7673-7_6

2023). In reality, the special characteristics of nanoscale materials make them appropriate candidates for the creation of new technologies to help sustainable agriculture (Fraceto et al., 2016). Using nanopesticides and nanofertilizers to boost production, improving soil quality, stimulating plant development using nanomaterials, and managing plant and soil health are just a few uses of nanotechnology in agriculture sectors. Novel nanomaterials based on inorganic and polymer improve plant productivity by immobilizing nutrients and their release as intelligent nanosystems. Zinc (Zn) is an essential micronutrient required for plant growth and development that performs vital metabolic reactions (Awan et al., 2021). In order to boost agricultural output and efficiency and minimize environmental impact, Zn can be administered to the soil in various ways, including zinc and zinc oxide nanoparticles (ZnONPs) as nanofertilizers (Mittal et al., 2020). ZnONPs are non-toxic, environmentally friendly, biologically safe and biocompatible, soluble, and sensitive compared to conventional ZnO fertilizers due to their nanoscale size and precise surface area. These nanofertilizers significantly impact corn's fatty acid profiles, yield, and plant development (Taheri et al., 2016).

ZnONPs improve soybean plants' total plant growth and root biomass (De la Rosa et al., 2013). In lowland rice, using ZnONPs at sowing dramatically boosted plant height and demonstrated improved responsiveness to shoot length at maturity (Hosseini & Maftoun, 2008). Additionally, exposure to ZnONPs improves germination of seed, maize plant root length, and oat and berseem plant shoot length (Meena et al., 2017). Foliage spray of Zinc NPs can remarkably increase the leaf area and dry mass of maize (Taheri et al., 2016). ZnONPs promote Zn bioavailability to plants and gradually release fertilizer, which raises plant growth and dry matter and fresh and dry matter in the roots. This enhanced plant development by enhancing elemental absorption and nutrient utilization. In order to boost Zn availability for plant nutritional demands in Zn-deficient soils, traditional Zinc fertilizer can be replaced with ZnONPs. This increases fertilizer usage efficiency and can benefit the environment (Adil et al., 2022).

ZnONPs can modify the food and agriculture sector by enhancing productivity globally in reversing oxidative stress symptoms even under stressful conditions. Foliar application of conventional and ZnO nanofertilizers induces morpho-physiological characteristics and harvest index in sunflowers under water deficit stress, and seed treatment with such nanoparticles reduces salt stress in lupine plants (Abdel Latef et al., 2016; Asadazade et al., 2015). Under salt stress, it accelerates plant growth, yield, and antioxidant enzymes in wheat (*Triticum aestivum*) as foliar nanofertilizers (Babaeia et al., 2017). The positive effects of ZnONPs on the secretion of phosphorus mobilizing enzymes and gum concentrations in beans and the sustainable production of *Atriplex halimus* in saline habitats were demonstrated (Raliya & Tarafdar, 2013; Tawfik et al., 2017). Furthermore, it alleviates chilling stress and adverse effects of cadmium in *Oryza sativa* by regulating the antioxidant defense system and chilling response transcription factors, even greatly influencing the rhizospheric environment (Faizan et al., 2021a; Song et al., 2021). Hence, ZnONPs seem to modify interkingdom cell signaling processes relevant to crop performance and production (Anderson et al., 2017). However, the higher concentration of these nanoparticles adversely affects the plants suggesting the toxic effect on plant growth and metabolism in different periods of development (Faizan et al., 2021b; Liu et al., 2022). The high solubility of the particles, which led to cytotoxicity, oxidative stress, and mitochondrial dysfunction, was blamed for these harmful consequences. The degree of ZnONPs' impact is influenced by increased nanoscale reactivity, dosage, plant species and age, exposure route and duration, and environmental factors, including pH and surface interactions with other soil components (Dimkpa et al., 2019a). Taken together, the present chapter explores the pragmatic and negative effect of ZnO nanofertilizer, its properties, and role in the antioxidant systems, and their efficient utilization in agriculture for future sustainability.

2 Synthesis Methods of Zn and ZnONPs

ZnONPs may be produced using various preparation techniques, including spray pyrolysis, vapor transfer, sol–gel, deposition, and hydrothermal processes. The biogenic production of ZnONPs currently frequently uses diverse plant extracts. Through this procedure, NPs of various sizes and shapes may be produced. Various synthesis methods and their properties are delineated in Fig. 1.

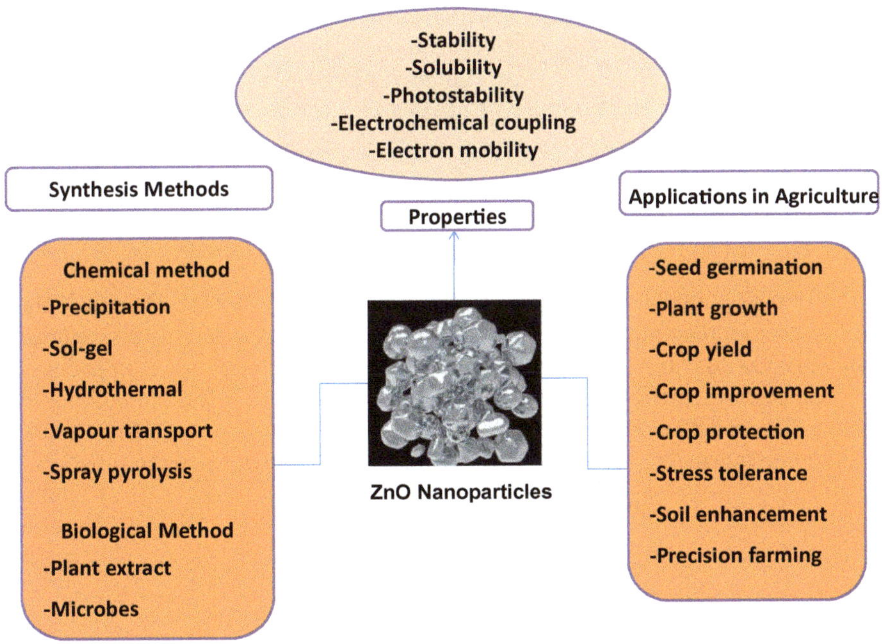

Fig. 1 Synthesis methods and properties of ZnO nanoparticles

2.1 Precipitation Method

Through the direct precipitation method, ZnONPs can be synthesized using zinc nitrate/acetate (0.2 M) and KOH/NaOH (0.4 M) as precursors (Ghorbani et al., 2015). In this method, zinc acetate solution in a beaker is placed on a magnetic stirrer with a hot plate at 30 °C for 40 min, after which NaOH (0.2 M) is added dropwise to the solution with constant stirring. Once the solution became white from the precipitation of ZnO and $Zn(OH)_2$, it was agitated for an additional hour and then allowed to stand for 50 min. The previously synthesized zinc acetate dihydrate (0.1 M) can also be refluxed at 180 or 220 °C in diethylene glycol or trimethylene glycol to produce ZnONPs (Mahamuni et al., 2018). This technique has excellent yield, purity, surface effect, no requirement for organic solvents, simple repeatability, and low cost as its benefits.

2.2 Sol–Gel Method

Chelating chemicals are added after the metal alkoxide has been dissolved in an organic solvent to create a homogenous solution. Coating the substrate, followed by drying, thermal breakdown, and annealing, produces inorganic thin films. By employing zinc acetate dehydrate as a precursor, ethanol as a solvent, sodium hydroxide, and distilled water as a media, ZnONPs can be created (Hasnidawani et al., 2016). High purity and uniformity may be attained by using ethanol at a low temperature. In this liquid-phase synthesis, the dispersion can be stabilized by covering the particles with the appropriate ligands.

2.3 Hydrothermal Process

One of the most popular techniques for creating nanomaterials is this one. This is a solution-reaction strategy. Nanomaterials may be created at various temperatures, from very low to extremely high, regarding hydrothermal synthesis. Depending on the vapor pressure of the base composition in the reaction, either low or high-pressure conditions can influence the materials' morphology (Bulcha et al., 2021). ZnONPs by hydrothermal method contain $Zn(NO_3)_2{\cdot}6H_2O$ and NaOH precursors. 1 M NaOH aqueous ethanol solution is added dropwise to 0.6 M ethanol $Zn(NO_3)_2{\cdot}6H_2O$ aqueous solution, which is placed in a magnetic stirrer for 45 min (Yong et al., 2020).

2.4 *Vapor Transport Method*

The following steps make up the synthesis process. First, when the precursor (ZnO:C) and substrate were suitably positioned in the tube furnace, the furnace temperature was increased from 22 °C (room temperature) to 900–1000 °C (growth temperature) over 15 min while maintaining a continuous flow of Ar (carrier gas) of 70–200 cm. The specific growth temperature was then maintained for reaction times varying from 30 to 120 min. The substrates were taken out when the tube furnace had cooled to room temperature. Growth, temperature, reaction time, and Ar flow rate all impact the creation of a product (Yu et al., 2010).

2.5 *Spray Pyrolysis Method*

When a solution is sprayed over a heated surface, the contents react to produce a chemical product, leaving behind a thin film. This process is known as spray pyrolysis. The chemical reactants are selected to make the target molecule and any additional products volatile at the deposition temperature. Spray pyrolysis may create ZnONPs at different concentrations between 5 and 25 weight percent. Under various atomizing pressures, precursor solution breakdown occurs at 800, 1000, and 1200 °C (Ghaffariana et al., 2011).

2.6 *Green Synthesis*

The bioreduction of metal ions into their elemental form in the range of 1–100 nm size is accomplished by using plants or plant parts in ZnONPs green synthesis, which is an environmentally benign process. Zinc nitrate/acetate is added to the produced plant extracts using a magnetic stirrer for 120 min, and then dropwise additions of NaOH solution are made (Haripriya et al., 2018). For the manufacture of biogenic ZnONPs, the metabolites present in the aqueous plant extract serve as oxidizing, reducing, and capping agents (Godeto et al., 2023). By adjusting the reaction conditions, it is possible to optimize the size and form of NPs during their production using microorganisms. *Aspergillus aeneus*, *Pichia kudriavzevii* yeast strain, and Lactic acid bacteria are heavily utilized for the microbial-mediated production of ZnONPs for their antibacterial properties (Mohd Yusof et al., 2019, 2020).

3 Properties

ZnONPs are often found in clusters and come in a variety of forms, including rod-, star-, and isometric ones. Among the exceptional qualities are high chemical stability and solubility, high photostability, high electrochemical coupling coefficient, and a broad spectrum of radiation absorption. ZnO will concurrently be present in a formulation as a combination of solubilized Zn ions and a significant portion of undissolved ZnONPs, each exhibiting distinct diffusion and uptake properties. The dissolution rate describes the solubilization of ZnONPs in a certain fluid matrix over time, which is also relevant to nanoparticles. The dissolving rate will vary significantly depending on the fluid composition. ZnONPs have a variety of advantageous optoelectronic characteristics, including strong electron mobility, a broadband gap, and superior transparency. With a broadband gap of 3.37 eV and an exciton binding energy of up to 60 meV even at ambient temperature, ZnONPs are also known as n-type multi-functional semiconductor materials (Haque et al., 2020).

3.1 ZnONPs in Plant Growth, Development, and Productivity

Improving yield was necessary to feed a growing population and, at the same time, reduce the environmental impact of food production. Plant productivity hinges on environmental factors, the availability of nutrients in the soil and water, and the photosynthetic capacity of plants. Inappropriate use of fertilizers degrades the soil and results in poor yield. Replacing traditional Zn fertilizer with ZnONPs can help increase Zn availability for plant nutrient needs in Zn-deficient soils. Nanofertilizer provides the plant with nutrients and restores the soil to an organic state without the harmful factors of chemical fertilizer.

ZnONPs can be used as a fertilizer that is released in a controlled manner so that nutrients only reach plants and are not lost to unwanted targets such as soil, water, and microbes. A nanoscale zinc oxide particle promotes germination, growth, and yield in peanut and wheat plants (Prasad et al., 2012; Rizwan et al., 2019). Zinc nanoparticle delivery in maize boosts the plant's grain development, yield, and zinc content (Subbaiah et al., 2016). Different concentrations ZnONPs were applied to peanut seeds to promote germination of seed, seedling vigor, and plant growth. These ZnONPs also effectively promote stem and root growth in peanuts. The average particle size was 25 nm at 1000 ppm concentration.

Nanoparticles are introduced into the pores of the seed coat to increase the penetration of water molecules and stimulate the activity of starch-degrading and ROS-producing enzymes, which physiologically increase the germination of seeds (Javed et al., 2022). An organic synthesized cow dung-based nanomaterial has recently been reported for future good agricultural practices and seed preparation applications in the agricultural seed sector. Enhancing development and vigor depends heavily on NMs' capacity to pierce the tough seed and let water inside. Additionally, a promising technique is nanotechnology's seed priming technology, which boosts the potential of high-yielding plants before seeding (Acharya et al., 2020; Anand et al., 2020). An antioxidant system that boosts phytohormones like indoleacetic acid (IAA) in roots as a result of foliar application of ZnONPs to *Cicer arietinum* promotes growth and root growth (Burman et al., 2013; Pandey et al., 2010). Similarly, applying ZnONPs to rice and pearl millet plants' leaves boosted plant height, root length, and dry biomass (Song et al., 2021; Tarafdar et al., 2014).

According to a study on *Capsicum annuum* seeds treated with ZnONPs, germination rates, root length, and stem length, all increased (Afrayeem & Chaurasia, 2017). Researchers (Atteya et al., 2018; Gheith et al., 2018) discovered that jojoba and maize plants' growth and yield metrics were improved by zinc treatment. The soil components may be impacted by the movement of ZnONPs, which will then impact plant structure. This study demonstrated the synergistic effect of compost and nanoparticles on the growth and yield of flax plants by increasing flax chlorophyll and related molecules, free amino acids, and carbohydrates (Sadak & Bakry, 2020). Treatment of flax plants with ZnO and nano-ZnO improved flax productivity growth, quantity, and quality. The rate of photosynthesis would rise with a more significant concentration of overall photosynthetic pigments and, boosting the process of output, which enhanced growth of plants (Tawfik et al., 2017).

Another study on ZnO-nanorod's impact on the symbiotic connection between *P. indica* and *B. oleracea* demonstrates a synergistic effect of *P. indica* and this nanorod on the growth of *B. oleracea* as well as the biomass of *P. indica.* Investigation on *Arachis hypogaea* has exhibited that exposure of seeds with ZnONPs enhances seed germination percentage and seedling vigor, causing earlier flowering and pod production by 34%. ZnONPs may enter plant cells due to their small size, which helps in seed germination and growth. In order to improve germination of seed and seedling growth, ZnONPs at low dosages can function as a seed priming agent (Srivastav et al., 2021). The ability to use ZnONPs as nanofertilizer in two Brassicaceae family species such as *B. napus* and *Camelina sativa* showed a positive effect on seed germination and improves vigor index, rootlet, and plumule growth (Sarkhosh et al., 2022). Venkatachalam et al., (2017a, 2017b) reported the plant growth-promoting role of ZnONPs coated with molecules in cotton with P supplementation. These nanoparticles increase wheat yield, plant height, root fresh and dry weight, and grain weight, indicating that applying ZnNPs is a more effective method of raising agricultural production (Adil et al., 2022). After adding ZnONPs in low and high doses to *H. sativum* grown in hydroponic systems, Voloshina et al. (2022) looked at the physiological and biochemical responses. The dose, plant species, age, exposure route, length of exposure, and environmental conditions, such as pH and surface interactions with other soil components, all influence the extent of the ZnONPs impact. Exposure route and duration have an impact on nanoscale reactivity as well (Dimkpa et al., 2020a, 2020b). Table 1 shows the effects of zinc and ZnONPs on several plant development metrics.

Table 1 Impact of zinc and zinc oxide nanoparticles in plant growth and development

Nanoparticle	Plant	Response	References
ZnONPs	Wheat, Maize	Increased root and shoot length, antioxidant enzymes (superoxide dismutase, ascorbate peroxidase, guaiacol peroxidase, and catalase)	Srivastav et al. (2021)
ZnO NPs	Tomato	Changes in photosynthetic efficiency and antioxidant system	Faizan et al. (2018)
Phycomolecules coated zinc oxide nanoparticles	Cotton (*Gossypium hirsutum* L.)	Enhanced plant growth	Venkatachalam et al. (2017a, 2017b)
Hybrid ZnONPs	Green pea (*Pisum sativum* L.)	Plant growth	Mukherjee et al. (2016)
ZnONPs	Tomato plants	Growth, photosynthetic traits, and antioxidative enzymes	Wang et al. (2018)
Zinc nanoparticles	Wheat	Plant growth	Du et al. (2019)
Nano-CuO + nano-ZnO	Plants	Plant development	Dimkpa et al. (2015)
ZnONPs and salicylic acid	Rice	Modulation of cellular redox status and antioxidant defense system	Faizan et al. (2021b)
ZnONPs and its bulk form	*Hordeum vulgare* L	Growth, antioxidant defense system, and expression of oxidative stress-related genes	Azarin et al. (2022)
ZnONPs	Spring barley	Physiological and anatomical indices	Rajput et al. (2021)
Nanoscale ZnONPs	Peanut	Germination, growth, and yield	Prasad et al. (2012)
Organic fertilizer and zinc oxide nanoscale	Wheat	Wheat performance and grain nutrient accumulation	Dimkpa et al. (2020a, 2020b)
ZnONPs	Tomato	Physiological and molecular changes	Pejam et al. (2021)
ZnONPs	*Lactuca sativa*	Seedling growth	Galindo-Guzmán et al. (2022)
ZnO nano and ZnO bulk particles	Tobacco	Growth responses, photosynthetic pigment content, and gene expression pattern	Mardi et al. (2022)

(continued)

Table 1 (continued)

Nanoparticle	Plant	Response	References
ZnONPs	*Pleioblastus pygmaeus*	Stimulating antioxidant defense and reducing the metal accumulation and translocation	Emamverdian et al. (2022)
ZnONPs	Soybean plants	Induced shoot and root, net photosynthetic rate, transpiration, stomatal conductance, photochemical yield	Ahmad et al. (2020)
ZnONPs and bacteria	Rice (*Oryza sativa* L.)	Enhanced plant growth and reduced heavy metal toxicity	Akhtar et al. (2021)
Green synthesized (ZnONPs) + *Bacillus cereus* + *Lysinibacillus macroides*)	Rice (*Oryza sativa* L.)	Plant growth and germination	Akhtar et al. (2022)
ZnONPs	*Brassica oleracea* var *italic*	Plant growth	Awan et al. (2021)
ZnONPs	*Chenopodium murale* L	Increased proline content and catalase (CAT), guaiacol peroxidase (GPX), and superoxide dismutase (SOD)	Zoufan et al. (2020)
ZnONPs-based seed priming	Fragrant rice	Modulates early growth and enhances physio-biochemical and metabolic profiles	Li et al. (2020)
Zinc and ZnONPs	*Allium cepa* L	Germination and seedling growth	Tymoszuk and Wojnarowicz (2020)
ZnONPs	*Leucaena leucocephala* seedlings	Plant growth and alleviate heavy metal toxicity	Venkatachalam et al. (2017a, 2017b)
Zinc and iron oxide nanoparticles	Wheat	Improved the plant growth and reduced the oxidative stress	Javed et al. (2022)
Biogenic zinc nanoparticles	*Brassica napus* L	Growth and development	Sohail et al. (2022)
ZnONPs	Alfalfa, tomato, and cucumber	Germination, increased root biomass	De la Rosa et al. (2013)

(continued)

Table 1 (continued)

Nanoparticle	Plant	Response	References
Nano-priming of zinc oxide	rapeseed (*Brassica napus* L.)	Modulation of salinity impact on early seedling stage	El-Badri et al. (2021)
ZnONPs	Onion	Growth, flowering, and seed productivity	Laware and Raskar (2014)
Biosynthesized nanoscale ZnO	*Brassica juncea*	Morphological, biochemical, and molecular aspects	Mazumder et al. (2020)
ZnONPs	Maize	Seed germination	Meena et al. (2017)
Iron-doped zinc oxide nanoparticles	Green peas (*Pisum sativum* L.)	Plant growth	Mukherjee et al. (2014)
ZnONPs and AM fungi	Wheat	Upregulate antioxidant system	Raghib et al. (2020)
ZnONPs	Clusterbean (*Cyamopsis tetragonoloba* L.)	Gum contents	Raliya and Tarafdar (2013)
Titanium dioxide and ZnONPs	Tomato (*Solanum lycopersicum* L.)	Physiological impact	Raliya et al. (2015)
Zinc nanoparticles	*B. napus* L	In vitro germination and biochemical profiling	Sohail et al. (2019)
Nano zinc	Rapeseed	Nutritional enhancement and antioxidant system	Sohail et al. (2020)
ZnONPs	Corn	Plant growth, increased leaf area, and dry biomass	Taheri et al. (2016)
ZnONPs	Sunflower cultivars	Growth and ion concentration	Torabian et al. (2016)
ZnONPs	Brassica nigra seedlings	Growth and antioxidative response	Zafar et al. (2016)
ZnONPs and biofertilizer	Safflower	Regulating ion homeostasis and antioxidant defense responses	Yasmin et al. (2021)
ZnONPs	Wheat (Triticum aestivum L.)	Improved the growth and decreased cadmium concentration in grains	Adrees et al. (2021)
ZnONPs	Fenugreek (*Trigonella foenum-graecum*) plants	Improved seed germination and root development	Shaik et al. (2020)

3.2 *Impact of ZnONPs on Antioxidant Enzymes*

Based on the plant type, the dose, and duration, the antioxidant activity of enzymes in plants subjected to the harmful effects of ZnONPs varies substantially. Depending on the formation of ZnONPs, the comparative activity of antioxidant enzymes may cause the plant to begin a robust antioxidant response. According to studies, there is an increase in low-molecular-weight antioxidants. ZnONPs exhibit antioxidant capabilities because of the electron density transfer at the O atoms, which depends on the atomic arrangement (Zeghoud et al., 2022). Under stress, the concentration of proline, free amino acids, and total soluble sugars is essential for osmotic adjustment, protecting the structure of macromolecules and cell membranes. Increased levels of H_2O_2, MDA, and proline, as well as antioxidant enzyme activity (superoxide dismutase, catalase, and peroxidase) in rice plants, resulted in significantly higher levels of superoxide dismutase, catalase, and peroxidase gene expression (Song et al., 2021). The formation of O_2 and OH radicals from hydrogen peroxide is catalyzed by superoxide dismutase (SOD), significantly reducing the toxicity caused by superoxide. It regulates ROS damage in addition to controlling ROS signaling. The direct or indirect effects of NPs on the SOD gene expression or ROS level might be due to an increase or reduction in superoxide dismutase activity (Voloshina et al., 2022). Because of its well-known function as a cofactor of SOD, zinc serves as an antioxidant and aids plants in quenching ROS. This element functions as a crucial cofactor or a part of the oxidative reactions in several enzymes. ZnONPs have been shown to affect tomato plant development, photosynthetic properties, and antioxidative enzymes, according to Wang et al. (2018).

The ascorbate–glutathione detoxification cycle for H_2O_2 and metal chelation include Glutathione (GSH), an essential non-enzymatic antioxidant. The quantity of synthesis and degradation of plants' balance throughout development scenarios make up GSH's component part, which is a critical redox buffer for plant activity. *Hordeum vulgare* cultivated in hydroponic systems showed higher glutathione in roots and shoots by nanozinc oxide controlling plant growth indices (Voloshina et al., 2022). High Zn concentrations, however, primarily cause plants to impede root development, thicken, and disturb cell division. Superoxide dismutase, ascorbate peroxidase, and guaiacol peroxidase are antioxidant enzymes that help wheat and maize plants develop more successfully (Srivastav et al., 2021). High Zn concentrations, however, primarily cause plants to impede root development, thicken, and disturb cell division. Superoxide dismutase, ascorbate peroxidase, and guaiacol peroxidase are antioxidant enzymes that help wheat and maize plants develop more successfully.

3.3 Impact of ZnONPs Under Stress Situations

By modifying important physiological parameters, ZnONPs have been discovered to operate as a natural regulator for plants in both stressed and non-stressed environments, hence promoting plant growth and development. When developing fertilizers using nanotechnology for soil application, product effectiveness in field crop production may be impacted by unavoidable occurrences like dryness, which affects nutrient mobility in soil and, as a result, plant absorption. Indeed, the impact of drought on soil nutrient availability and agricultural yield continue to be disastrous in some parts of the world (Lesk et al., 2016; Moreno-Jiménez et al., 2019). Due to Zn's part in metabolic mechanisms that control water dynamics, it can mechanically minimize the effects of drought on crops (Dimkpa et al., 2017; Karim et al., 2012). For instance, plants create more abscisic acid (ABA) under water stress to maximize stomatal closure and preserve water. Zn is known to boost abscisic acid production in plants, improving its ability to regulate stomata when water is scarce (Karim & Rahman, 2015; Yang et al., 2018; Zengin, 2006).

Adding ZnONPs increased soybean percentage of seed germination and rate, while decreasing seed fresh and dry weight under drought stress. Additionally, when exposed to salt stress, ZnONPs significantly enhance sunflower and wheat plant growth and development (Song et al., 2021). By promoting wheat growth and Zn concentrations while lowering plant Cd concentrations, ZnONPs also reduced wheat's cadmium (Cd) toxicity.

Flag leaf and grain head were delayed by drought; however, Dimpka et al. (2019a) examined the acceleration of sorghum growth by ZnONPs, finding that ZnONPs reduced this delay. As a result, under drought conditions, the start of reproductive development in wheat was sped up in the presence of ZnONPs. The use of nanoscale micronutrients in field applications may be made more accessible, the problem of smaller and larger nutrient particle segregation in bulk fertilizer blends may be solved, and one-time Zn-urea application may be made easier by coating urea or other N-fertilizers with nano-scale micronutrients like Zn. Coating might not have a bigger impact on output than separate Zn and urea treatments. Therefore, increasing the coating effectiveness of ZnONPs by modifying the urea coating method will further enhance crop performance and Zn uptake. Also, Dimpka and coworkers (2019a) exhibited the positive influence of ZnONPs on shoot/root length, plant height, biomass, chlorophyll, grain yield, and uptake in wheat crops, under 40% field capacity of moisture.

The physiological and metabolic processes of rice are negatively impacted by cooling stress, which lowers the yield. However, by controlling the gene expression of transcription factors involved in the chilling reaction, ZnONPs applied topically to rice may effectively reduce the toxicity of chilling stress. This suggests that ZnONPs were crucial in controlling the chilling response. Additionally, ZnONPs exhibited significant and varied impacts on plant development and chlorophyll production, ultimately boosting antioxidant capability and ROS scavenging capabilities under freezing stress (Song et al., 2021).

Increased cropping system resilience, continued food/feed and nutrition security for humans and animals, and a reduction in nutrient losses and environmental pollution brought on by N-fertilizers are all strongly impacted by ZnONPs' capacity to speed up plant development, increase yield, fortify edible grains with crucial nutrients like Zn, and improve N acquisition under drought stress. The induction of drought tolerance genes by ZnONPs, which effectively facilitated wheat deficiency tolerance, mitigates drought stress's undesirable effects, as reported by Sadati et al. (2022).

3.4 ZnONPs in Crop Protection

Crop plant diseases caused by microorganisms significantly affect productivity and yield loss. Zinc and copper nanoformulations have shown to be the top performers in creating various commercially available agricultural bio-/pesticides used to control weeds and plant diseases. Applying ZnONPs prevents *Sclerospora graminicola* from causing downy mildew in pearl millet, which is reduced by 35%. The study of defense enzymes revealed that the treatment with NPs noticeably increased the defense enzyme activities, namely, peroxidase, phenylalanine ammonia-lyase, lipoxygenase, and polyphenol oxidase. This showed that the defense enzyme genes were overexpressed in seedlings that had received treatment, indicating that ZnONPs could boost growth and generate systemic resistance in pearl millet against *S. gramini* (Nandhini et al., 2019). They function similarly to chemical pesticides and transport bioactive pesticide components, host defense-inducing chemicals, to the target pathogens (Khan et al., 2019). Nanoformulated zinc oxide was known for its potential to control citrus canker on sweet orange (*Citrus sinensis*) and 'Ruby red' grapefruit trees. Citrus scab and melanose, two fungal diseases that affect grapefruit, were also successfully treated by it. This success may be attributable to the translaminar transport of Zinkicide (Graham et al., 2016).

A new platform for environmentally friendly and efficient ways to manage plant diseases has been created by the green synthesis of nanoparticles, which enables the creation of nanoparticles containing bioactive compounds from plants and bacteria (Bachheti et al., 2023; Sabir et al., 2014). According to Wagner et al. (2016), zinc and ZnONPs are potent suppressors of spore germination of *Peronospora tabacina* and infectiousness on tobacco leaves. The growth of *Hordeum vulgare* L., the antioxidant defense system, and the expression of genes related to oxidative stress are all impacted by ZnONPs and their bulk form (Azarin et al., 2022). According to Thunugunta et al. (2018), ZnONPs can effectively increase aubergine development under greenhouse circumstances by enhancing seed germination, photosynthetic pigments, carbohydrates, protein and raising the activity of antioxidant enzymes. In place of synthetic fungicides, ZnONPs synthesized by *Penicillium expansum* prevented *Fusarium* wilt disease in grown aubergine (*Solanum melongena*), enhanced growth characteristics, and improved metabolic functions (Abdelaziz et al., 2022).

ZnONPs have been used to treat various plant diseases and acceleration of plant growth (Khan & Siddiqui, 2018; Siddiqui et al., 2018). The unique use of *Syzygium aromaticum* flower bud extract in the fabrication of ZnONPs allows for the control of *Fusarium graminearum* development and mycotoxins. The effectiveness of ZnONPs as an antifungal and anti-mycotoxin against Fusarium sp. may be attributed to elevated levels of lipid peroxidation, reactive oxygen species (ROS), and shifting ergosterol concentration, which changed the membrane integrity and shape of macroconidia (Lakshmeesha et al., 2019). The synthesis of metal nanoparticles using plant-derived saponins as efficient capping agents and their potential use as antibacterial and anticancer agents are reported in many publications (Nandhini et al., 2019). In order to prevent *Tetranychus urticae* oviposition, Pavela et al. (2017) created silver nanoparticles with saponin caps made from *Saponaria officinalis* root extract.

3.5 Impact of ZnONPs on the Environment

It is significant to note that the process utilized to create nanoparticles affects how dangerous they are. It has been demonstrated that ZnONPs have beneficial and adverse impacts on plant growth and metabolism at various developmental stages. They are absorbed, transported, and accumulated by plants depending on their anatomy and the characteristics of the NPs (Singh et al., 2018). The transfer of water and nutrients in plants' above-ground tissues is halted by the root tissue's ability to absorb nanoparticles, pierce plant cell walls, enter the apoplast, and proceed to the aerial section. By causing cell cycle arrest and inducing apoptosis, which severely damages DNA and protects plants, the accumulation of such ZnONPs in leaf tissues also triggers a plant defense mechanism. Nanofertilizers enter the xylem vessels through the root epidermis and endodermis, then go to the aerial parts of the plant.

Furthermore, the phloem and leaf stomata can carry these nanoparticle nutrients to various plant parts. Arbuscular mycorrhizae have been shown by Wang et al. (2016) to reduce the detrimental effects of ZnONPs and zinc buildup in maize plants. Despite the benefits, studies show that nanoparticles pose potential environmental harm. ZnONPs dramatically decreased the biomass of ryegrass, shrank the root tip, and caused the root epidermal and cortical cells to become extensively vacuolated and collapse, according to toxicology tests. The bulk of ZnONPs remained attached to the root's surface, and particular NPs were also discovered in the root endodermis, stele, and apoplast. Toxic effects of ZnONPs on bacteria, *Daphnia magna*, and freshwater microalga were also reported (Sabir et al., 2014). The photosynthesis features may be significantly hampered by a highly high NP concentration, which might stunt or even kill plant development. Therefore, there has to be a thorough reexamination of NP release into the environment, NP contact with plants, and NP effects on plants, ecosystems, and the overall environment.

4 Conclusion

The early availability of nutrients to roots and, subsequently, an improvement in crop yield are made possible by metal and metal oxide nano-fertilizers, which offer enormous potential in sustainable agriculture. Excessive amounts of nanoparticles are inadequate for plants, although their traces might benefit them. Therefore, there is a need for further study to comprehend the molecular mechanism of plant nanoparticle interaction in order for nanotechnology to improve agricultural productivity in the future and assure food security.

References

Abdel Latef, A. A. H., Abu Alhmad, M. F., & Abdelfattah, K. E. (2016). The possible roles of priming with ZnO nanoparticles in mitigation of salinity stress in lupine (*Lupinus termis*) plants. *Journal of Plant Growth Regulation*. https://doi.org/10.1007/s00344-016-9618-x

Abdelaziz, A. M., Salem, S. S., Khalil, A. M. A., El-Wakil, D. A., Fouda, H. M., & Hashem, A. H. (2022). Potential of biosynthesized zinc oxide nanoparticles to control *Fusarium* wilt disease in eggplant (*Solanum melongena*) and promote plant growth. *BioMetals, 35*, 601–616.

Acharya, P., Jayaprakasha, G. K., Crosby, K. M., Jifon, J. L., & Patil, B. S. (2020). Nanoparticle-mediated seed priming improves germination, growth, yield, and quality of watermelons (*Citrullus lanatus*) at multi-locations in texas. *Science and Reports, 10*, 5037.

Adil, M., Bashir, S., Bashir, S., Aslam, Z., Ahmad, N., Younas, T., Asghar, R. M. A., Alkahtani, J., Dwiningsih, Y., & Elshikh, M. S. (2022). Zinc oxide nanoparticles improved chlorophyll contents, physical parameters, and wheat yield under salt stress. *Frontiers in Plant Science, 13*, 932861.

Adrees, M., Khan, Z. S., Hafeez, M., Rizwan, M., Hussain, K., Asrar, M., Alyemeni, N., Wijaya, L., & Ali, S. (2021). Foliar exposure of zinc oxide nanoparticles improved the growth of wheat (*Triticum aestivum* L.) and decreased cadmium concentration in grains under simultaneous Cd and water deficient stress. *Ecotoxicology and Environmental Safety, 208*, 111627.

Afrayeem, S. M., & Chaurasia, A. (2017). Effect of zinc oxide nanoparticles on seed germination and seed vigour in chilli (*Capsicum annuum* L.). *Journal of Pharmacognosy and Phytochemistry, 6*, 1564–1566.

Ahmad, P., Alyemeni, M. N., Al-Huqail, A. A., Alqahtani, M. A., Wijaya, L., Ashraf, M., Kaya, C., & Bajguz, A. (2020). Zinc oxide nanoparticles application Alleviates Arsenic (As) toxicity in soybean plants by restricting the uptake of as and modulating key biochemical attributes, antioxidant enzymes, ascorbate-glutathione cycle and glyoxalase system. *Plants, 9*, 825.

Akhtar, N., Khan, S., Réhman, S. U., Rehman, Z. U., Khatoon, A., Rha, E. S., & Jamil, M. (2021). Synergistic effects of zinc oxide nanoparticles and bacteria reduce heavy metals toxicity in rice (*Oryza sativa* L.). *Plant Toxics, 9*(5),113.

Akhtar, N., Khan, S., Rehman, S. U., Rha, E. S., & Jamil, M. (2022). Combined effect of zinc oxide nanoparticles and bacteria on osmolytes and antioxidative parameters of rice (*Oryza sativa* L.). *Plant Grown in Heavy Metal-Contaminated Water*, 4148765. https://doi.org/10.1155/2022/4148765

Anand, K. V., Anugraga, A. R., Kannan, M., Singaravelu, G., & Govindaraju, K. (2020). Bio-engineered magnesium oxide nanoparticles as nano-priming agent for enhancing seed germination and seedling vigor of green gram (*Vigna radiata* L.). *Materials Letters, 8*, 127792.

Anderson, A. J., McLean, J. E., Jacobson, A. R., & Britt, D. W. (2017). CuO and ZnO nanoparticles modify interkingdom cell signaling processes relevant to crop production. *Journal of Agricultural and Food Chemistry, 66*(26), 6513–6524.

Asadazade, N., Moosavi, S. G., & Seghatoleslami, M. J. (2015). Effect of low irrigation and Zn and SiO2 nano-fertilizers and conventional fertilizers on morphophysiological traits and seed yield of sunflower. *Biological Forum, 7*(1), 357–364.

Atteya, A. K., Esmail, G., Genaidy, G. E., & Hamdy, Z. A. (2018). Chemical constituents and yield of Simmondsia chinensis plants as affected by foliar application of gibberellic acid and zinc sulphate. *Bioscience Research, 15*(3), 1528–1541.

Awan, S. J., Shahzadi, K., Javad, S., Tariq, A., Ahmad, A. H., & Ilyas, S. (2021). A preliminary study of the influence of zinc oxide nanoparticles on growth parameters of *Brassica oleracea* var *italic*. *Journal of the Saudi Society of Agricultural Sciences, 20*, 18–24.

Azarin, K., Usatov, A., Minkina, T., Plotnikov, A., Kasyanova, A., Fedorenko, A., Duplii, N., Vechkanov, E., Rajput, V. D., & Mandzhieva, S. (2022). Effects of ZnO nanoparticles and its bulk form on growth, antioxidant defense system and expression of oxidative stress-related genes in *Hordeum vulgare* L. *Chemosphere, 287*, 132167. https://doi.org/10.1016/j.chemosphere.2021.132167

Babaeia, K., Sharifia, R. S., Pirzadb, A., & Khalilzadeha, R. (2017). Effects of biofertilizer and nano Zn-Fe oxide on physiological traits, antioxidant enzymes activity and yield of wheat (*Triticum aestivum* L.) under salinity stress. *Journal of Plant Interactions, 12*(1), 381–389.

Bachheti, R. K., & Bachheti, A. (Eds.). (2023). *Secondary metabolites from medicinal plants: Nanoparticles synthesis and their applications*. CRC Press. https://doi.org/10.1201/9781003213727

Bulcha, B., Tesfaye, J. L., Anatol, D., Shanmugam, R., Priyanka, L., Dwarampudi, N., Nagaprasad, V. L., Bhargavi, N., & Krishnaraj, R. (2021). Synthesis of zinc oxide nanoparticles by hydrothermal methods and spectroscopic investigation of ultraviolet radiation protective properties. *Journal of Nanomaterials, 2021*, 8617290.

Burman, U., Saini, M., & Kumar, P. (2013). Effect of zinc oxide nanoparticles on growth and antioxidant system of chickpea seedlings. *Toxicological and Environmental Chemistry, 95*(4), 605–612.

Dimkpa, C. O., Andrews, J., Fugice, J., Singh, U., Bindraban, P. S., Elmer, W. H., Gardea-Torresdey, J. L., & White, J. C. (2020a). Facile coating of urea with low-dose ZnO nanoparticles promotes wheat performance and enhances Zn uptake under drought stress. *Frontiers in Plant Science, 11*, 168.

Dimkpa, C. O., Andrews, J., Sanabria, J., Bindraban, P. S., Singh, U., Elmer, W. H., & White, J. C. (2020b). Interactive effects of drought, organic fertilizer, and zinc oxide nanoscale and bulk particles on wheat performance and grain nutrient accumulation. *Science of the Total Environment, 722*, 137808.

Dimkpa, C., Bindraban, P., Fugice, J., Agyin-Birikorang, S., Singh, U., & Hellums, D. (2017). Composite micronutrient nanoparticles and salts decrease drought stress in soybean. *Agronomy for Sustainable Development, 37*, 5.

Dimkpa, C. O., McLean, J. E., Britt, D. W., & Anderson, A. J. (2015). Nano-CuO and interaction with nano-ZnO or soil bacterium provide evidence for the interference of nanoparticles in the metal nutrition of plants. *Ecotoxicol, 24*, 119–129.

Dimkpa, C. O., Singh, U., Bindraban, P. S., Adisa, I. O., Elmer, W. H., Gardea-Torresdey, J. L., & White, J. C. (2019a). Addition-omission of zinc, copper, and boron nano and bulk particles demonstrate element and size-specific response of soybean to micronutrients exposure. *Science of the Total Environment, 665*, 606–616.

Dimkpa, C. O., Singh, U., Bindraban, P. S., Elmer, W. H., Gardea-Torresdey, J. L., & White, J. C. (2019b). Zinc oxide nanoparticles alleviate drought-induced alterations in sorghum performance, nutrient acquisition, and grain fortification. *Science of the Total Environment, 688*, 926–934.

Du, W., Yang, J., Peng, Q., Liang, X., & Mao, H. (2019). Comparison study of zinc nanoparticles and zinc sulfate on wheat growth: From toxicity and zinc biofortification. *Chemosphere, 227*, 109–116.

El-Badri, A. M., Batool, M., Mohamed, I. A., Khatab, A., Sherif, A., Wang, Z., Salah, A., Nishawy, E., Ayaad, M., Kuai, J., Wang, B., & Zhou, G. (2021). Modulation of salinity impact on early seedling stage via nano-priming application of zinc oxide on rapeseed (*Brassica napus* L.). *Plant Physiology and Biochemistry, 166*, 376–392.

Emamverdian, A., Hasanuzzaman, M., Ding, Y., Barker, J., Mokhberdoran, F., & Liu, G. (2022). Zinc oxide nanoparticles improve *Pleioblastus pygmaeus* plant tolerance to arsenic and mercury by stimulating antioxidant defense and reducing the metal accumulation and translocation. *Frontiers in Plant Science, 13*, 841501. https://doi.org/10.3389/fpls.2022.841501

Faizan, M., Faraz, A., Yusuf, M., Khan, S. T., & Hayat, S. (2018). Zinc oxide nanoparticle-mediated changes in photosynthetic efficiency and antioxidant system of tomato plants. *Photosynthetica, 56*, 678–686.

Faizan, M., Bhat, J. A., Hessini, K., Yu, F., Ahmad, P. (2021a). Zinc oxide nanoparticles alleviate the adverse effects of cadmium stress on *Oryza sativa* via modulation of the photosynthesis and antioxidant defense system. *Ecotoxicology and Environmental Safety, 1*(220), 112401.

Faizan, M., Sehar, S., Rajput, V. D., Faraz, A., Afzal, S., Minkina, T., Sushkova, S., Adil, M. F., Yu, F., Alatar, A. A., Akhter, F., & Faisal, M. (2021b). Modulation of cellular redox status and antioxidant defense system after synergistic application of zinc oxide nanoparticles and salicylic acid in rice (*Oryza sativa*) plant under arsenic stress. *Plants, 10*, 2254.

Fraceto, L. F., Grillo, R., de Medeiros, G. A., Scognamiglio, V., Rea, G., & Bartolucci, C. (2016). Nanotechnology in Agriculture: Which Innovation Potential Does It Have? *Frontiers in Environmental Science, 4*, 20.

Galindo-Guzmán, A. P., Fortis-Hernández, M., Rosa-Reta, C. V., Zermeño-González, H., & Galindo-Guzmán, M. (2022). Chemical synthesis of zinc oxide nanoparticles and their evaluation in *Lactuca sativa* seedlings. *Remexca*. https://doi.org/10.29312/remexca.v13i28.3284

Ghaffariana, H. R., Saiedi, M., Sayyadnejad, M. A., & Rashidi, A. M. (2011). Synthesis of ZnO nanoparticles by spray pyrolysis method. *Iranian Journal of Chemistry and Chemical Engineering, 30*(1), 1–6.

Gheith, E. M. S., Shafik, M. M., El-Badry, O. Z., Abdul Kareem, B. M. (2018). Growth and productivity of maize (Zea mays L.) as affected by nitrogen and zinc fertilizer levels: Growth analysis. *Bioscience Research, 15*(1), 54–59.

Ghorbani, H. R., Mehr, F. P., Pazoki, H., & Rahmani, B. M. (2015). Synthesis of ZnO nanoparticles by precipitation method. *Oriental Journal of Chemistry, 31*(2), 1219–1221.

Godeto, Y. G., Ayele, A., Ahmed, I. N., Husen, A., & Bachheti, R. K. (2023). Medicinal plant-based metabolites in nanoparticles synthesis and their cutting-edge applications: An overview. In *Secondary metabolites from medicinal plants* (pp. 1–34). CRC Press.

Graham, J. H., Johnson, E. G., Myers, M. E., Young, M., Rajasekaran, P., Das, S., & Santra, S. (2016). Potential of nano-formulated zinc oxide for control of citrus canker on grapefruit trees. *Plant Disease, 100*(12), 2442–2447.

Haque, M. J., Bellah, M. M., Hassan, M. R., & Rahman, S. (2020). Synthesis of ZnO nanoparticles by two different methods & comparison of their structural, antibacterial, photocatalytic and optical properties. *Nano Ex., 1*, 010007.

Haripriya, P., Stella, P. M., Anusuya, S. (2018). Foliar spray of zinc oxide nanoparticles improves salt tolerance in finger millet crops under glasshouse condition. *SCIOL Biotechnology*, 20–29.

Hasnidawani, J. N., Azlina, H. N., Norita, H., Bonnia, N. N., Ratim, S., & Ali, E. S. (2016). Synthesis of ZnO Nanostructures Using Sol-Gel Method. *Procedia Chemistry, 19*, 211–216.

Hosseini, Y., & Maftoun, M. (2008). Effects of nitrogen levels, nitrogen sources, and zinc rates on the growth and mineral composition of lowland rice. *JAST, 10*, 307–316.

Husen, A., & Siddiqi, K. S. (2023). *Advances in smart nanomaterials and their applications*. Elsevier Inc. https://doi.org/10.1016/C2021-0-02202-1

Husen, A. (2022). *Engineered nanomaterials for sustainable agricultural production, soil improvement and stress management*. Elsevier Inc. https://doi.org/10.1016/C2021-0-00054-7

Husen, A. (2023). *Nanomaterials and nanocomposites exposures to plants (Response, Interaction, Phytotoxicity and Defense Mechanisms)*. Springer Nature Singapore Pte Ltd. https://doi.org/10.1007/978-981-99-2419-6

Javed, Z., Tripathi, G. D., Mishra, M., Meghana, G., & Kavya, D. (2022). Cow dung extract mediated green synthesis of zinc oxide nanoparticles for agricultural applications. *Science and Reports, 12*, 20371.

Karim, M. R., & Rahman, M. A. (2015). Drought risk management for increased cereal production in Asian least developed countries. *Weather and Climate Extremes, 7*, 24–35.

Karim, M. R., Zhang, Y. Q., Zhao, R. R., Chen, X. P., Zhang, F. S., & Zou, C. Q. (2012). Alleviation of drought stress in winter wheat by late foliar application of zinc, boron, and manganese. *Journal of Soil Science and Plant Nutrition, 175*, 142–151.

Khan, M., & Siddiqui, Z. A. (2018). Zinc Oxide nanoparticles for the management of *Ralstonia solanacearum, Phomopsis vexans* and *Meloidogyne incognita* incited disease complex of eggplant. *Indian Phytopathol, 71*(3), 355–364.

Khan, M. R., Rizvi, T. F., & Ahamad, F. (2019). Application of nanomaterials in plant disease diagnosis and management. *Nanobiotechnology Applications in Plant Protection,* 19–33

Lakshmeesha, T. R., Kalagatur, N. K., Mudili, V., Mohan, C. D., Rangappa, S., Prasad, B. D., Ashwini, B. S., Hashem, A., Alqarawi, A. A., Malik, J. A., Abd_Allah, E. F., Gupta, V. K., Siddaiah, C. N., & Niranjana, S. R. (2019). Biofabrication of zinc oxide nanoparticles with *Syzygium aromaticum* flower buds extract and finding its novel application in controlling the growth and mycotoxins of *Fusarium graminearum. Frontiers in Microbiology, 10*, 1244

Laware, S. L., & Raskar, S. (2014). Influence of zinc oxide nanoparticles on growth, flowering and seed productivity in onion. *International Journal of Current Microbiology and Applied Sciences, 3*, 874–881.

Lesk, C., Rowhani, P., & Ramankutty, N. (2016). Influence of extreme weather disasters on global crop production. *Nature, 529*, 84–87.

Li, Y., Liang, L., Li, W., Ashraf, U., Ma, L., Tang, X., Pan, S., Tian, H., & Mo, Z. (2020). ZnO nanoparticle-based seed priming modulates early growth and enhances physio-biochemical and metabolic profiles of fragrant rice against cadmium toxicity. *Journal of Nanbiotechnology, 19*, 75.

Liu, L., Nian, H., & Lian, T. (2022). Plants and rhizospheric environment: Affected by zinc oxide nanoparticles (ZnO NPs) A review. *Plant Physiology and Biochemistry, 185*, 91–100.

Mahamuni, P. P., Patil, P. M., Dhanavade, M. J., Badiger, M. V., Shadija, P. G., Lokhande, A. C., & Bohara, R. A. (2018). Synthesis and characterization of zinc oxide nanoparticles by using polyol chemistry for their antimicrobial and antibiofilm activity. *Biochemistry and Biophysics Report, 17*, 71–80.

Mardi, A., Mohajjel Shoja, H., & Kazemi, M. E. (2022) Comparative study of growth responses, photosynthetic pigment content, and gene expression pattern in tobacco plants treated with ZnO nano and ZnO bulk particles. *Journal of Nano Research, 24*(10). https://doi.org/10.1007/s11051-022-05583-4.

Mazumder, J. A., Khan, E., Perwez, M., Gupta, M., Kumar, S., Raza, K., & Sardar, M. (2020). Exposure of biosynthesized nanoscale ZnO to *Brassica juncea* crop plant: Morphological, biochemical and molecular aspects. *Science and Reports, 10*, 8531.

Meena, D. S., Jayadeva, H. M., Gautam, C., & Meena, H. M. (2017). Effects of nano zinc oxide (ZnO) particles on germination of maize (*Zea mays* L.) seeds. *International Journal of Plant and Soil Science, 16*, 1–5.

Mittal, D., Kaur, G., Singh, P., Yadav, K., & Ali, S. A. (2020). Nanoparticle-based sustainable agriculture and food science: Recent advances and future outlook. *Frontiers in Nanotechnology, 2*, 579954.

Mohd Yusof, H., Abdul Rahman, N., Mohamad, R., Zaidan, U. H., & Samsudin, A. A. (2020). Biosynthesis of zinc oxide nanoparticles by cell-biomass and supernatant of *Lactobacillus plantarum* TA4 and its antibacterial and biocompatibility properties. *Science and Reports, 10*, 19996.

Mohd Yusof, H., Mohamad, R., Zaidan, U. H., & Abdul Rahman, N. A. (2019). Microbial synthesis of zinc oxide nanoparticles and their potential application as an antimicrobial agent and a feed supplement in animal industry: A review. *Journal of Animal Science and Biotechnology, 10*, 57.

Moreno-Jiménez, E., Plaza, C., Saiz, H., Manzano, R., Flagmeier, M., & Maestre, F. T. (2019). Aridity and reduced soil micronutrient availability in global drylands. *Nature Sustainability, 2*, 371–377.

Mukherjee, A., Pokhrel, S., Bandyopadhyay, S., Madler, L., Peralta-Videa, J. R., & Gardea-Torresdey, J. L. (2014). A soil mediated phyto-toxicological study of iron-doped zinc oxide nanoparticles (Fe@ ZnO) in green peas (*Pisum sativum* L.). *Chemical Engineering, 258*, 394–401.

Mukherjee, A., Sun, Y., Morelius, E., Tamez, C., Bandyopadhyay, S., Niu, G., & Gardea-Torresdey, J. L. (2016) Differential toxicity of bare and hybrid ZnO nanoparticles in green pea (*Pisum sativum* L.): A life cycle study. *Frontiers in Plant Science, 6*, 1242.

Nandhini, M., Rajini, S. B., Udayashankar, A. C., Niranjana, S. R., Lund, O. S., Shetty, H. S., & Prakash, H. S. (2019). Biofabricated zinc oxide nanoparticles as an eco-friendly alternative for growth promotion and management of downy mildew of pearl millet. *Crop Protection, 121*, 103–112.

Pandey, A. C., Sanjay, S. S., & Yadav, R. S. (2010). Application of ZnO nanoparticles in influencing the growth rate of *Cicer arietinum*. *Journal of Experimental Nanoscience, 5*(6), 488–497.

Pavela, R., Angelo, C., Giovanni, B., & Murugan, K. (2017). *Saponaria officinalis*-synthesized silver nanocrystals as effective biopesticides and oviposition inhibitors against *Tetranychus urticae* Koch. *Industrial Crops and Products, 97*, 338–344.

Pejam, F., Ardebili, Z. O., Ladan-Moghadam, A., & Danaee, E. (2021). Zinc oxide nanoparticles mediated substantial physiological and molecular changes in tomato. *PLoS ONE, 16*. https://doi.org/10.1371/journal.pone.0248778

Prasad, T. N. V. K. V., Sudhakar, P., Sreenivasulu, Y., Latha, P., Munaswamy, V., Reddy, K. R., Sree Prasad, T. S., Sanjanla, P. R., & Pradeep, T. (2012). Effect of nanoscale zinc oxide particles on the germination, growth and yield of peanut. *Journal of Plant Nutrition, 35*, 905–927.

Raghib, F., Naikoo, M. I., Khan, F. A., Alyemeni, M. N., & Ahmad, P. (2020). Interaction of ZnO nanoparticle and AM fungi mitigates Pb toxicity in wheat by upregulating antioxidants and restricted uptake of Pb. *Journal of Biotechnology, 323*, 254–263.

Rajput, V. D., Minkina, T., Fedorenko, A., Chernikova, N., Hassan, T., Mandzhieva, S., Sushkova, S., Lysenko, V., Soldatov, M. A., & Burachevskaya, M. (2021). Effects of zinc oxide nanoparticles on physiological and anatomical indices in spring barley tissues. *Nanomater, 30*, 1722.

Raliya, R., Nair, R., Chavalmane, S., Wang, W. N., & Biswas, P. (2015). Mechanistic evaluation of translocation and physiological impact of titanium dioxide and zinc oxide nanoparticles on the tomato (*Solanum lycopersicum* L.) plant. *Metallomics, 7*, 1584–1594.

Raliya, R., & Tarafdar, J. C. (2013). ZnO nanoparticle biosynthesis and its effect on phosphorous-mobilizing enzyme secretion and gum contents in cluster bean (*Cyamopsis tetragonoloba* L.). *Agricultural Research, 2*, 48–57.

Rizwan, M., Ali, S., Ali, B., Adrees, M., Arshad, M., Hussain, A., Ur Rahman, M. Z., & Waris, A. A. (2019). Zinc and iron oxide nanoparticles improved the plant growth and reduced the oxidative stress and cadmium concentration in wheat. *Chemosphere, 214*, 269–277.

De la Rosa, G., Lopez-Moreno, M. L., de Haro, D., Botez, C. E., Peralta-Videa, J. R., & Gardea-Torresdey, J. L. (2013). Effects of ZnO nanoparticles in alfalfa, tomato, and cucumber at the germination stage: Root development and X-ray absorption spectroscopy studies. *Pure and Applied Chemistry, 85*, 2161–2174.

Sabir, S., Arshad, M., & Chaudhari, S. K. (2014). Zinc oxide nanoparticles for revolutionizing agriculture: synthesis and applications. *The Scientific World Journal, 2014*, 925494.

Sadak, M. S., & Bakry, B. A. (2020). Zinc-oxide and nano ZnO oxide effects on growth, some biochemical aspects, yield quantity and quality of flax (*Linum uitatissimum* L.) in absence and presence of compost under sandy soil. *Bulletin of the National Research Centre, 44*, 98.

Sadati, R. S. Y., Godehkahriz, J. S., Ebadi, A., & Sedghi, M. (2022). Zinc oxide nanoparticles enhance drought tolerance in wheat via physio-biochemical changes and stress genes expression. *Iranian Journal of Biotechnology, 20*(1), e3027.

Sarkhosh, S., Kahrizi, D., Darvishi, E., Tourang, M., Haghighi-Mood, S., Vahedi, P., & Ercisli, S. (2022). Effect of zinc oxide nanoparticles (ZnO-NPs) on seed germination characteristics in two Brassicaceae family species: *Camelina sativa* and *Brassica napus* L. *Journal of Nanomaterials,* 1892759. https://doi.org/10.1155/2022/1892759

Shaik, A. M., David Raju, M., & Rama Sekhara Reddy, D. (2020). Green synthesis of zinc oxide nanoparticles using aqueous root extract of *Sphagneticola trilobat*e Lin and investigate its role in toxic metal removal, sowing germination and fostering of plant growth. *Inorganic and Nano-Metal Chemistry, 50*, 569–579.

Siddiqui, Z. A., Khan, A., Khan, M. R., Abd-Allah, E. F. (2018). Effects of Zinc Oxide Nanoparticles (ZnO NPs) and some plant pathogens on the growth and nodulation of lentil (*Lens culinaris* Medik.). *Acta Phytopathol et Entomol Hungarica, 53*(2), 195.

Singh, A., Singh, N. B., Afzal, S., Afzal, S., Singh, T., & Hussain, I. (2018). Zinc oxide nanoparticles: A review of their biological synthesis, antimicrobial activity, uptake, translocation and biotransformation in plants. *Journal of Materials Science, 53*, 185–201.

Sohail, K. K., Kemmerling, B., Shutaywi, M., & Mashwani, Z. U. R. (2020). Nano zinc elicited biochemical characterization, nutritional assessment, antioxidant enzymes, and fatty acid profiling of rapeseed. *PLoS ONE, 15*, e0241568.

Sohail, A. U., Shad, S., Ilyas, N., Manaf, A., Raja, N. I., & Mashwani, Z. U. (2019). In vitro germination and biochemical profiling of *B. napus* L. in response to biosynthesized zinc nanoparticles. *IET Nanobiotechnology, 13*, 46–51.

Sohail, S. L., Ferrari, E., Stierhof, Y. D., Kemmerling, B., & Mashwani, Z. R. (2022). Molecular effects of biogenic zinc nanoparticles on the growth and development of *Brassica napus* L. Revealed by proteomics and transcriptomics. *Frontiers in Plant Science, 13*, 798751.

Song, Y., Jiang, M., Zhang, H., Li, R. (2021). Zinc oxide nanoparticles alleviate chilling stress in rice (*Oryza Sativa* L.) by regulating antioxidative system and chilling response transcription factors. *Molecules, 26*(8), 2196.

Srivastav, A., Ganjewala, D., Singhal, R. K., Rajput, V. D., Minkina, T., Voloshina, M., Srivastava, S., & Shrivastava, M. (2021). Effect of ZnO nanoparticles on growth and biochemical responses of wheat and maize. *Plants, 10*(12), 2556.

Subbaiah, L. V., Prasad, T. N., Krishna, T. G., Sudhakar, P., Reddy, B. R., & Pradeep, T. (2016). Novel effects of nanoparticulate delivery of zinc on growth, productivity, and zinc biofortification in maize (*Zea mays* L.). *Journal of Agricultural and Food Chemistry, 64*(19), 3778–3788.

Taheri, M., Qarache, H. A., Qarache, A. A., & Yoosefi, M. (2016). The effects of zinc-oxide nanoparticles on growth parameters of corn (SC704). *STEM Fellowship J, 1*, 17–20.

Tarafdar, J. C., Raliya, R., Mahawar, H., & Rathore, I. (2014). Development of zinc nanofertilizer to enhance crop production in pearl millet (*Pennisetum americanum*). *Agricultural Research, 3*(3), 257–262.

Tawfik, M. M., Bakhoum, G. S., Sadak Mervat, S., & Kabesh, M. O. (2017). Application of ZnO nanoparticles for sustainable production of *Atriplex halimus* in saline habitats. *Bulletin of the NRC, 41*(2), 286–305.

Thunugunta, T., Reddy, A. C., Seetharamaiah, S. K., Ramanna Hunashikatti, L., Gowdra Chandrappa, S., Cherukatu Kalathil, N., Dhoranapalli Chinnappa Reddy, L. R. (2018) Impact of zinc oxide nanoparticles on eggplant (*S. melongena*): studies on growth and the accumulation of nanoparticles. *IET Nanobiotechnol, 12*(6), 706–713.

Torabian, S., Zahedi, M., & Khoshgoftar, A. H. (2016). Effects of foliar spray of two kinds of zinc oxide on the growth and ion concentration of sunflower cultivars under salt stress. *Journal of Plant Nutrition, 39*, 172–180.

Tymoszuk, A., & Wojnarowicz, J. (2020). Zinc oxide and zinc oxide nanoparticles impact on in vitro germination and seedling growth in *Allium cepa* L. *Mater, 13*(12), 2784.

Venkatachalam, P., Jayaraj, M., Manikandan, R., Geetha, N., Rene, E. R., Sharma, N. C., Sahi, S. V. (2017). Zinc oxide nanoparticles (ZnONPs) alleviate heavy metal-induced toxicity in *Leucaena leucocephala* seedlings: A physiochemical analysis. *Plant Physiology and Biochemistry, 110*, 59–69.

Venkatachalam, P., Priyanka, N., Manikandan, K., Ganeshbabu, I., Indiraarulselvi, P., Geetha, N., & Sahi, S. V. (2017b). Enhanced plant growth promoting role of phycomolecules coated zinc oxide nanoparticles with P supplementation in cotton (*Gossypium hirsutum* L.) *Plant Physiol Biochem, 110*, 118–127.

Voloshina, M., Rajput, V. D., Minkina, T., Vechkanov, E., Mandzhieva, S., Mazarji, M., Churyukina, E., Plotnikov, A., Krepakova, M., & Wong, M. H. (2022). Zinc oxide nanoparticles: physiological and biochemical responses in barley (*Hordeum vulgare* L.). *Plants, 11*, 2759.

Wagner, G., Korenkov, V., Judy, J. D., & Bertsch, P. M. (2016). Nanoparticles composed of Zn and ZnO inhibit *Peronospora tabacina* spore germination in vitro and *P. tabacina* infectivity on tobacco leaves. *Nanomater (Basel), 6*(3), 50.

Wang, X. P., Li, Q. Q., Pei, Z. M., & Wang, S. C. (2018). Effects of zinc oxide nanoparticles on the growth, photosynthetic traits, and antioxidative enzymes in tomato plants. *Biologia Plantarum, 62*, 801–808.

Wang, F., Liu, X., Shi, Z., Tong, R., Adams, C. A., & Shi, X. (2016). Arbuscular mycorrhizae alleviate negative effects of zinc oxide nanoparticle and zinc accumulation in maize plants–a soil microcosm experiment. *Chemosphere, 147*, 88–97.

Yang, K. Y., Doxey, S., McLean, J. E., Britt, D., Watson, A., Al Qassy, D., Jacobson, A., & Anderson, A. J. (2018). Remodeling of root morphology by CuO and ZnO nanoparticles: Effects on drought tolerance for plants colonized by a beneficial pseudomonad. *Botany, 96*, 175–186.

Yasmin, H., Mazher, J., Azmat, A., Nosheen, A., Naz, R., Hassan, M. N., Noureldeen, A., & Ahmad, P. (2021). Combined application of zinc oxide nanoparticles and biofertilizer to induce salt resistance in safflower by regulating ion homeostasis and antioxidant defense responses. *Ecotoxicology and Environmental Safety, 218*, 112262.

Yong, X. G., Jayatissa, A. H., Yu, Z., Chen, X., & Li, M. (2020). Hydrothermal synthesis of nanomaterials. *Journal of Nanomaterials*, 8917013.

Yu, D., Trad, T., McLeskey, J. T., Craciun, V., & Taylor, C. R. (2010). ZnO nanowires synthesized by vapor phase transport deposition on transparent oxide substrates. *Nanoscale Research Letters, 5*, 1333.

Zafar, H., Ali, A., Ali, J. S., Haq, I. U., & Zia, M. (2016). Effect of ZnO nanoparticles on *Brassica nigra* seedlings and stem explants: Growth dynamics and antioxidative response. *Frontiers in Plant Science, 7*, 535.

Zeghoud, S., Hemmami, H., Seghir, B. B., Amor, I. B., Kouadri, I., Rebiai, A., Messaoudi, M., Ahmed, S., Pohl, P., & Simal-Gandara, J. (2022). A review on biogenic green synthesis of ZnO nanoparticles by plant biomass and their applications. *Materials Today Communications, 33*, 104747.

Zengin, F. K. (2006). The effects of Co_{2+} and Zn_{2+} on the contents of protein, abscisic acid, proline and chlorophyll in bean (*Phaseolus vulgaris* cv. *strike*) seedlings. *Journal of Environmental Biology, 27*, 441–448.

Zoufan, P., Baroonian, M., & Zargar, B. (2020). ZnO nanoparticles-induced oxidative stress in *Chenopodium murale* L, Zn uptake, and accumulation under hydroponic culture. *Environmental Science and Pollution Research, 27*, 11066–11078.

Recent Updates on the Use of Smart Nanostructures for Food Packaging Applications

Noopur Khare, Anil Patani, Pragati Khare, and Sachidanand Singh

Abstract The food packaging sector is undergoing different scientific developments in response to consumer demands and trends. Current trends emphasize both sophisticated food protection and consumer-friendly food presentation. Non-biodegradable materials are replaced with eco-friendly biodegradable composite materials that have good public awareness in the present scenario. It has also addressed the shortcomings of the current approaches; nanotechnology is thriving and has shown to be a terrific addition to the packaging sector. Different nanomaterials have been investigated and put to the test for making packaging materials. In this chapter, we will talk about the contemporary development in food packing and the implementation of sustainable packaging. Nanotechnology is also emerging as a potential and widely utilized option, besides the nanomaterials previously used in food packaging, such as zinc nanoparticles, silver, and clay of polymers. This chapter will add potential understanding and issues related to the use of nanotechnology in food packaging.

Keywords Nanotechnology · Nanostructure · Biodegradable · Packaging · Polymers

N. Khare
Bhai Gurdas Institute of Engineering and Technology, Sangrur, Punjab, India

P. Khare
Department of Pharmacy, Bhagwant University, Ajmer, Rajasthan, India

A. Patani
Department of Biotechnology, Smt. S.S. Patel Nootan Science and Commerce College, Sankalchand Patel University, Visnagar, Mehsana 384315, Gujarat, India

S. Singh (✉)
Department of Biotechnology, Faculty of Energy and Technology, Pandit Deendayal Energy University, Gandhinagar 382426, Gujarat, India
e-mail: drsachinbioinfo@gmail.com

R. K. Bachheti et al. (eds.), *Metal and Metal-Oxide Based Nanomaterials*, Smart Nanomaterials Technology, https://doi.org/10.1007/978-981-99-7673-7_7

1 Introduction

Food packaging is one of the most rapid fluxes and a sector under constant pressure to production. The material that is used in packing food serves as a fence against outside influences and makes it possible to deliver food to consumers in a safe and reliable manner. Therefore, packaging is increasingly important in the processed food industry (Akelah, 2013). Creating novel packaging materials with properties like appearance, durability, hygiene, aesthetics, and Eco-friendliness became more crucial with the advent of the industrial revolution (Ketelsen et al., 2020). Furthermore, the primary goal of food packaging is to give the best quality, protection, and good shelf life to the food so that it may fit flawlessly into the hectic lifestyle of today. Changes in the food, such as fragrance sorption, flavor owing to aroma sorption, and flavor transfer from packing materials, are significant factors in the deterioration of packed food (Coles et al., 2003). Due to their affordability, lightweight, clarity, low gas and moisture permeability, and low cost, plastic is one of the most used packaging materials in the food industry. However, it has been reported that the material used in food packaging is degradable, includes harmful softeners, and has low heat resistance (Ncube et al., 2020). These factors contribute to environmental contamination. As a result, the food sector has a pressing need for the creation of vigorous, intelligent packaging that overcomes the drawbacks of prior packaging. Following the widespread usage of non-biodegradable plastic, the development of sustainable packaging prompted the creation of several technologies and the identification of numerous materials with the necessary characteristics to serve as plastic replacements. Another major source of improvement for the existing problems with food security and maintainability is nanotechnology (Bajpai et al., 2018).

Nanotechnology is a branch of development, characterization, manufacturing, administration, and use of nanostructured materials for a variety of purposes. Nanostructures are materials that can be altered in a range of sizes from 1 to 100 nm. The size reduction to the nanoscale range improves the surface-to-volume ratio, biological effectiveness, and adsorption capacity (Abate et al., 2023; Godeto et al., 2023; Pathakoti et al., 2017). Additionally, nanomaterials' chemical and physical properties, including optical, magnetic, and thermodynamic properties as well as their diffusivity, strength, color, and solubility, have been improved. Other unique characteristics of nanoparticles include their decreased density and chemical, mechanical, and kinetic stability. According to Cerqueira et al. (2018), nanostructure materials and nanocomposites have a more comprehensive range of applications and offer better performance than their macro and bulk analogs. One of the fastest-emerging areas of research in the agri-food sector is nanotechnology. Agri-food production has increased significantly worldwide thanks to advances in nanotechnology, which have improved food quality, nutritional content, and safety (Sekhon, 2014). With the rapidly expanding nanotechnology business, the need for nanomaterials has dramatically expanded in recent years. Market revenue for nanomaterials worldwide was \$38.5 billion in 2020, and from 2021 to 2026, it is projected to increase by 12.2% CAGR (Chausali et al., 2022).

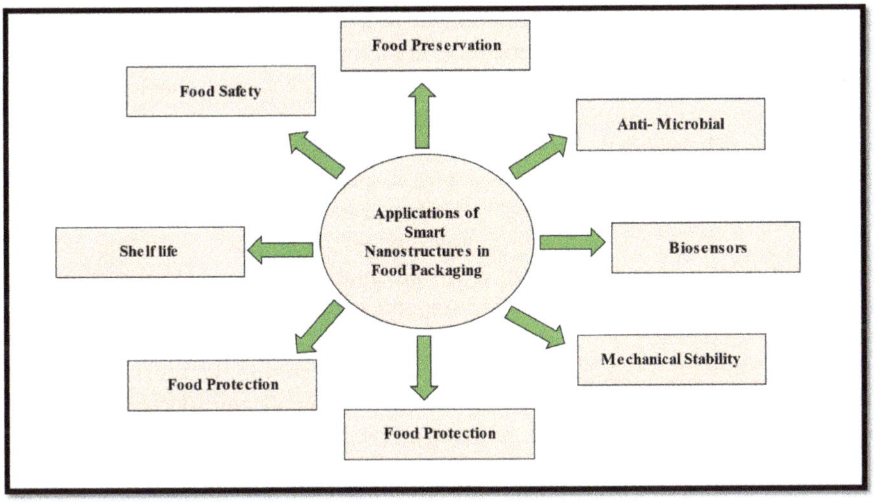

Fig. 1 Applications of smart nanostructures in food packaging

Food nanosensing and food nanostructured ingredients are two categories in which nanotechnology is used in the food business. Both of these categories are subdivided further into subcategories. Nanostructured components cover food packaging and food processing, while nanosensing in food provides food safety and quality (Sahani & Sharma, 2021). Combining nanoparticles and nanocomposites, the most cutting-edge features of nanotechnology provides enhanced, active, smart/intelligent, and bio-based packaging. Modern packaging based on nanotechnology contains a variety of nanomaterials, such as metal oxides, nanofillers, nanometals, nanosensors, bioactive chemicals, mixed polymers, oxygen scavengers, and antioxidants. Thus, the food packaging not only enhances the quality, and shelf life of the food but also guards against infection inside the container (Alfei et al., 2020) (Fig. 1).

2 Classification of Food Packaging

Numerous studies have examined new materials and packaging methods (Husen & Siddiqi, 2023; Husen et al., 2023). Nanotechnology-based nanoreinforcement, nanocomposite active, nanocomposite smart, and bio-based packing are all popular right now (Kumar et al., 2017, 2021a, 2021b). Food containment and confinement, as well as the provision of graphical or textual composition, production, and expiration information, are some of the main functions of packaging. Food is a plentiful source of nutrients., making it vulnerable to deterioration and contamination by the environment's microbes (Tripathi & Khare, 2017). Because of this, food packaging must guarantee the safety of the products within and preserve their quality and freshness for as long as feasible. As a result, polymers have been produced with nanosized or nanostructured elements inserted into them to suit contemporary packaging demands (Rossi et al., 2017). Therefore, using nanosensors, intelligent packaging enables monitoring environmental conditions such as temperature, moisture, gas, or surroundings and food spoiling. The two types of sensors most frequently employed for food deterioration detection are gas-specific and pathogen-specific sensors (Madhusudan et al., 2018). The final type of packaging is bio-based packaging, which uses biological components instead of traditional or non-biodegradable ones, including proteins, lipids, and polysaccharides. It is an environmentally friendly, biodegradable alternative that might provide a sustainable contender for packaging material because it uses these ingredients. Food packaging has various responsibilities in preserving the quality and safety of food (Martins et al., 2022). You can stop surface microbes from contaminating food by putting together active ingredients and food packing materials. Active packaging integrates nutrients and nanoparticles to assure food quality and safety (Fang et al., 2017).

2.1 *Types of Nanomaterials*

2.1.1 Polymer Nanomaterials

Polymer nanomaterials are particles that may include or have active materials on their surface. Food packaging made of polymer nanomaterials is used because it promotes food quality while also being safe, affordable, increasing shelf life, improving mechanical and microbiological resistance qualities, and being very good for the environment (McClements, 2020). Due to its high product yield in industrialized countries, it is most frequently utilized in the manufacturing and packaging industries. Compared to commercial packaging of gases such as O_2 and CO_2, polymer nanomaterial can improve performance against barriers such as UV radiation in addition to stiffness, strength, and resistance to heat. According to Chaudhary et al. (2020) research, using polymer nanoparticles in the food packaging sector may increase barrier properties in food packaging, intelligent packaging, active packaging, and biopolymer degradation. Commercial packaging for a range of goods, including alcoholic beverages, carbonated beverages, bakery goods, and meat products, can use this. Prepolymerization and postpolymerization are the two processes used to create polymer nanoparticles for food packaging (PNFP). PNFPs have many benefits, but they also have drawbacks. For example, they can be harmful when inhaled, absorbed via the skin, or consumed (Chadha et al., 2022).

2.1.2 Biopolymer Nanocomposites

Synthetic polymers and most of the polymer nanocomposites threaten the environment. Biopolymers can thus be utilized in place of synthetic polymers. Biopolymers are synthetic polymers that decompose quickly in soil and are favorable to the environment. However, biopolymers' mechanical and thermal characteristics are poor (Gowthaman et al., 2021). Clay's organic structure must be altered to improve its qualities. This will improve the biopolymer nanocomposites' mechanical characteristics (Awad et al., 2019). The biopolymer becomes mechanically robust when montmorillonite clay is converted into organic ammonium salts. This makes a structure cheap, simple to use, renewable, and degradable. The most prevalent food-borne bacteria, *S. aureus* and *E. coli* are particularly susceptible to this changed organic structure's antibacterial capabilities, which are then observed and examined using SEM, Fourier spectroscopy, and X-ray diffraction (Shehabeldine et al., 2023).

2.1.3 Antimicrobial Nanomaterials

Active packaging is used in antimicrobial nanomaterials packaging. The use of antimicrobial packaging materials reduces microbial activity, preventing food spoiling and foodborne illnesses (Ahmed et al., 2022). Nanomaterials such as TiO_2, Ag, ZnO, Fe_3O_4, and CuO are antibacterial examples (Worku et al., 2023a). Because antimicrobial nanostructures work in various ways and varying degrees, they are frequently used in food packaging. Despite the promise of nanotechnology in food packaging, many issues still need to be resolved, including their toxicity levels, migration, and storage, while keeping the human race's environment and wellness in mind. This should be stated before the product is authorized for sale on the market. It should also change to maintain the caliber of the packing material. Some active packages preserve food for longer by dispensing specific components; other intelligent packages alert consumers to the product's safety. Additionally, there is probiotic-infused antimicrobial packaging, a relatively recent biopreservation technique (Swaminathan & Sharma, 2019).

2.1.4 Gelatin-Based Nanomaterials

According to Nerus (2020) fish, prawns, chickens, and their gills can be used to synthesize protein hydrolysate, gelatin, collagen, and bioactive peptides (Nerus, 2020). The high protein content, even molecular weight distribution, and amino acid content of the gelatin derived from poultry feces make it widely utilized. The hydrophobic and hydrophilic configurations of amino acids and the functional groups COOH, –OH, and NH_2 are all present in gelatin. This can be combined with additional polymeric functional compounds in the future. Food packaging nanocomposites based on gelatin can be made by engineering gelatin with nanocomposites (Chadha et al., 2022). Nanomaterial for packaging made from gelatin has several advantageous qualities, including high physicochemical flexibility, nature, brittleness, biodegradability, and natural abundance; it also has no negative effects on people or the environment (Taherimehr et al., 2021).

2.1.5 Cellulose-Based Nanomaterials

Biobased materials are developed because of their special properties in many different industries. They have features that make them simple to bind and are safe for the environment (Bedian et al., 2017). The nanocomposites are also combined with biobased materials to create nanocellulose material structures, which have excellent mechanical and physical characteristics used in the packaging of the food industry (Blanco et al., 2018).

2.1.6 Electrospinning of Nanofibers

An easy method to create nanofibers from the viscoelastic fluid using an electric field is electrospinning, an electrohydrodynamic process. It can be used to assist in the creation of secondary structures and functionalities in electrospun nanofibers. The spinneret at the syringe's tip spun a hard droplet of viscoelastic fluid. The droplet took on a conical shape because there was enough force to break through the surface tension. Upon leaving a cone, a jet extends partially in the shape of an expanding helix (Subbiah et al., 2005). Finally, a polymer solution droplet is stretched to receive the nanofiber on the grounded collector. Nanomaterials were functionalized onto the surface of nanofibers during the electrospinning process. Applications for the electrospun nanofibers include medicine delivery, environmental protection, food packaging, and air and water purification. The heat-sensitive chemicals can improve food stability throughout processing, handling, and storage thanks to electrospinning's nonthermal procedure (Charles et al., 2021).

2.1.7 Microstructure of Nanomaterials

Nisin is the sole antibacterial bacteriocin employed in green products, also known as antimicrobial peptide (AMP). It is created by the Lactococcus lactis strain's 3.5 kDa cationic polypeptide, which is used in more than 40 countries to enhance food biopreservation (Paiva & Breukink, 2012). Biodegradable films and coating materials were created by encapsulating the biobased nisin microstructure with nanoliposomes. This substance will eventually raise the thickness of the film and grease resistance; it also keeps the moisture level and permeability to oxygen constant, reduces CO_2; and increases transparency's resistance to ultraviolet (UV) radiation. Due to the liposome's ability to connect to certain regions and form a stable structure, unlike some other structures. Additionally, this is less hazardous and more environmentally friendly (Chawla et al., 2021). The fluorinated silica's nanoparticle's fundamental composition is created by the combination of microstructure and nanoparticle deposition (Mehdizadeh et al., 2022).

2.2 *Use of Nanomaterials in Food Packaging*

Zinc oxide, clay, silver, titanium dioxide, carbon nanotubes, copper oxides, and copper are examples of nanomaterials that are frequently utilized in food packaging. These can improve physical and chemical characteristics (color, taste, flavor, weight, and physical texture measured moisture content), resistance to microbes, particularly for food-related illnesses, and environmental factors (Awuchi & Dendegh, 2022).

2.2.1 Clay Nanoparticles

Clay nanoparticles serve a variety of functions in the body as vitamin and antioxidant carriers. Clay nanoparticles can reduce food contamination, alleviate stomach ulcers, clean the blood, and even take the role of anti-diarrhea drugs (Brandelli et al., 2017). Tetrahedral and octahedral sheets, which are present in the single layer of nanoparticles, organize themselves like book pages. Nanoclays cut permeability to moisture and CO_2 by 35 and 55%, respectively (Farzadnia et al., 2013).

2.2.2 Silver

The ability of silver to inhibit microbial growth is widely established. Lipopolysaccharides can break down if silver attaches to the cell's surface and produces pitting in the cell's outer layer. According to a report, low-density polyethylene, which maintains and extends the shelf life, is present in silver (Dallas et al., 2011). By utilizing or omitting the standard covering liquid, silver may have a synergistic impact and atmosphere packaging changed with 55% N_2 and another 55% CO_2 to increase the shelf life of the cheese. The best material for making packaging films is silver nanoparticles. They exhibit outstanding mechanical and thermal capabilities. There is always a chance that the food's nanoparticles will move and disrupt the balance of biomass used to produce meals (Ogunsona et al., 2020).

2.2.3 Zinc Oxide

The primary application of zinc oxide as a photocatalytic agent is the degradation of bacteria and other organic compounds. By adding nutrients, zinc oxide improves the nutritional value of food products. It functions like a fortifier and helps to enhance the food's flavor, preservation capabilities, appearance, and texture. Zinc oxide characteristics are governed by growth, nucleation, and aging processes. ZnO nanoparticles aggregate and bind together, losing some of their useful characteristics. Melanin is a biologically active substance with a wide range of functions and functions as an agent of stabilization. Numerous mineral components can be found in zinc oxide. Zinc oxide has potent antimicrobial properties. Hydrophobicity and barrier qualities are improved by zinc oxide. On the skin, zinc oxide appears as an opaque white coating (Ifijen et al., 2022).

2.2.4 Gold

The gold nanoparticle has shown its benefits in food packaging in the present scenario. Ionic silsesquioxane is added to the mixture as a stabilizer to help minimize the gold salts. As a result, gold nanoparticles are created. This stabilizer is required for nanoparticle production (Pagno et al., 2015). Additionally, it contains quaternary ammonium groups known for their potent antibacterial effects and may be dissolved in water. The ability to create bioactive films using this technique is quite promising. Gold nanoparticles improve biofilms mechanically, optically, and morphologically. It demonstrated significant antibacterial activity against food-borne infections. The loss of red color and irreversible aggregation of gold nanoparticles is because of the tendency of temperature integrators to freeze over time. They cost more than the other types of nanomaterials (Hoffmann et al., 2019).

2.2.5 Titanium Dioxide

The food-grade version of titanium dioxide enhances and brightens white foods like dairy goods. Due to its ability to increase food sensitivity and prevent food from rotting, titanium dioxide is an ingredient that is used in the food industries to ensure that food is kept safe (Peters et al., 2014). Additionally, because of its high melting and boiling points of 1,843 °C and 2,972 °C, this causes it to be a solid at room temperature and also makes it insoluble in water at the particle level. It serves as an insulator as well. Food's shelf life is extended due to this food rotting prevention. It serves as a UV filter. TiO_2 is nontoxic and inert. Biodegradable films are nontoxic, affordable, and photostable, and they can change their qualities (Modugu et al., 2022).

3 Packaging for Food with Organic, Inorganic, and Hybrid Nanoparticles

The creation of effective food packaging uses a wide variety of nanoparticles, some of which are organic, some of which are inorganic, and some of which are a mixture of the two. This is done to prevent both the quantitative and qualitative loss of food products (Ashfaq et al., 2022).

3.1 Nanocellulose

A polysaccharide, cellulose is the most common type of polymer and gives plant cell walls their strength. Nanocellulose has a diameter of 100 nm and a few micrometers. Nanocellulose can take the form of nanocrystals or nanofibers. The capacity to form strong, thick surfaces impermeable to molecules thanks to good hydrogen bonding makes nanocelluloses, harvested from plants via mechanical and chemical methods, an ideal barrier material. In order to increase its tensile strength, mechanical characteristics, homogeneity, and biodegradability, nanocellulose is utilized in the paper and composite industries (Jonoobi et al., 2015; Worku et al., 2023a, 2023b).

3.2 Nanostarch

Amylose and amylopectin are the two polymers that make up the complex polysaccharide known as starch. In order to produce nanostarch, starch granules must first be reduced in size by means of many different chemical and physical methods. Nanoparticles made of starch have one of its dimensions which is less than 350 nm and a large surface area compared to their volume. It has been discovered that incorporating starch nanoparticles into composite materials as nanofillers can improve strength, biodegradability, flexibility, heat, and water impermeability (Ge et al., 2017).

3.3 Chitosan Nanoparticles

Chitosan nanoparticles are a form of polysaccharide produced from chitin and majorly seen in the exoskeletons of crustaceans, arthropods, and fungi. Because chitosan is nontoxic, safe for the environment, and possesses excellent antibacterial properties, it is considered a potential material for use in the packaging sector. When chitosan is combined with a biodegradable polymer, such as polylactic acid films, the properties of the gas and moisture barriers are enhanced. Chitosan is a type of polysaccharide. The physical and mechanical properties of the film may be significantly enhanced by adding chitosan nanoparticles when combined with biocomposites. In addition, they improve the antibacterial capabilities and the impermeability to moisture (Babaei-Ghazvini et al., 2021).

3.4 Carbon Nanotubes

Carbon nanotubes are alternative forms of carbon that have been rolled up into cylindrical shapes and have a diameter that falls somewhere in the region of nanometers.

It is possible that they are single-walled nanotubes or multi-walled nanotubes that are made up of a number of cylinders arranged in a concentric pattern. Using carbon nanotubes in the production of the polymer used for packaging results in improvements to the material's mechanistic and antibacterial properties (Gupta et al., 2019). Additionally, oxygen sensors are made with carbon nanotubes, which are used to monitor the amount of oxygen present in modified atmospheric packaging. These types of nanotubes are embedded within an artificial polymer matrix that is used for packaging food. These nanotubes give the matrix antibacterial properties and smart sensors that can detect when food has gone bad (Chausali et al., 2022).

3.5 Silver Nanoparticles

Silver nanoparticles (AgNPs) are used as an antibacterial agent in food packing and are becoming more prevalent in the interest of increasing the life of packaged foods. In order to manufacture silver nanoparticles, which exhibit antibacterial action against a wide range of gram-negative and gram-positive bacteria as well as antioxidant activity, it is essential to develop methods that are user-friendly, inexpensive, and kind to the environment (Gonfa et al., 2023; Kumar et al., 2021a, 2021b). These polymers can either be biodegradable or not. Because of the toxicity features of AgNPs, which might act against bacteria by slowly releasing food material from packaging material, the passage must be taken into consideration (Fahmy et al., 2020). Active packaging films and coatings loaded with AgNPs moderately release nanoparticles on food surfaces, inhibiting microbial growth and increasing the life of the food (Omerović et al., 2021).

3.6 Zinc Oxide Nanoparticles

Zinc can be added to food and supplements to boost nutrition. Zinc oxide nanoparticles create Zn^{2+} ions and reactive oxygen species, which destroy organelles of the cell and kill bacteria like *E. coli*, *Listeria monocytes*, *Staphylococcus aureus*, *Salmonella enteritidis*, and others. (Krishnamoorthy et al., 2022). When added to the polymer matrix, ZnO NPs improve the composite film barrier and antibacterial and mechanical characteristics. Polylactic acid (PLA), low-density polyethylene (LDPE), linear low-density polyethylene (LLDPE), poly 3-hydroxybutyrate (PHB), PHBV were easily incorporated and studied for their effects (Silva-Leyton et al., 2019).

3.7 Titanium Dioxide Nanoparticles (TiO_2 NPs)

Food additives and packaging nanocomposites contain TiO_2 NPs, and white metal oxides. TiO_2 NPs limit UV radiation in food packaging polymers and improve film's chemical, mechanical, and barrier properties. Due to their high index of refraction, stability, and brightness, TiO_2 NPs are used to color many processed foods (Chaudhary et al., 2020). Titanium dioxide creates ROS and free radicals that destroy microorganisms. TiO_2, a potent photocatalyst, is chemically stable, nontoxic, affordable, and environmentally friendly. In addition, TiO_2 is discovered to be an oxygen and ethylene scavenger since ethylene photodegrades when exposed to sunshine (Canaparo et al., 2020).

3.8 Nanoclay

The term "nanoclay" is typically used to describe layered crystalline silicates, composed of octahedral or tetrahedral sheets. In order to enhance the physical and barrier qualities of plastic materials, nanoclay is frequently utilized in food packaging. The most common form of nanoclays is a platelet with a blistering and soft surface (Majeed et al., 2013). The smectite subclass of naturally occurring montmorillonites (MMT) is a hydrophilic nanoclay in the phyllosilicate family that is widely used to improve protein-based film mechanical properties. It's critical to comprehend how nanocomposites incorporated in the packaging material migrate and are harmful. The interface of the nanoclay with the polymer, contact with food, temperature, and exposure period all affect how the nanoclay migrates. Starch thermoplastic's mechanical and barrier qualities, as well as the biodegradability of synthetic polymers, are improved by nanoclay (Saba et al., 2016) (Table 1).

4 Techniques of Nanoparticle Synthesis

The creation of nanoparticles can be accomplished in one of two broad ways: top-down or bottom-up.

4.1 Top-Down Method

The top-down methodology reduces bigger materials to nanoscale materials by employing mechanical and chemical forces in dismantling. Top-down approaches to

Table 1 Different types of nanostructured materials, their activities, and applications

Nanostructured materials	Activity	Applications	References
Silicon dioxide	Demonstrates hygroscopic applicability by soaking up water molecules in food; moisture leakage is reduced; acts as a food coloring, drying, and anti-caking agent; typical particle size, large surface area, stability, biocompatibility, low toxicity, poor heat conductivity, and exceptional insulation	Active food preservation packaging extends the shelf life of the food and reduces harmful and spoilage germs	Alam et al. (2022)
Nanoclay and silicate	Higher levels of organic acids, antioxidants, and volatiles overall; antibacterial activity	Active food preservation packaging extends the shelf life of the food and reduces harmful and spoilage germs	Saba et al. (2016)
Nanocomposites with zinc oxide, pediocin, and silver coating	Breakdown of lipopolysaccharides, irreversible bacterial DNA damage, and assistance in the fight against microbes	Composition of enhanced food packaging with distinguishing features (antimicrobial agent)	Ogunsona et al., (2020)
Nano biosensors	Identification of viruses and bacteria	Smart (intelligent) food packaging that controls and identifies harmful and spoilage germs to extend shelf life	Thakur et al. (2022)
Silver-based	Greater antioxidant activity; antibacterial activity that is effective against both gram-positive and gram-negative bacteria; better mechanical and barrier properties; yellowness, poor transparency, and heat stability	Active food preservation packaging extends the shelf life of the food and reduces harmful and spoilage germs	Al-Tayyar et al. (2020)

(continued)

Table 1 (continued)

Nanostructured materials	Activity	Applications	References
Zinc oxide	Strong antibacterial agent; UV-A irradiation had no effect on the mechanical properties of the nanomaterial generated; oxygen flow within packaging containers is prevented by using activated oxygen scavenging materials	The antibacterial properties of packaging used for food preservation are highlighted, and it is used to increase the shelf life of fresh items while preventing them from sticking together	Ifijen et al. (2022)
Titanium dioxide	It possesses a number of advantages, including being affordable, nontoxic, and photo-stable; gaining popularity as a better photocatalyst particle for economic and power applications (water splitting, air or gas and water decontamination, antibacterial, and surfaces that clean themselves); antibacterial activity; polymer nanocomposites' mechanical characteristics have been improved; and being used as food whiteners in milk, cheese, and other various products	Active food preservation packaging extends the shelf life of the food and reduces harmful and spoilage germs	Modugu et al. (2022)

the production of nanoparticles include mechanical milling, electrospinning, lithography, sputtering, etching, and laser ablation. These are the most common of these approaches.

4.1.1 Mechanical Milling

The mechanical milling process involves placing the components into a high-power mill with or without the medium. This can be done to reduce the elements' particle size. As a result of the rolling ball imparting some of its kinetic energy to the elements, those elements can now be machined to dimensions on the nanoscale. The kinetic energy transfer depends on various elements, such as the kind of mill, the speed of the milling process, the ball packing, the milling method, the milling length, and the milling temperature. Conventional milling techniques offer fewer benefits and less reliability than ball milling techniques due to the fact that ball milling techniques may be used on both dry and wet materials, function in an enclosed environment, are feasible from an economic aspect, and can be used at room temperature. Additionally, conventional milling procedures do not give the advantages that ball milling techniques offer. Ball milling techniques can be utilized effectively to manufacture nanomaterials on a wide scale due to their low cost and relatively straightforward processes (Mattos et al., 2016).

4.1.2 Nanolithography

A nanofabrication technique called nanolithography is used to create nanopatterns between 1 and 100 nm in size. Lithography can be divided into three categories: nanoimprint, soft, and photolithography. Masks, molds, or templates are employed to create nanopatterns in masked lithography, whereas maskless lithography is not used at all when creating nanopatterns (Liu et al., 2000).

4.2 Bottom-Up Method

The bottom-up strategy entails utilizing a wide range of methods to fabricate the nanomaterial by starting with atomic or molecular species. The production of nanoparticles can be accomplished by the use of a wide variety of processes and methods, including chemical vapor deposition, solvothermal, sol–gel, hydrothermal processes, and reverse micelle approaches.

4.2.1 Chemical Vapor Deposition (CVD)

Chemical vapor deposition (CVD) is a well-known approach, i.e., bottom-up process used to deposit nanosized material and thin films on the substrate of choice. In this process, a chemical reaction takes place at high temperatures between a precursor (which can be a gas or a vapor) and the substrate that has been chosen in advance. As a consequence of this reaction, the desired product will be deposited on the surface that has been selected. The drawbacks of CVD are its high cost of production and the toxicity of its gaseous byproducts, both of which make it difficult to produce nanocrystals with a high degree of purity, quality and few flaws on the substrate. Despite these drawbacks, CVD is still widely used as a method for producing nanocrystals (Abid et al., 2022).

4.2.2 Sol–gel Method

The sol–gel method is a flexible and popular approach for creating nanoparticles. It entails the creation of a sol, a colloidal solution, which is then turned into a solid network that resembles gel through the gelation process. One benefit of the sol–gel

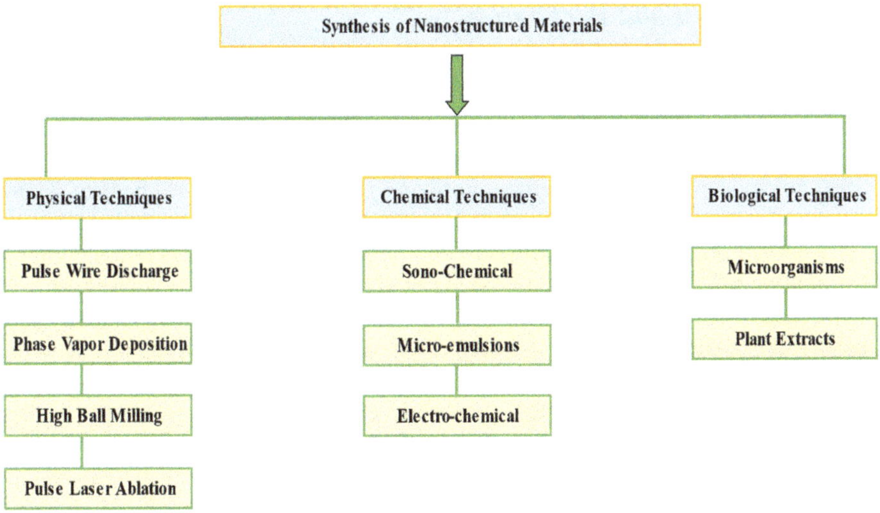

Fig. 2 Synthesis of nanostructured materials

process is the ability to customize particle size, shape, composition, and surface qualities. The sector of food packaging has also used the sol–gel technique. It is a promising method for improving the performance and functionality of packaging materials since it provides special advantages for manufacturing thin coatings and films with specialized features (Balbyshev et al., 2003) (Fig. 2).

5 Nanomaterial-Based Food Packaging

Food packaging, a key element of the food industry, enables food preservation with retained nutritional and organoleptic properties. Along with preventing food from going bad, food packaging protects delicate bioactive molecules from harmful physical and environmental elements. Numerous studies are being conducted on nanotechnology to improve food packaging (Rout et al., 2022). Functional nanoparticles improve packaging material barrier qualities, strength, thermal stability, and durability, extending food product shelf life. Nano-based packaging materials can be made by mixing nanoparticles into films and containers or by developing multilayer nanocomposite packaging materials and organic, inorganic, and combined nanocoatings by immersion, spraying, or rubbing (Qin et al., 2020).

5.1 Improved Food Packaging

Nanoparticles' special qualities and capabilities have demonstrated significant promise for enhancing food packaging. In food packaging, nanoparticles can be used as sensors or indications. For instance, nanoparticles may be programmed to glow or change color in response to temperature, gas composition, or pH changes. These nanoparticles serve as sensors and offer visual clues to show if the packed food is fresh or of a certain standard. These sensing capabilities assist customers and merchants in evaluating the state of the food (Ashfaq et al., 2022). In food packaging, active-release systems can be made utilizing nanoparticles. For instance, active compounds (such as antioxidants and antimicrobials) can be enclosed within nanoparticles using nanoencapsulation processes before being incorporated into the packaging material. These nanoparticles gradually release the active ingredients, giving the packed food antioxidant protection, antibacterial properties, or other desired capabilities (Akhila & Badwaik, 2022).

5.2 Active Packaging

Active packaging prevents moisture gain, microbial invasion, oxidation, over-ripening, etc., extending food shelf life. Oxygen scavengers, ethylene absorbers, carbon dioxide emitters, antioxidants, and antibacterial agents are added to packaging material or package headspace to improve packing polymer qualities. Antimicrobial agents either weaken microorganisms' cellular structure or disrupt their metabolic route to stop future growth. Reactive oxygen species, which are produced in part by nanoparticles and stop DNA replication and ATP generation, harm or kill cells.

In order to interact with the food, active packaging includes adding active substances or nanoparticles to the packaging materials. To stop the growth of microbes and improve food safety, for instance, packaging films can be infused with nanoscale antimicrobial agents like silver nanoparticles. Similar to this, antioxidants, antibacterial substances, or flavor enhancers can be included in nanosized capsules within the packaging material using nanoencapsulation processes. These active ingredients can be gradually released to lengthen the food's shelf life or improve its sensory qualities (Ahmed et al., 2017) (Fig. 3).

5.3 Intelligent Packaging

The creation of intelligent or smart packaging solutions can benefit from nanoparticles. Packaging materials can be created to detect and indicate changes in temperature, light, or other environmental conditions by integrating nanoparticles having responsive qualities, such as stimuli-responsive polymers or quantum dots. This

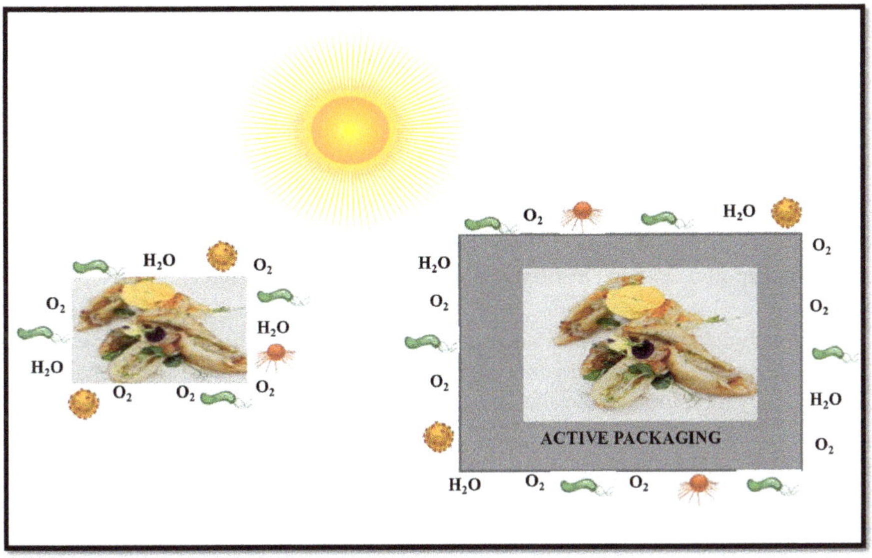

Fig. 3 Comparison between normal packaging and active packaging

makes it possible to monitor food quality and conditions in real time while they are being transported and stored. It is crucial to remember that nanoparticles in food packaging must pass stringent safety tests to guarantee compliance with laws governing food contact and to analyze any potential concerns posed by nanoparticle migration into food. The proper rules and regulations must be in place to regulate the use of nanoparticles in food packaging and assure consumer safety (Fuertes et al., 2016).

6 Role of Nanotechnology in Food Microbiology and Science

The topic of food technology has a wide variety of potential nanotechnology applications. Nanoparticles are typically put to use in one of two different ways: first, they can be directly incorporated into food in order to create nutraceutical delivery systems and modify the optical and rheological properties of food products; second, they can be used in the sanitization of food manufacturing facilities, as well as in packaging materials, sensors, and other applications. In the industry that deals with the preparation of food, nanoparticles are utilized not only as nanocarriers, preservatives, and nanocapsules but also as transporters of bioactive substances found in food items, flavor additives, color additives, and anticaking agents (Fathima et al., 2022).

Foodborne viruses and pollutants can be quickly and accurately detected using nanotechnology-based technologies. When combined with specialized probes, nanoparticles like gold nanoparticles or quantum dots can be used in colorimetric tests or fluorescence-based procedures to identify the presence of infections. Additionally, high-sensitivity and specificity nanosensors and nanobiosensors are being developed for the detection of toxins, allergens, and other pollutants. Foodborne disease risk can be decreased, and nanotechnology can control foodborne pathogens. In order to get rid of or reduce microbial contamination in food processing facilities, water treatment systems, or food preparation surfaces, nano-based disinfection techniques can be used. Innovative preservation methods, including nanoemulsions, films with nanoparticles, and nanoencapsulation, which can stop pathogens and bacteria that cause food spoiling from growing, are developed in part because of nanotechnology. Overall, techniques, materials, and tactics for advancing food microbiology and science are provided by nanotechnology. It provides better food microbe detection, management, and monitoring, which improves food quality, safety, and preservation. However, in order to ensure the safe and responsible application of nanotechnology in the food business, it is crucial to carry out exhaustive risk assessments and follow regulatory requirements (Spirescu et al., 2021).

7 Conclusion

Smart nanostructures have exciting chances to transform food packaging. These cutting-edge materials and technologies offer special qualities and functions that solve important issues in the food industry, such as sustainability, preservation, and safety. There are various advantages to using smart nanostructures in food packaging. Smart nanostructures can significantly lower the permeability of gases, moisture, and pollutants while improving the barrier qualities of packaging materials. This results in increased freshness, decreased food waste, and increased shelf life. Smart nanostructures' antibacterial qualities aid in preventing the growth of microbes, improving food safety, and lowering the risk of contracting foodborne illnesses. To continuously guard against microbial contamination, packaging materials might integrate nanoparticles with antimicrobial capabilities, such as silver or copper nanoparticles.

Smart nanostructures also aid in the creation of innovative packaging techniques. Real-time monitoring of food quality characteristics is made possible by nanosensors and indicators included in packing materials, which provide important details about the state of the food throughout storage and transportation. Overall, smart nanostructures can completely change how food is packaged by enhancing food quality, safety, and preservation. The application of intelligent nanostructures in the food sector will advance with continued study and development, supporting creative and sustainable packaging solutions.

References

Abate, L., Tadesse, M. G., Bachheti, A., Husen, A., & Bachheti, R. K. (2023). Medicinal plant-based lignin and its role in nanoparticles synthesis and applications. In R. K. Bachheti & A. Bachheti (Eds.), *Secondary metabolites from medicinal plants* (pp. 87–110). CRC Press.

Abid, N., Khan, A. M., Shujait, S., Chaudhary, K., Ikram, M., Imran, M., et al. (2022). Synthesis of nanomaterials using various top-down and bottom-up approaches, influencing factors, advantages, and disadvantages: A review. *Advances in Colloid and Interface Science, 300*, 102597.

Ahmed, S., Sameen, D. E., Lu, R., Li, R., Dai, J., Qin, W., & Liu, Y. (2022). Research progress on antimicrobial materials for food packaging. *Critical Reviews in Food Science and Nutrition, 62*(11), 3088–3102.

Ahmed, I., Lin, H., Zou, L., Brody, A. L., Li, Z., Qazi, I. M., et al. (2017). A comprehensive review of the application of active packaging technologies to muscle foods. *Food Control, 82*, 163–178.

Akelah, A. (2013). Functionalized polymeric materials in agriculture and the food industry.

Akhila, V., & Badwaik, L. S. (2022). Recent advancement in improvement of properties of polysaccharides and proteins based packaging film with added nanoparticles: A review. *International Journal of Biological Macromolecules.*

Al-Tayyar, N. A., Youssef, A. M., & Al-Hindi, R. (2020). Antimicrobial food packaging based on sustainable Bio-based materials for reducing foodborne Pathogens: A review. *Food Chemistry, 310*, 125915.

Alam, M. A., Khatoon, R., Huda, S., Ahmad, N., & Sharma, P. K. (2022). Applications of nano-biocomposite materials as antimicrobial and antioxidant finishing agents for packaged food products. In *Biomaterials in food packaging* (pp. 233–255). Jenny Stanford Publishing.

Alfei, S., Marengo, B., & Zuccari, G. (2020). Nanotechnology application in food packaging: A plethora of opportunities versus pending risks assessment and public concerns. *Food Research International, 137*, 109664.

Ashfaq, A., Khursheed, N., Fatima, S., Anjum, Z., & Younis, K. (2022). Application of nanotechnology in food packaging: Pros and Cons. *Journal of Agriculture and Food Research, 7*, 100270.

Awad, A. M., Shaikh, S. M., Jalab, R., Gulied, M. H., Nasser, M. S., Benamor, A., & Adham, S. (2019). Adsorption of organic pollutants by natural and modified clays: A comprehensive review. *Separation and Purification Technology, 228*, 115719.

Awuchi, C. G., & Dendegh, T. A. (2022). Active, smart, intelligent, and improved packaging. *Application of nanotechnology in food science, processing and packaging* (pp. 189–202). Springer International Publishing.

Babaei-Ghazvini, A., Acharya, B., & Korber, D. R. (2021). Antimicrobial biodegradable food packaging based on chitosan and metal/metal-oxide bio-nanocomposites: A review. *Polymers, 13*(16), 2790.

Bajpai, V. K., Kamle, M., Shukla, S., Mahato, D. K., Chandra, P., Hwang, S. K., et al. (2018). Prospects of using nanotechnology for food preservation, safety, and security. *Journal of Food and Drug Analysis, 26*(4), 1201–1214.

Balbyshev, V. N., Anderson, K. L., Sinsawat, A., Farmer, B. L., & Donley, M. S. (2003). Modeling of nanosized macromolecules in silane-based self-assembled nano-phase particle coatings. *Progress in Organic Coatings, 47*(3–4), 337–341.

Bedian, L., Villalba-Rodríguez, A. M., Hernández-Vargas, G., Parra-Saldivar, R., & Iqbal, H. M. (2017). Bio-based materials with novel characteristics for tissue engineering applications–A review. *International Journal of Biological Macromolecules, 98*, 837–846.

Blanco, A., Monte, M. C., Campano, C., Balea, A., Merayo, N., & Negro, C. (2018). Nanocellulose for industrial use: cellulose nanofibers (CNF), cellulose nanocrystals (CNC), and bacterial cellulose (BC). In *Handbook of nanomaterials for industrial applications* (pp. 74–126). Elsevier.

Brandelli, A., Brum, L. F. W., & dos Santos, J. H. Z. (2017). Nanostructured bioactive compounds for ecological food packaging. *Environmental Chemistry Letters, 15*, 193–204.

Canaparo, R., Foglietta, F., Limongi, T., & Serpe, L. (2020). Biomedical applications of reactive oxygen species generation by metal nanoparticles. *Materials, 14*(1), 53.
Cerqueira, M. A., Vicente, A. A., & Pastrana, L. M. (2018). Nanotechnology in food packaging: opportunities and challenges. *Nanomaterials for Food Packaging*, 1–11.
Chadha, U., Bhardwaj, P., Selvaraj, S. K., Arasu, K., Praveena, S., Pavan, A., et al. (2022). Current trends and future perspectives of nanomaterials in food packaging application. *Journal of Nanomaterials*, 1–32.
Charles, A. P. R., Jin, T. Z., Mu, R., & Wu, Y. (2021). Electrohydrodynamic processing of natural polymers for active food packaging: A comprehensive review. *Comprehensive Reviews in Food Science and Food Safety, 20*(6), 6027–6056.
Chaudhary, P., Fatima, F., & Kumar, A. (2020). Relevance of nanomaterials in food packaging and its advanced future prospects. *Journal of Inorganic and Organometallic Polymers and Materials, 30*, 5180–5192.
Chausali, N., Saxena, J., & Prasad, R. (2022). Recent trends in nanotechnology applications of bio-based packaging. *Journal of Agriculture and Food Research, 7*, 100257.
Chawla, R., Sivakumar, S., & Kaur, H. (2021). Antimicrobial edible films in food packaging: Current scenario and recent nanotechnological advancements-a review. *Carbohydrate Polymer Technologies and Applications, 2*, 100024.
Coles, R., McDowell, D., & Kirwan, M. J. (Eds.). (2003). *Food packaging technology* (Vol. 5). CRC Press.
Dallas, P., Sharma, V. K., & Zboril, R. (2011). Silver polymeric nanocomposites as advanced antimicrobial agents: Classification, synthetic paths, applications, and perspectives. *Advances in Colloid and Interface Science, 166*(1–2), 119–135.
Fahmy, H. M., Eldin, R. E. S., Serea, E. S. A., Gomaa, N. M., AboElmagd, G. M., Salem, S. A., et al. (2020). Advances in nanotechnology and antibacterial properties of biodegradable food packaging materials. *RSC Advances, 10*(35), 20467–20484.
Fang, Z., Zhao, Y., Warner, R. D., & Johnson, S. K. (2017). Active and intelligent packaging in the meat industry. *Trends in Food Science and Technology, 61*, 60–71.
Farzadnia, N., Ali, A. A. A., Demirboga, R., & Anwar, M. P. (2013). Effect of halloysite nano clay on mechanical properties, thermal behavior and microstructure of cement mortars. *Cement and Concrete Research, 48*, 97–104.
Fathima, A., Chiome, T. J., Catherine, A. A., Egbuna, C., Achar, R. R., & Srinivasan, A. (2022). Smart use of nanomaterials as sensors for detection and monitoring of food spoilage. *Application of nanotechnology in food science, processing and packaging* (pp. 169–188). Springer International Publishing.
Fuertes, G., Soto, I., Carrasco, R., Vargas, M., Sabattin, J., & Lagos, C. (2016). Intelligent packaging systems: sensors and nanosensors to monitor food quality and safety. *Journal of Sensors*.
Ge, S., Xiong, L., Li, M., Liu, J., Yang, J., Chang, R., et al. (2017). Characterizations of pickering emulsions stabilized by starch nanoparticles: Influence of starch variety and particle size. *Food Chemistry, 234*, 339–347.
Godeto, Y. G., Ayele, A., Ahmed, I. N., Husen, A., & Bachheti, R. K. (2023). Medicinal plant-based metabolites in nanoparticles synthesis and their cutting-edge applications: An overview. In: R. K. Bachheti, A. Bachheti (Eds.), *Secondary metabolites from medicinal plants* (pp. 1–34). CRC Press.
Gonfa, Y. H., Gelagle, A. A., Hailegnaw, B., Kabeto, S. A., Workeneh, G. A., Tessema, F. B., Tadesse, M. G., Wabaidur, S. M., Dahlous, K. A., Abou Fayssal, S., & Kumar, P. (2023). Optimization, characterization, and biological applications of silver nanoparticles synthesized using essential oil of aerial part of *Laggera tomentosa*. *Sustainability, 15*(1), 797.
Gowthaman, N. S. K., Lim, H. N., Sreeraj, T. R., Amalraj, A., & Gopi, S. (2021). Advantages of biopolymers over synthetic polymers: social, economic, and environmental aspects. In *Biopolymers and Their Industrial Applications* (pp. 351–372). Elsevier.
Gupta, N., Gupta, S. M., & Sharma, S. K. (2019). Carbon nanotubes: Synthesis, properties and engineering applications. *Carbon Letters, 29*, 419–447.

Hoffmann, T., Peters, D., Angioletti, B., Bertoli, S., Péres, L., Reiter, M., & De Souza, C. (2019). Potential nanocomposites in food packaging. *Chemical Engineering Transactions, 75*.
Husen, A., & Siddiqi, K. S. (2023). *Advances in smart nanomaterials and their applications*. Elsevier Inc. https://doi.org/10.1016/C2021-0-02202-1
Husen, A., Bachheti, R. K., & Bachheti, A. (2023). *Current trends in green nano-emulsions (Food, Agriculture and Biomedical Sectors)*. Springer Nature Singapore Pte Ltd.
Ifijen, I. H., Maliki, M., & Anegbe, B. (2022). Synthesis, photocatalytic degradation and antibacterial properties of selenium or silver doped zinc oxide nanoparticles: A detailed review. *OpenNano*, 100082.
Jonoobi, M., Oladi, R., Davoudpour, Y., Oksman, K., Dufresne, A., Hamzeh, Y., & Davoodi, R. (2015). Different preparation methods and properties of nanostructured cellulose from various natural resources and residues: A review. *Cellulose, 22*, 935–969.
Ketelsen, M., Janssen, M., & Hamm, U. (2020). Consumers' response to environmentally-friendly food packaging-A systematic review. *Journal of Cleaner Production, 254*, 120123.
Krishnamoorthy, R., Athinarayanan, J., Periyasamy, V. S., Alshuniaber, M. A., Alshammari, G., Hakeem, M. J., et al. (2022). Antibacterial mechanisms of zinc oxide nanoparticle against bacterial food pathogens resistant to beta-lactam antibiotics. *Molecules, 27*(8), 2489.
Kumar, S., Basumatary, I. B., Sudhani, H. P., Bajpai, V. K., Chen, L., Shukla, S., & Mukherjee, A. (2021a). Plant extract mediated silver nanoparticles and their applications as antimicrobials and in sustainable food packaging: A state-of-the-art review. *Trends in Food Science and Technology, 112*, 651–666.
Kumar, A., Choudhary, A., Kaur, H., Mehta, S., & Husen, A. (2021b). Metal-based nanoparticles, sensors and their multifaceted application in food packaging. *Journal of Nanobiotechnology, 19*(256), 1–25. https://doi.org/10.1186/s12951-021-00996-0
Kumar, S., Rani, R., Dilbaghi, N., Tankeshwar, K., & Kim, K. H. (2017). Carbon nanotubes: A novel material for multifaceted applications in human healthcare. *Chemical Society Reviews, 46*(1), 158–196.
Liu, G. Y., Xu, S., & Qian, Y. (2000). Nanofabrication of self-assembled monolayers using scanning probe lithography. *Accounts of Chemical Research, 33*(7), 457–466.
Madhusudan, P., Chellukuri, N., & Shivakumar, N. (2018). Smart packaging of food for the 21st century–A review with futuristic trends, their feasibility and economics. *Materials Today: Proceedings, 5*(10), 21018–21022.
Majeed, K., Jawaid, M., Hassan, A. A. B. A. A., Bakar, A. A., Khalil, H. A., Salema, A. A., & Inuwa, I. (2013). Potential materials for food packaging from nano clay/natural fibers filled hybrid composites. *Materials and Design, 46*, 391–410.
Martins, B. A., de Albuquerque, P. B. S., & de Souza, M. P. (2022). Bio-based films and coatings: Sustainable polysaccharide packaging alternatives for the food industry. *Journal of Polymers and the Environment, 30*(10), 4023–4039.
Mattos, B. D., Rojas, O. J., & Magalhães, W. L. (2016). Biogenic SiO2 in colloidal dispersions via ball milling and ultrasonication. *Powder Technology, 301*, 58–64.
McClements, D. J. (2020). Nanotechnology approaches for improving the healthiness and sustainability of the modern food supply. *ACS Omega, 5*(46), 29623–29630.
Mehdizadeh, A., Shahidi, S. A., Shariatifar, N., Shiran, M., & Ghorbani-Hasan Saraei, A. (2022). Physicochemical characteristics and antioxidant activity of the chitosan/zein films incorporated with Pulicariagnaphalodes L. extract-loaded nanoliposomes. *Journal of Food Measurement and Characterization*, 1–11.
Modugu, S. C. R., Schuster, J., & Shaik, Y. P. (2022). Synthesis and characterization of carbon fiber nanocomposite using titanium dioxide and silicon carbide nanomaterials. *Journal of Composites Science, 6*(10), 312.
Ncube, L. K., Ude, A. U., Ogunmuyiwa, E. N., Zulkifli, R., & Beas, I. N. (2020). Environmental impact of food packaging materials: A review of contemporary development from conventional plastics to polylactic acid-based materials. *Materials, 13*(21), 4994.

Nerus, T. (2020). Emerging of bio-nano composite gelatine-based film as bio-degradable food packaging: A review. *Food Research, 4*(4), 944–956.
Ogunsona, E. O., Muthuraj, R., Ojogbo, E., Valerio, O., & Mekonnen, T. H. (2020). Engineered nanomaterials for antimicrobial applications: A review. *Applied Materials Today, 18*, 100473.
Omerović, N., Djisalov, M., Živojević, K., Mladenović, M., Vunduk, J., Milenković, I., et al. (2021). Antimicrobial nanoparticles and biodegradable polymer composites for active food packaging applications. *Comprehensive Reviews in Food Science and Food Safety, 20*(3), 2428–2454.
Pagno, C. H., Costa, T. M., de Menezes, E. W., Benvenutti, E. V., Hertz, P. F., Matte, C. R., et al. (2015). Development of active biofilms of quinoa (Chenopodium quinoa W.) starch containing gold nanoparticles and evaluation of antimicrobial activity. *Food Chemistry, 173*, 755–762.
Paiva, A. D., & Breukink, E. (2012). Antimicrobial peptides are produced by microorganisms. In *Antimicrobial peptides and innate immunity* (pp. 53–95). Springer.
Pathakoti, K., Manubolu, M., & Hwang, H. M. (2017). Nanostructures: Current uses and future applications in food science. *Journal of Food and Drug Analysis, 25*(2), 245–253.
Peters, R. J., van Bemmel, G., Herrera-Rivera, Z., Helsper, H. P., Marvin, H. J., Weigel, S., et al. (2014). Characterization of titanium dioxide nanoparticles in food products: analytical methods to define nanoparticles. *Journal of Agricultural and Food Chemistry, 62*(27), 6285–6293.
Qin, Y., Liu, Y., Zhang, X., & Liu, J. (2020). Development of active and intelligent packaging by incorporating betalains from red pitaya (Hylocereuspolyrhizus) peel into starch/polyvinyl alcohol films. *Food Hydrocolloids, 100*, 105410.
Rossi, M., Passeri, D., Sinibaldi, A., Angjellari, M., Tamburri, E., Sorbo, A., et al. (2017). Nanotechnology for food packaging and food quality assessment. *Advances in Food and Nutrition Research, 82*, 149–204.
Rout, S., Tambe, S., Deshmukh, R. K., Mali, S., Cruz, J., Srivastav, P. P., et al. (2022). Recent trends in the application of essential oils: The next generation of food preservation and food packaging. *Trends in Food Science and Technology.*
Saba, N., Jawaid, M., & Asim, M. (2016). Recent advances in nano clay/natural fibers hybrid composites. *Nanoclay Reinforced Polymer Composites: Natural Fiber/Nanoclay Hybrid Composites*, 1–28.
Sahani, S., & Sharma, Y. C. (2021). Advancements in applications of nanotechnology in the global food industry. *Food Chemistry, 342*, 128318.
Sekhon, B. S. (2014). Nanotechnology in agri-food production: an overview. *Nanotechnology, Science, and Applications*, 31–53.
Shehabeldine, A. M., Amin, B. H., Hagras, F. A., Ramadan, A. A., Kamel, M. R., Ahmed, M. A., et al. (2023). Potential antimicrobial and antibiofilm properties of copper oxide nanoparticles: time-kill kinetic essay and ultrastructure of pathogenic bacterial cells. *Applied Biochemistry and Biotechnology, 195*(1), 467–485.
Silva-Leyton, R., Quijada, R., Bastías, R., Zamora, N., Olate-Moya, F., & Palza, H. (2019). Polyethylene/graphene oxide composites toward multifunctional active packaging films. *Composites Science and Technology, 184*, 107888.
Spirescu, V. A., Chircov, C., Grumezescu, A. M., Vasile, B. Ş, & Andronescu, E. (2021). Inorganic nanoparticles and composite films for antimicrobial therapies. *International Journal of Molecular Sciences, 22*(9), 4595.
Subbiah, T., Bhat, G. S., Tock, R. W., Parameswaran, S., & Ramkumar, S. S. (2005). Electrospinning of nanofibers. *Journal of Applied Polymer Science, 96*(2), 557–569.
Swaminathan, M., & Sharma, N. K. (2019). *Antimicrobial activity of the engineered nanoparticles used as coating agents* (pp. 549–563). Springer International Publishing.
Taherimehr, M., YousefniaPasha, H., Tabatabaeekoloor, R., & Pesaranhajiabbas, E. (2021). Trends and challenges of biopolymer-based nanocomposites in food packaging. *Comprehensive Reviews in Food Science and Food Safety, 20*(6), 5321–5344.
Thakur, M., Wang, B., & Verma, M. L. (2022). Development and applications of nano biosensors for sustainable agricultural and food industries: Recent developments, challenges, and perspectives. *Environmental Technology and Innovation*, 102371.

Tripathi, A., & Khare, N. (2017). Applications of nanobiotechnology in food packaging. *International Journal of Trend in Scientific Research and Development, 2*(1), 662–666.

Worku, L. A., Nigussie, Y., Bachheti, A., Bachheti, R. K., & Husen, A. (2023b). Antimicrobial activities of nanomaterials. In A. Husen & K. S. Siddiqi (Eds.), *Advances in smart nanomaterials and their applications* (pp. 127–148). Elsevier.

Worku, L. A., Bachheti, R. K., & Tadesse, M. G. (2023). Preparation and characterization of carboxylated cellulose nanocrystals from *Oxytenanthera abyssinica* (Ethiopian lowland bamboo) cellulose via citric acid anhydrous hydrolysis catalyzed by sulfuric acid. *Biomass Conversion and Biorefinery*. https://doi.org/10.1007/s13399-022-03718-0

Potential of Silver and Zinc Nanoparticles in Mediating Abiotic Stress Tolerance in Crop Plants

Gouranga Upadhyaya and Aryadeep Roychoudhury

Abstract Being rooted, plants are highly vulnerable to extreme abiotic stress environment like high temperature, drought, and salinity, severely affecting the quality and yield of crops. Among the multiple approaches to fortifying plants, nanoparticles (NPs) have attracted global attention. In particular, silver (Ag) and zinc (Zn) NPs are very popular owing to their biosolubility, small size, and large surface area. Once NPs reach the outer membrane of plant cells, they move in through the plasmodesmata or endocytic pathways and are translocated to other organs via apoplastic and symplastic routes. NPs, upon reaching the xylem, are transferred to the shoots, and some flow back to the roots through the phloem. These NPs show notable stress protection in enhanced growth, seed germination, chlorophyll content, stomatal conductance, transpiration rate, and decreased reactive oxygen species (ROS) and membrane damage. Specifically, Ag and Zn NPs protect from mitochondrial dysfunction, ribosome disassembly, and DNA and protein damage. However, beyond a certain concentration, Ag and Zn, like other NPs, are cytotoxic and genotoxic, thereby causing excess ROS accumulation, DNA damage, chromosomal aberrations, and membrane damage leading to decreased leaf area, pollen viability, chlorophyll degradation, vacuole shrinkage, etc. Hence, only regulated exposure to NPs can help us achieve sustainable nano-agriculture in the near future. This chapter will probe into the role of Ag and Zn NPs in protecting plants from stressful abiotic conditions, along with insights into the cytotoxic nature and activity of NPs.

Keywords Nanotechnology · Silver Nanoparticles (AgNPs) · Zinc Nanoparticles (ZnNPs) · Abiotic stress · Cereal crops · Crop yield · Sustainable agriculture

---

G. Upadhyaya
Indian Institute of Science Education and Research Kolkata, Mohanpur 741246, West Bengal, India

A. Roychoudhury (✉)
Discipline of Life Sciences, School of Sciences, Indira Gandhi National Open University, Maidan Garhi, New Delhi 110068, India
e-mail: aryadeep.rc@gmail.com

R. K. Bachheti et al. (eds.), *Metal and Metal-Oxide Based Nanomaterials*, Smart Nanomaterials Technology, https://doi.org/10.1007/978-981-99-7673-7_8

1 Nanoagriculture: Boon or Bane to Crops Exposed to Abiotic Stresses?

Scientists have been referring to the agricultural application of nanoparticles (NPs) to protect crops from environmental stresses and boost their production as "nanoagriculture." It is developed opposite to the conventional agricultural practice, which relies on the use of bulky fertilizers with heavy impacts on the human and animal world. Nanoagriculture using bare minimum quantity perfectly fits into the concept of sustainable agriculture, where low production expenditure leads to higher returns. Hence, NPs have been probed globally, and researchers have developed nanoforms of various plant growth promoters, pesticides, fertilizers, herbicides, agrochemical encapsulated nanocarrier systems, for an enriched agricultural practice (Husen, 2022, 2023; Husen & Iqbal, 2019; Husen & Siddiqi, 2023; Singh et al., 2021). As understood from the terms, these nanomaterials have significant applications, including higher crop productivity and stress protection with lesser cost, energy, and waste production. Focusing on stress protection, NPs have been immensely researched and reported for their role in stress alleviation from heavy metals, salinity, and drought (El-Saadony et al., 2022; Singh & Husen, 2019, 2020).

NPs facilitate present agriculture transformation into precision agriculture by targeting maximum output from minimum resources. However, these multitalented NPs come with specific agroecological toxicity concerns, such as the synthesis, uptake, penetration, and internalization inside the plant cells, which still need elaborate investigation (Singh et al., 2021). Thus, risk assessment and management is an essential zone in the field of nanoagriculture that needs attention.

According to the definition by the International Standardization Organization (ISO) Technical Committee 229 (in 2010), nanotechnology refers to the "study/use of objects and organisms at nanoscale of 1–100 nm." Many physiochemical properties, such as size, surface charge, shape, and potential interaction of these NPs with plants, determine their potential application. Many different forms of NPs have been reported for their application in agriculture, but this chapter will solely focus on silver (Ag) and zinc (Zn) NPs. Concerning Ag and Zn NPs, both of these NPs have shown tremendous potential in protecting plants from abiotic stress-mediated damage. They have shown a beneficial impact on many physiological processes of plants, including seed germination, photosynthesis, growth and development, and reactive oxygen species (ROS) scavenging (Banerjee & Roychoudhury, 2022). However, this chapter is focused on the ability of these NPs to protect the plants from abiotic stresses.

Abiotic stresses are often interrelated, whether they are individual or in combination. High temperature, cold, salinity, and drought are the major stress forms that induce damage to plant species, including crop plants. Since the origin of the life forms in this world, temperature has continuously risen due to global warming. Increasing temperature is the primary source of heat/high-temperature stress and the most frequent version of the abiotic stressor. Heat stress varies for different plants, depending upon their ambient growth temperature. Most crop plants grow between

28 and 32 °C; hence, above 35 °C is generally considered high-temperature stress (Jha et al., 2014).

On the other hand, plants often face cold stress at variable degrees (0–15 °C). Plants apart from tropical and subtropical regions do not have mechanisms to acclimate to cold conditions under low temperatures. They suffer from cold stress. Cold and high-temperature stress cumulatively generates drought stress by limiting the groundwater level and moisture content (Krasensky & Jonak, 2012). Another form of abiotic stress, i.e., salinity stress, is also generated from poor-quality water with higher salt concentrations. Since higher salt content in water also creates osmotic imbalance and hinders water absorption through the root, salt stress shares almost identical features to drought stress (Navarro et al., 2008). Continuous development of urbanization and industrialization are the key factors for another form of abiotic stress, i.e., heavy metal stress. To serve food to the growing population, excess fertilizers and pesticides cause heavy metal stress directly (Khalid et al., 2022). The toxic heavy metal affects the cellular composition of a plant, and it has a direct impact on human uptake (Angulo-Bejarano et al., 2021). These abiotic stressors follow a common path to imbalance the redox signaling. These primary stressors further culminate into secondary stress, i.e., oxidative stress through the overproduction/generation of ROS (Upadhyaya et al., 2021). Hence, it causes severe damage to protein, lipids, carbohydrates, DNA, and other molecules and finally impairs cellular homeostasis. The effect of different abiotic stresses in plants, starting from molecular to physiological aspects, has been well documented and elaborated in several studies. Altogether they restrict plant growth and development, resulting in a significant loss of crop productivity and quality (He et al., 2018). How NPs can help to survive against such stress factors by enhancing tolerance levels at different stages has been illustrated here. This chapter aims to highlight the characteristics, application, uptake, and translocation method, site, and mode of action along with the toxic properties of Ag and Zn NPs in the present-day agricultural practice.

2 Ag and Zn NPs: Uptake, Translocation, and Site of Action

2.1 Source of AgNPs and ZnNPs

Although AgNPs are mainly introduced into the soil by anthropogenic sources, geogenic sources also exist, majorly as chemically inert Ag_2S. Biotic or abiotic transformation of Ag_2S often releases Ag^+ and leads to the generation of AgNPs. Several environmental factors like light, humidity and soil conditions govern abiotic transformation. On the other hand, biotic transformation to form AgNPs by microorganisms, fungi, and protozoa can be enzymatic or non-enzymatic. AgNPs can be synthesized by enzymes like nitrate reductase. Interestingly, root exudates of plants can also boost in vitro AgNP formation around the root surface zone (Guo et al., 2021). ZnNPs exist as ZnS and ZnSe, or quantum dots CdSe/ZnS, or the most widespread type of zinc oxide (nano-ZnO) (Ali et al., 2018).

2.2 *Uptake of AgNPs and ZnNPs*

The uptake of NPs in plants is a significant scientific interest because it is regarded as the major entry point into the food chain. There can be two major entry points: foliar uptake and root uptake pathways. Petiole feeding and trunk injection have also been attempted but have lost attention because of their limited applicability (normally trees) (Su et al., 2019).

In foliar exposure, plant leaves prefer cuticular and stomatal pathways for AgNP uptake (Wu & Li, 2022). NPs below 2 nm in size are taken up by cuticular openings directly, while larger particles choose the stomatal route (Chichiriccò & Poma, 2015). Even hydathode pores and trichomes might serve as additional pathways for NP uptake, although they have not yet been experimentally validated. Literature reports also observe AgNPs within leaf stomata and other tissues through high-resolution microscopy technologies (He et al., 2022). AgNPs ranging from 10 to 40 nm can be absorbed by leaves. The absorption of ZnO NPs is linked with the stomatal diameter. Zhu et al. (2020) reported that when the stomatal aperture diameter reduced from ~9 μm to ~5 μm, Zn concentrations in wheat leaf apoplast and cytoplasm decreased by 33.2 and 8.3%. In the case of root uptake, AgNPs smaller than 40 nm can directly pass through the almost impermeable cell wall. However, larger AgNPs require endocytic uptake, pore formation, or wounds for cellular entry, although direct evidence is yet to be reported (Zhu et al., 2020). Yang et al. (2018b) confirmed the direct uptake of AgNPs and reported AgNPs as the dominating Ag species in rice roots. The authors proposed that Ag^+ ions mainly resided on the root surface, possibly due to the adsorption of free Ag^+ ions with ligand residues on the cell wall, thus limiting their internalization. Wu et al. (2020) showed that the uptake of PVP-coated Ag nanowires is responsible for more than 85% of accumulated Ag in lettuce for 18-day exposure. Similarly, Dang et al. (2020) also stated that direct uptake of AgNPs by 15 day wheat seedlings was prominent compared to dissolved Ag uptake.

Most importantly, the uptake of NPs is affected by multiple environmental factors like physicochemical properties of soil, heavy metals, or organic pollutants. How various symbiotic microorganisms play a significant role in the transformation and uptake of AgNPs in plants still needs to be discovered. Also, there exists variation in the uptake and translocation rate between different plant species based on differences in root surface area, number of lateral roots, leaf area, stomatal size, and cuticle thickness.

2.3 Translocation of AgNPs and ZnNPs

The NPs taken in by the roots or leaves are translocated to other plant parts through the apoplastic or the symplastic pathway. AgNPs have been reported to be present in the epidermal cells, columella cells, or root meristem cells. AgNPs have been commonly observed in the cell wall. From the cell wall onwards, the NPs move through the pores in the cell wall into the intercellular space or the middle lamella (i.e., the region between the cell wall and plasma membrane) using the apoplastic route. Following the epidermal and the cortical region, most of the NPs get impeded by the Casparian strip in the endodermis. However, some NPs can infiltrate through the plasma membrane of the endodermal cells via the symplastic path. Once NPs reach the cell interior, AgNPs can move from one cell to the other via plasmodesmata. This plasmodesmata connection helps AgNPs reach the vascular bundles, from where they can travel throughout the plant body. In the vascular bundles, xylem helps AgNPs travel upward from roots to stems or leaves, while phloem aids downward movement from leaves to roots. In the phloem tissues, the perforation plate of vessels and the sieve plate hold a critical role in the translocation of NPs. They were observed to prevent the passage of most of the PVP-AgNPs and citrate-Ag NPs but not GA-Ag NPs.

Translocation factors ($TF_{shoot/root}$, which is calculated as the ratio of AgNP amount present in the shoots to that in roots) showed wide differences when the particle properties and plant species varied. Smaller Ag particles showed a greater tendency to be transported upward in tomatoes exposed to 10 mg/L Ag for 7 days (Noori et al., 2020). Even Souza et al., (2021) reported that the smaller AgNPs (30 nm) translocated easily from roots to the leaves of *Lemna minor* compared to the larger Ag NPs. Even the surface charge of NPs regulates the ability to translocate. Many studies have reported that negatively charged NPs bear higher $TF_{shoot/root}$, i.e., showing the tendency to be transferred from the roots to the shoots (Rajput et al., 2020). Ag content in citrus leaves was maximum in the case of branch-feeding GA–AgNPs as compared to PVP–AgNPs and citrate-AgNPs. Thus, GA coating enables better transport competency of NPs (Spielman-Sun et al., 2019). The movement of AgNPs within plants is also governed by the inner atmosphere of plants. For example, alfalfa (*Medicago sativa*) roots, and leaves of soybean (*Glycine max*) were found to restrict the movement of internalized AgNPs (Stegemeier et al., 2015). On the other hand, many studies showed that internalized AgNPs aggregated into larger NPs in plants like soybean, rice, wheat, lettuce, tomato, and Arabidopsis. AgNPs have also accumulated into many transformed products such as Ag_2O, AgCl, Ag_2S, and Ag-thiolate complexes.

Although ZnO NP is predicted to be the source of Zn^{2+} within tissues, spectroscopy studies failed to detect Zn^{2+} (Doolette et al., 2018; Rosa et al., 2013). Hence, ZnO NPs were concluded to be ionized by the root exudates on the root surface. Similarly, in *Lolium perenne*, ZnO NPs existed in scattered form within the apoplast, cytoplasm, nuclei of the endodermal cells and the vascular cylinder (Lin & Xing, 2008). In the seeds of *Vigna unguiculata*, Zn was present abundantly in the outer layer of cotyledon and hypocotyl, while very less Zn was noted in the testa and inner cotyledon (Wang et al., 2013). Zn was principally linked with citrate, histidine, malate, phytate, and polygalacturonic acid and precipitated in zinc phosphate. ZnO NPs moved in through the root surfaces are transported into the root cortex via the apoplastic pathway and were perceived in the epidermis, cortex, vascular system, and root tip cells. Through the gaps of the Casparian strips at the location of the primary and lateral root junction, ZnO NPs enter the vascular system. The route through the endodermis was proposed to be symplastic. However, to date, no ZnO NPs could be translocated into shoots, possibly due to the dissolution and biotransformation of ZnO NPs from ZnO NPs to $ZnPO_4$, which limits their long-distance transport.

3 Mode of Action of Ag and Zn NPs: Cellular Targets, Site, and Mode of Action

NPs have been found to exert a stress-protective umbrella on the plant cells against multiple abiotic stress factors. Different NPs have come up with different types of stress-mitigation strategies against different stressors (Fig. 1). For example, NPs have positively affected temperature stress mitigation and crop yield protection. NPs have been reported to play a pivotal role under high-temperature stress by maintaining the hydration state and growth rate. Heat shock protein (Hsp70, 90) and chaperones are the classical members of temperature stress alleviator of plants. Upregulation of these molecular chaperones and *Hsp* genes and their biosynthesis pathway members have been noted in plants treated with NPs to maintain stability of cellular components under stressful conditions (Shafqat et al., 2021; Wang et al., 2022). NPs help plants overcome temperature-mediated oxidative stress by exhibiting antioxidant properties. Under temperature stress, the application of ZnONP helps to maintain photosynthetic activity, stomatal movement, and activity of stress-responsive proteins that further reflects through normal grain yield and nutrition (Kareem et al., 2022; Song et al., 2021). AgNPs can provide membrane integrity, osmotic regulation, and antioxidant defense. Through these processes, Zn and Ag NPs can modulate plant metabolites like soluble sugar, proline, and IAA and increase shoot and root length and plant biomass to fight against heat stress (Azmat et al., 2022). Zn is the most effective NP for cold tolerance in crops like rice. Applying Zn can even combat cold stress by reducing photosynthetic injury and oxidative stress, thus helping maintain plant biomass (Song et al., 2021). Similarly, applications of ZnO NPs to sugarcane also help to combat cold stress by increasing chlorophyll synthesis and photosynthetic efficiency and impeding the gas exchange parameters and oxidative activity (Elsheery et al., 2020b).

To date, most of the studies have focused on the role of NPs in drought stress tolerance. It is the most common stress form, which is frequently raised when the water demand of the plant cannot be fully met. For survival, plants reduce water loss,

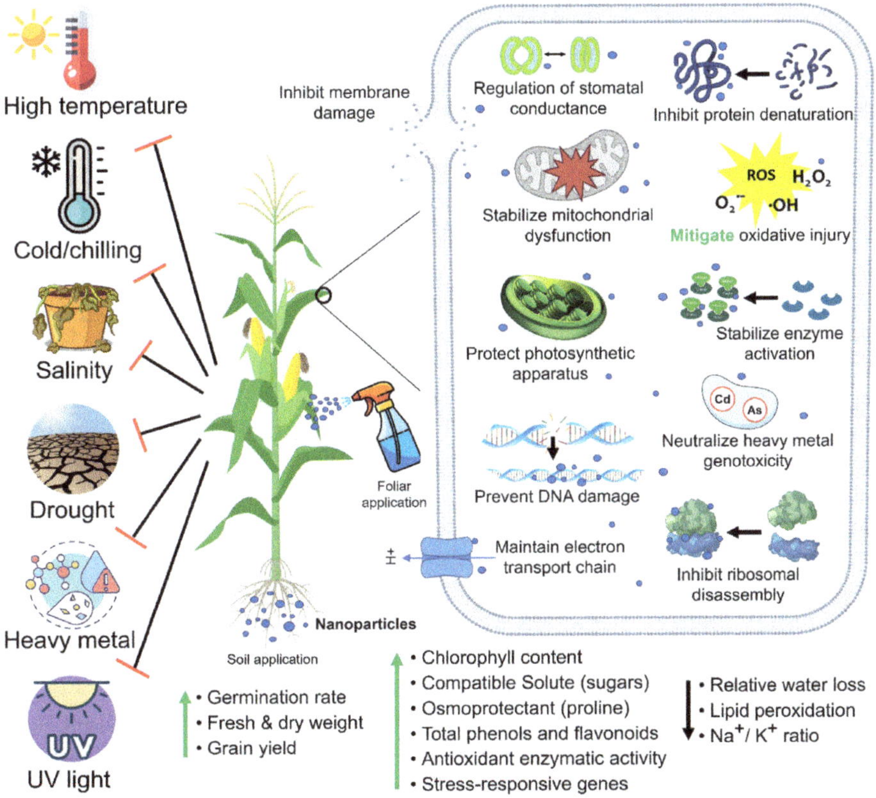

Fig. 1 Schematic diagram of Zinc (Zn) and Silver (Ag) NP-mediated defense responses in crop plants against abiotic stress

adjust morphological and anatomical adaptation, and specifically their biochemical processes. Zn and Ag NPs were found to render protection to multiple crop plants against drought stress through improvement in growth, physiochemical and biochemical characteristics (Allabdalah et al., 2021; Dimpka et al., 2020; Hojjat & Hojjat, 2016). Exogenous application of Zn and Ag can promote RuBisCo activase activity, photosynthetic index, and chlorophyll synthesis in cucumber (Ghani et al., 2022). Foliar application of NPs can induce photosynthetic rate and yield in soybean plants (Linh et al., 2020). Zn NPs help plants to expand lateral root formation, long root hairs, and water uptake ability in wheat (Yang et al., 2018a). Studies suggest that these NPs can also facilitate intracellular water binding, generating more water reserves to adapt to drought stress. AgNPs are the most used NPs in research experiments. Formulation of nano-sized ZnNPs and AgNPs as a fertilizer increased plant dry mass (Khalid et al., 2022). These NPs have been reported to alleviate drought-induced oxidative injury in crops in various ways, like enhancement in the activity of antioxidant enzymes, i.e., SOD, CAT, and POD, elevation in the expression of genes related to drought response, upregulation of aquaporins, improvement of metabolism, accumulation of compatible solutes, reduction in electrolyte leakage and ionic homeostasis (Allabdalah et al., 2021; Semida et al., 2021).

Plants often face salinity stress, resulting in lower water availability, nutritional disparities, and an upsurge in ionic toxicity (salt stress). Physiological, biochemical, and molecular pathways protect crop plants from salinity stress-mediated damage. Crop plants have a very complex mode of action to mitigate salt stress. The adaptive mechanism starts from osmotic regulation, compartmentalization of ions, and inhibition of toxic ion uptake until cellular ROS generation and electron transport chain is maintained. Additionally, the accumulation of compatible solutes, nitric oxide regulation, and hormonal directives to alter gene expression are the effective mechanisms for salt stress resistance (Ahmed et al., 2021; Khalid et al., 2020, 2022). NPs are a practical tool to improve plant growth and development subjected to salinity stress. Ag and Zn NPs can significantly increase the chlorophyll content as well as the photosynthetic rate in salinity-stressed plants (Adil et al., 2022; Mohamed et al., 2017; Zafar et al., 2021). The foliar application of such NPs helps to induce more light-harvesting complexes and photosynthesis (Ali et al., 2021). Ag NPs help increase plant growth characteristics and increase oxidative stress marker genes (CAT, SOD, GPX) in millet under 150 mM NaCl.

Interestingly, plants are more tolerant when they are primed with 10–30 mM Ag NPs (Khan et al., 2020). Salt stress negatively alters gene expression of different cellular processes, which NPs can neutralize. *Brassica napus,* upon foliar application of Zn NPs, also showed improved salt tolerance. It could suppress the expression of genes like *SKRD2*, *MYC*, and *MPK4* and increase the expression of other genes such as *ARP* and *MPK*. Zn NPs could stimulate genes associated with hormonal response and key transcription factors like ARP, considerably decreasing the expression of *MYC*, *MPK4*, and *SKRD2* genes (Hezaveh et al., 2019).

Among multiple stress forms, the deposition of heavy metals like cadmium, nickel, copper, lead, and mercury in the soil is highly toxic to plants. Heavy metals pass through the plasma membrane by partnering with other required nutrients with the help of metal carriers in the plant cells. They are non-degradable and are easily transferable to different parts (leaves, seed, stem) of plants, followed by human consumption. Hence, they have been considered as most toxic to the environment (Ghori et al., 2019; Sassykova et al., 2020). Mainly their highly oxidative states change the redox balance at the cellular level, resulting in protein denaturation, enzyme activity loss, membrane destruction, and imbalanced plant physiology (Manoj et al., 2020). The use of NPs as nano-fertilizer and nano-pesticides in agriculture is flourishing. Zn NPs have been reported to alleviate Cd stress in wheat (Hussain et al., 2018). It neutralizes Cd toxicity with enhanced POD and SOD activity.

Similarly, ZnO NPs were observed to mitigate As and Cd stress in rice (Ma et al., 2020). Venkatachalam et al. (2017) suggested a potential role of ZnONPs in *Leucaena leucocephala* exposed to Pb and Cd stress (Venkatachalam et al., 2017). Ag NPs have been found to mitigate lead toxicity in *Vigna radiata* by balancing the uptake of minerals and substantially enhancing plant growth (Chen et al., 2022).

Interestingly, Ag NPs also show a potential impact on the chelation of wastewater heavy metals (Zhou et al., 2021). Tables 1 and 2 list the possible activity modes of Ag NPs and Zn NPs against different types of abiotic stress forms.

4 Adverse Effects of Ag and Zn NPs

Despite the beneficial effect of NPs, these NPs have also been attributed with certain cytotoxic and genotoxic effects. Most importantly, reports suggest that NPs might accumulate through the food chain and cause severe toxicity. Plants being the primary producers in the chain are the first target of such bioaccumulation. Although the exact mechanism of toxicity induced by NPs is yet to be discovered, the possibility suggests that they depend on the chemical composition, structure, particle size, and surface area of the NPs (Albanese et al., 2012). It can be of the following types: (1) chemical toxicity due to chemical composition, e.g., the release of (toxic) ions; or (2) stress or stimuli caused by the surface, size, and shape of the particles. Arruda et al. (2015) stated that only some of the studies have assessed the effects of the ions released from the NPs, raising doubts about the data interpretation.

The phytotoxicity of ZnO NP (2 g/L) has been investigated by Lin and Xang (2008), where seed germination and root growth were found to be severely affected in six higher plant species. Even Ag NPs have been reported for their toxicity in *Phaseolus radiatus* and *Sorghum bicolor*, where authors suggest that the media (i.e., the soil in a real-life environment) and solubility of NPs is an important factor determining the toxicity (Lee et al., 2012). Even Gubbins et al., (2011) reported the phytotoxicity of AgNPs in terms of growth inhibition at 5 mg/L with size 20–100 nm. This study validated the hypothesis that the phytotoxicity of the NPs is related to the size, shape, exposure time, and concentration. Although certain authors have argued that proof of toxicity of Ag NPs is debatable, it is clear that Ag NP poses a great environmental risk. The toxicity of Ag NPs has also been verified in the seedlings of *Arabidopsis thaliana* at a low dose of 1 mg/L (Geisler-Lee et al., 2012). AgNPs ranging from 20 to 80 nm certainly arrested the growth, where the phytotoxicity relied on the concentration and particle size of the NPs.

The root tip (the cap and columella) turned light brown, perhaps attributed to the adsorption of the AgNPs with the cell wall materials or secondary metabolites produced by the root tips. The precise mechanism of phytotoxicity remains unclear. A recent study by Ke et al. (2020) showed that this toxicity of AgNPs is transferred to the offspring of *Arabidopsis thaliana* plants. Genes related to flowering and the development of floral organs were found to be significantly downregulated, causing a noticeable delay in flowering. A study of the phytotoxicity of ZnO NPs in *Lolium perenne* by Lin and Xing (2007) indicated that the phytotoxicity could not be clarified by the dissolution of ZnO NPs from bulk materials alone. The cytotoxic and genotoxic effects of ZnO NPs on root cells of *Allium cepa* were reported based on the mitotic inhibition index and chromosomal aberrations. The most common type of aberration was the stick chromosomes at the anaphase–telophase stages, probably due to the

Table 1 Variable mode of action of AgNPs

Organism	Type of stress	Nanoparticle (Amount)	Mode of action	References
Cuminum cyminum	Salinity stress	20 mg/kg	Enhanced germination percentage and vigor	Ekhtiyar and Moraghebi
Lens culinaris	Drought stress	20 μg/ml	Improved germination, shoot length, fresh and dry weight	Hojjat and Hojjat (2016)
Phaseolus vulgaris	Freezing stress	0.25, 1.25 mg/dm	Increased seedling height, fresh and dry weight, net photosynthesis	Prazak et al. (2020)
Crocus sativus	Flooding stress	40, 80 ppm	Enhanced root and leaf fresh and dry weight	Rezvani et al. (2012)
Satureja hortensis	Salinity stress	60 ppm	Enhanced germination and growth parameters	Nejatzadeh (2021)
Solanum lycopersicum	Salinity stress	0.5 mg/L	Increased germination, fresh and dry weight	Almutairi (2016)
Triticum aestivum	Salinity stress	300 ppm	Protected from oxidative damage by upregulating antioxidative enzymes	Wahid et al. (2020)
		5 mM	Improved growth parameters and decreased lipid peroxidation and ROS formation (H_2O_2)	Mohamed et al. (2017)
		50 mg/L	Balanced relative water content and improved chlorophyll	Iqbal et al. (2019a, 2019b)
	Heat stress	0.1 μmol	Restricted proline accumulation and reduced H_2O_2 and MDA content by up-regulating the antioxidant enzymes	Iqbal et al.
Solanum melongena	Drought stress	10–40 ppm	Promoted shoot and root growth, increased chlorophyll and carotenoid contents, reduced ROS, lipid peroxidation, elevated catalase (CAT), ascorbate peroxidase (APX) and glutathione reductase (GR) activities and gene expressions	Allabdalah et al. (2021)
Oryza sativa L., cv. Swarna	Oxidative stress	100 mg/L	Improved root length and weight, chlorophyll, and carotenoids	Gupta et al. (2018)

(continued)

Table 1 (continued)

Organism	Type of stress	Nanoparticle (Amount)	Mode of action	References
Thymus vulgaris	UV-B stress	75 ppm	Enhanced proline, total, and reduced glutathione, glyoxalase I and II activities, and expression of pyrroline-5-carboxylate synthetase gene	Azadi et al. (2021)
Cajanus cajan	Fluoride-induced oxidative stress	0–30 mM	Improved plant growth attributes, relative water content, proline content, fresh and dry weight, decreased oxidative damage by enhancing the antioxidant enzymes, lessened Na^+/K^+ ratio and increased total phenols and flavonoids	Yadu et al. (2017)

degradation or depolymerization of chromosomal DNA. Even micronucleus was observed in the root cell meristems at 100 μg/mL of ZnO NPs. When the toxicity of the ZnO NPs and Zn was compared, the authors concluded that the toxicity of the ZnO NPs was more significant. This shows that the toxicity of the ZnO NPs was not due to dissolved zinc ions alone. The toxic effects of ZnO NPs have already been reported in rice, maize, tomato, pea, *Medicago, and Brassica* (Rajput et al., 2018). Although several studies reported on the beneficial role of NPs are growing exponentially, the present-day scenario shows an exciting change. The report on toxicity studies of NPs has begun to gain pace. Hence, proper risk management is an absolute need to harvest the beneficial aspects of these NPs satisfactorily.

5 Conclusion

In a globe with constantly deteriorating environmental conditions, plants are being challenged daily with multiple environmental stressors, and frequently combined stresses also arise. This highlights the necessity of novel strategies to help plants cope with stress. Most importantly, crop plants cultivated to meet the food demand of the global population need to be protected. The discovery of NPs has given hope for protected sustainable agriculture. NPs, in a very minute amount, is turning up to be extremely beneficial as fertilizers/growth-promoters in the form of mineral/element(s) sources like Ag, Zn, etc. Many studies have reported how NPs like Ag and ZnO application in foliar spray, trunk injection, petiolar, or root application have helped plants face adverse abiotic stress factors, like salinity/drought/high temperature. Ag and ZnONPs often protect the plants by maintaining chlorophyll/carotenoid

Table 2 Variable mode of action of ZnO NPs

Organism	Type of stress	Nanoparticle (Amount)	Mode of action	References
Triticum asetivum	Cadmium stress	0–100 mg/L	Increase in plant height, spike length, dry weight, increased photosynthesis, reduced electrolyte leakage and SOD and POD activities, decreased Cd in roots, shoots, and grains	Khan et al. (2019)
	Drought stress	1%	Reduced time to panicle initiation, increased grain yield	Dimpka et al. (2020)
	Salinity stress	0.12 g/pot	Increased the physical parameters, chlorophyll content, plant height at vegetative and maturity stages, root fresh and dry weight, and grain yield	Adil et al. (2022)
	Cadmium + drought stress	100 mg/L	Improved growth and biomass, chlorophyll content, antioxidant enzyme activities by scavenging ROS and reducing Cd uptake	Bashir et al. (2021)
	Heat stress	10 ppm	Pigments, soluble sugars, protein, and indole acetic acid	Azmat et al. (2022)
Solanum melongena	Drought stress	50–100 ppm	Increased relative water content (RWC) and membrane stability associated with improved stem and leaf anatomical structures, enhanced photosynthetic efficiency, improved growth characteristics, and increased fruit yield	Semida et al. (2021)
Zea mays	Drought stress	100 mg/L	Enhanced melatonin synthesis and activated antioxidant enzymes, which alleviated the damage to mitochondria and chloroplast. Upregulation of Fe/Mn-SOD, Cu/Zn SOD, APX, CAT, TDC, SNAT, COMT, and ASMT genes	Sun et al. (2020)

(continued)

Table 2 (continued)

Organism	Type of stress	Nanoparticle (Amount)	Mode of action	References
	Cobalt stress	500 mg/L	Improved the plant growth, biomass, and photosynthetic machinery, reduced ROS and MDA accumulation, decreased Co uptake, conferred stability to plant ultra-cellular structures and photosynthetic apparatus, higher accumulation of nutrient content and antioxidant enzymes	Salam et al. (2022)
Sorghum bicolor	Drought stress	1, 3, and 5 mg/kg	Improved grain yield, grain N translocation, lowered total P acquisition, improved total K acquisition	Dimpka et al. (2019)
Glycine max	Arsenic stress		Improved growth, photosynthesis-related parameters, inhibited ROS, reversed MDA and H_2O_2 formation	Bhat et al. (2022)
Mangifera indica	Salinity stress	50, 100, and 150 mg/L	Improves resistance, annual crop load, and fruit quality	Elsheery et al. (2020a)
Abelmoschus esculentus	Salinity stress	0.1–0.3%	Increased shoot and root fresh and dry weight, chlorophyll content, and antioxidant activity	Zafar et al. (2021)
Lycopersicon esculentum	Cadmium stress	50 mg/L	Boosted plant height, fresh and dry weight, leaf area, photosynthetic attributes, stomatal conductance, increased protein content, activities of nitrate reductase and carbonic anhydrase, increase in stomatal aperture, decrease in MDA and superoxide radical (O^{2-})	Faizan et al.
	Salinity stress	15 and 30 mg/L	Callus induction, plant regeneration, improved element content, and antioxidant enzyme activity	Alharby et al. (2016)

(continued)

Table 2 (continued)

Organism	Type of stress	Nanoparticle (Amount)	Mode of action	References
Oryza sativa	Arsenic stress		Lessens the oxidative stress, lowers ROS, improves CAT, POX, and SOD, proline and total soluble protein content, and higher content of nitrogen, phosphorus, potassium, zinc, manganese, and iron	Faizan et al. (2021b)
	Cadmium stress	50 mg/L	Improved biomass, photosynthesis, protein, antioxidant enzyme activity, mineral nutrient content, and reduced Cd levels	Faizan et al. (2021a)
	Chilling stress	25, 50, and 100 mg/L	Restored chlorophyll accumulation and significantly ameliorated chilling-induced oxidative stress with reduced levels of H_2O_2, MDA, proline, and increased activities of major antioxidative enzymes; induced the chilling-responsive gene expression	Song et al. (2021)
Brassica napus	Salinity stress	100 mg/L	Improved germination, seed microstructure, and antioxidant enzyme activity decreased the expression of abscisic acid-related genes, BnCYP707A1, 3, and 4, and elevated the expression of BnGA20ox, BnGA3ox, and BnCPS genes	El-Badry et al. (2021)
Solanum lycopersicum	Salinity stress	20 and 40 mg/L	Increased genomic template stability, decreased DNA methylation	Hosseinpour et al. (2020)

(continued)

Table 2 (continued)

Organism	Type of stress	Nanoparticle (Amount)	Mode of action	References
Cucumis sativus	Drought stress	25 mg/L and 100 mg/L	Photosynthetic pigments and PSII activity enhanced, reduction in ROS and lipid peroxidation, reduction in oxidative damage manifested with the enhancement of enzymatic and non-enzymatic antioxidants, phenol and mineral contents were reduced, proline, glycine betaine, free amino acids, and sugars increased	Ghani et al. (2022)
Trigonella foenum-graecum	Salinity stress	1000 and 3000 ppm	Enhanced biochemical parameters, trigonelline, antioxidant enzymes, and elements such as Na, K, Ca, Zn, and Fe	Noohpisheh et al. (2020)
Vigna radiata	Heat stress	15, 30, 45, and 60 mg l^{-1}	Significantly upregulated the production of antioxidants and osmolytes; substantially abated the production of reactive oxygen species	Kareem et al. (2022)

content, improving photosynthesis, preventing water loss, preserving the membrane integrity, up-regulating the expression and activities of cellular antioxidants like SOD, CAT, APX. However, the study of NPs is still in its birth cradle and needs to be appropriately investigated for all the possibilities of toxicity, risk factor, and other side effects. Similar to the studies on the beneficial roles of Ag and ZnO NPs, multiple studies have also reported on the toxicity of these NPs in plants. How they may lead to oxidative stress is also being studied at present. Nevertheless, the multiple benefits of NPs must be addressed. Hence, proper risk management studies should be carried out to harvest the benefits of NPs completely. Applying these NPs might be a reliable tool to establish a protected sustainable agriculture for the future world.

References

Adil, M., Bashir, S., Bashir, S., Aslam, Z., Ahmad, N., & Younas, T., et al. (2022). Zinc oxide nanoparticles improved chlorophyll contents, physical parameters, and wheat yield under salt stress. *Frontiers in Plant Science, 13.*

Ahmed, T., AbouElezz, A., & Khalid, M. F. (2021). Hydropriming with moringa leaf extract mitigates salt stress in Wheat seedlings. *Agriculture, 11*(12), 1254.

Alabdallah, N. M., Hasan, M. M., Salih, A. M., Roushdy, S. S., Al-Shammari, A. S., Alsanie, S. I., & El-Zaidy, M. (2021). Silver nanoparticles improve growth and protect against oxidative damage in eggplant seedlings under drought stress. *Plant, Soil and Environment, 67*(11), 617–624.

Albanese, A., Tang, P. S., & Chan, W. C. (2012). The effect of nanoparticle size, shape, and surface chemistry on biological systems. *Annual Review of Biomedical Engineering, 14*(1), 1–16.

Alharby, H. F., Metwali, E. M., Fuller, M. P., & Aldhebiani, A. Y. (2016). The alteration of mRNA expression of SOD and GPX genes and proteins in tomato (*Lycopersicon esculentum* Mill) under stress of NaCl and/or ZnO nanoparticles. *Saudi Journal of Biological Sciences, 23*(6), 773–781.

Ali, A., Phull, A. R., & Zia, M. (2018). Elemental zinc to zinc nanoparticles: Are ZnO NPs crucial for life? Synthesis, toxicological, and environmental concerns. *Nanotechnology Reviews, 7*(5), 413–441.

Ali, S., Mehmood, A., & Khan, N. (2021). Uptake, translocation, and consequences of nanomaterials on plant growth and stress adaptation. *Journal of Nanomaterials.*

Almutairi, Z. M. (2016). Effect of nano-silicon application on the expression of salt tolerance genes in germinating tomato (*'Solanum lycopersicum'* L.) seedlings under salt stress. *Plant Omics, 9*(1), 106–114.

Angulo-Bejarano, P. I., Puente-Rivera, J., & Cruz-Ortega, R. (2021). Metal and metalloid toxicity in plants: An overview on molecular aspects. *Plants, 10*(4), 635.

Arruda, S. C. C., Silva, A. L. D., Galazzi, R. M., Azevedo, R. A., & Arruda, M. A. Z. (2015). Nanoparticles applied to plant science: A review. *Talanta, 131*, 693–705.

Azadi, M., Moghaddam, S. S., Rahimi, A., Pourakbar, L., & Popović-Djordjević, J. (2021). Biosynthesized silver nanoparticles ameliorate yield, leaf photosynthetic pigments, and essential oil composition of garden thyme (*Thymus vulgaris* L.) exposed to UV-B stress. *Journal of Environmental Chemical Engineering, 9*(5), 105919.

Azmat, A., Tanveer, Y., Yasmin, H., Hassan, M. N., Shahzad, A., Reddy, M., & Ahmad, A. (2022). Coactive role of zinc oxide nanoparticles and plant growth promoting rhizobacteria for mitigation of synchronized effects of heat and drought stress in wheat plants. *Chemosphere, 297*, 133982.

Banerjee, A., & Roychoudhury, A. (2022). Silver and zinc nanoparticles in the improvement of agricultural crops. In *Nano-enabled agrochemicals in agriculture* (pp. 199–209). Academic.

Bashir, A., ur Rehman, M. Z., Hussaini, K. M., Adrees, M., Qayyum, M. F., & Sayal, A. U., et al. (2021). Combined use of zinc nanoparticles and co-composted biochar enhanced wheat growth and decreased Cd concentration in grains under Cd and drought stress: A field study. *Environmental Technology and Innovation, 23*, 101518.

Bhat, J. A., Faizan, M., Bhat, M. A., Huang, F., Yu, D., & Ahmad, A., et al. (2022). Defense interplay of the zinc-oxide nanoparticles and melatonin in alleviating the arsenic stress in soybean (*Glycine max* L.). *Chemosphere, 288*, 132471.

Chen, F., Aqeel, M., Maqsood, M. F., Khalid, N., Irshad, M. K., & Ibrahim, M., et al. (2022). Mitigation of lead toxicity in *Vigna radiata* genotypes by silver nanoparticles. *Environmental Pollution, 308*, 119606.

Chichiriccò, G., & Poma, A. (2015). Penetration and toxicity of nanomaterials in higher plants. *Nanomaterials, 5*(2), 851–873.

Dang, F., Wang, Q., Cai, W., Zhou, D., & Xing, B. (2020). Uptake kinetics of silver nanoparticles by plant: Relative importance of particles and dissolved ions. *Nanotoxicology, 14*(5), 654–666.

de la Rosa, G., López-Moreno, M. L., de Haro, D., Botez, C. E., Peralta-Videa, J. R., & Gardea-Torresdey, J. L. (2013). Effects of ZnO nanoparticles in alfalfa, tomato, and cucumber at the germination stage: Root development and X-ray absorption spectroscopy studies. *Pure and Applied Chemistry, 85*(12), 2161–2174.

Dimkpa, C. O., Singh, U., Bindraban, P. S., Elmer, W. H., Gardea-Torresdey, J. L., & White, J. C. (2019). Zinc oxide nanoparticles alleviate drought-induced alterations in sorghum performance, nutrient acquisition, and grain fortification. *Science of the Total Environment, 688*, 926–934.

Dimkpa, C. O., Andrews, J., Sanabria, J., Bindraban, P. S., Singh, U., & Elmer, W. H., et al. (2020). Interactive effects of drought, organic fertilizer, and zinc oxide nanoscale and bulk particles on wheat performance and grain nutrient accumulation. *Science of the Total Environment, 722*, 137808.

Doolette, C. L., Read, T. L., Li, C., Scheckel, K. G., Donner, E., & Kopittke, P. M., et al. (2018). Foliar application of zinc sulfate and zinc EDTA to wheat leaves: differences in mobility, distribution, and speciation. *Journal of Experimental Botany, 69*(18), 4469–4481.

El-Badri, A. M., Batool, M., Mohamed, I. A., Khatib, A., Sherif, A., & Wang, Z., et al. (2021). Modulation of salinity impact on early seedling stage via nano-priming application of zinc oxide on rapeseed (*Brassica napus* L.). *Plant Physiology and Biochemistry, 166*, 376–392.

El-Saadony, M. T., Saad, A. M., Soliman, S. M., Salem, H. M., Desoky, E. S. M., & Babalghith, A. O., et al. (2022). Role of nanoparticles in enhancing crop tolerance to abiotic stress: A comprehensive review. *Frontiers in Plant Science, 13.*

Elsheery, N. I., Helaly, M. N., El-Hoseiny, H. M., & Alam-Eldein, S. M. (2020a). Zinc oxide and silicone nanoparticles to improve the resistance mechanism and annual productivity of salt-stressed mango trees. *Agronomy, 10*(4), 558.

Elsheery, N. I., Sunoj, V. S. J., Wen, Y., Zhu, J. J., Muralidharan, G., & Cao, K. F. (2020b). Foliar application of nanoparticles mitigates the chilling effect on photosynthesis and photoprotection in sugarcane. *Plant Physiology and Biochemistry, 149*, 50–60.

Faizan, M., Faraz, A., Mir, A. R., & Hayat, S. (2021a). Role of zinc oxide nanoparticles in countering negative effects generated by cadmium in *Lycopersicon esculentum*. *Journal of Plant Growth Regulation, 40*(1), 101–115.

Faizan, M., Sehar, S., Rajput, V. D., Faraz, A., Afzal, S., & Minkina, T., et al. (2021b). Modulation of cellular redox status and antioxidant defense system after synergistic application of zinc oxide nanoparticles and salicylic acid in rice (*Oryza sativa*) plant under arsenic stress. *Plants, 10*(11), 2254.

Geisler-Lee, J., Wang, Q., Yao, Y., Zhang, W., Geisler, M., Li, K., et al. (2012). Phytotoxicity, accumulation, and transport of silver nanoparticles by Arabidopsis thaliana. *Nanotoxicology, 7*(3), 323–337.

Ghani, M. I., Saleem, S., Rather, S. A., Rehmani, M. S., Alamri, S., Rajput, V. D., et al. (2022). Foliar application of zinc oxide nanoparticles: An effective strategy to mitigate drought stress in cucumber seedlings by modulating antioxidant defense system and osmolytes accumulation. *Chemosphere, 289*, 133202.

Ghori, N. H., Ghori, T., Hayat, M. Q., Imadi, S. R., Gul, A., Altay, V., & Ozturk, M. (2019). Heavy metal stress and responses in plants. *International Journal of Environmental Science and Technology, 16*(3), 1807–1828.

Gubbins, E. J., Batty, L. C., & Lead, J. R. (2011). Phytotoxicity of silver nanoparticles to Lemna minor L. *Environmental Pollution, 159*(6), 1551–1559.

Guo, H., Han, F., Shang, H., Xiong, S., Huynh, M., Thistle, L., et al. (2021). New insight into naturally formed nanosilver particles: Role of plant root exudates. *Environmental Science: Nano, 8*(6), 1580–1592.

Gupta, S. D., Agarwal, A., & Pradhan, S. (2018). Phytostimulatory effect of silver nanoparticles (AgNPs) on rice seedling growth: An insight from antioxidative enzyme activities and gene expression patterns. *Ecotoxicology and Environmental Safety, 161*, 624–633.

He, M., He, C. Q., & Ding, N. Z. (2018). Abiotic stresses: General defenses of land plants and chances for engineering multistress tolerance. *Frontiers in Plant Science, 9*, 1771.

He, J., Zhang, L., He, S. Y., Ryser, E. T., Li, H., & Zhang, W. (2022). Stomata facilitate foliar sorption of silver nanoparticles by *Arabidopsis thaliana*. *Environmental Pollution, 292*, 118448.

Hezaveh, T. A., Pourakbar, L., Rahmani, F., & Alipour, H. (2019). Interactive effects of salinity and ZnO nanoparticles on physiological and molecular parameters of rapeseed (*Brassica napus* L.). *Communications in Soil Science and Plant Analysis, 50*(6), 698–715.

Hojjat, S. S., & Hojjat, H. (2016). Effects of silver nanoparticle exposure on germination of Lentil (*Lens culinaris* Medik.). *International Journal of Farming and Allied Sciences, 5*(3), 248–252.

Hosseinpour, A., Haliloglu, K., TolgaCinisli, K., Ozkan, G., Ozturk, H. I., Pour-Aboughadareh, A., & Poczai, P. (2020). Application of zinc oxide nanoparticles and plant growth-promoting bacteria reduces genetic impairment under salt stress in tomatoes (*Solanum lycopersicum* L.'Linda'). *Agriculture, 10*(11), 521.

Husen, A., & Iqbal, M. (2019). *Nanomaterials and plant potential.* Springer International Publishing AG. https://doi.org/10.1007/978-3-030-05569-1

Husen, A., & Siddiqi, K. S. (2023). *Advances in smart nanomaterials and their applications.* Elsevier Inc. https://doi.org/10.1016/C2021-0-02202-1

Husen, A. (2022). *Engineered nanomaterials for sustainable agricultural production, soil improvement and stress management.* Elsevier Inc. https://doi.org/10.1016/C2021-0-00054-7

Husen, A. (2023). *Nanomaterials and nanocomposites exposures to plants (Response, Interaction, Phytotoxicity and Defense Mechanisms).* Springer Nature Singapore Pte Ltd. https://doi.org/10.1007/978-981-99-2419-6

Hussain, A., Ali, S., Rizwan, M., ur Rehman, M. Z., Javed, M. R., & Imran, M., et al. (2018). Zinc oxide nanoparticles alter the wheat physiological response and reduce the cadmium uptake by plants. *Environmental Pollution, 242*, 1518–1526.

Iqbal, M., Raja, N. I., Mashwani, Z. U. R., Hussain, M., Ejaz, M., & Yasmeen, F. (2019a). Effect of silver nanoparticles on growth of wheat under heat stress. *Iranian Journal of Science and Technology, Transactions a: Science, 43*(2), 387–395.

Iqbal, M., Raja, N. I., Mashwani, Z. U. R., Wattoo, F. H., Hussain, M., Ejaz, M., & Saira, H. (2019b). Assessment of AgNPs exposure on physiological and biochemical changes and antioxidative defense system in wheat (*Triticum aestivum* L) under heat stress. *IET Nanobiotechnology, 13*(2), 230–236.

Jha, U. C., Bohra, A., & Singh, N. P. (2014). Heat stress in crop plants: Its nature, impacts and integrated breeding strategies to improve heat tolerance. *Plant Breeding, 133*(6), 679–701.

Kareem, H. A., Saleem, M. F., Saleem, S., Rather, S. A., Wani, S. H., Siddiqui, M. H., et al. (2022). Zinc oxide nanoparticles interplay with physiological and biochemical attributes in terminal heat stress alleviation in mungbean (*Vigna radiata* L.). *Frontiers in Plant Science, 13*, 842349–842349.

Ke, M., Li, Y., Qu, Q., Ye, Y., Peijnenburg, W. J. G. M., & Zhang, Z., et al. (2020). Offspring toxicity of silver nanoparticles to *Arabidopsis thaliana* flowering and floral development. *Journal of Hazardous Materials, 386*, 121975.

Khalid, M. F., Hussain, S., Anjum, M. A., Ahmad, S., Ali, M. A., Ejaz, S., & Morillon, R. (2020). Better salinity tolerance in tetraploid versus diploid volkamer lemon seedlings is associated with robust antioxidant and osmotic adjustment mechanisms. *Journal of Plant Physiology, 244*, 153071.

Khalid, M. F., Iqbal Khan, R., Jawaid, M. Z., Shafqat, W., Hussain, S., Ahmed, T., et al. (2022). Nanoparticles: the plant savior under abiotic stresses. *Nanomaterials, 12*(21), 3915.

Khan, Z. S., Rizwan, M., Hafeez, M., Ali, S., Javed, M. R., & Adrees, M. (2019). The accumulation of cadmium in wheat (*Triticum aestivum*) is influenced by zinc oxide nanoparticles and soil moisture conditions. *Environmental Science and Pollution Research, 26*(19), 19859–19870.

Khan, I., Raza, M. A., Awan, S. A., Shah, G. A., Rizwan, M., Ali, B., et al. (2020). Amelioration of salt-induced toxicity in pearl millet by seed priming with silver nanoparticles (AgNPs): The oxidative damage, antioxidant enzymes, and ions uptake are major determinants of salt tolerant capacity. *Plant Physiology and Biochemistry, 156*, 221–232.

Krasensky, J., & Jonak, C. (2012). Drought, salt, and temperature stress-induced metabolic rearrangements and regulatory networks. *Journal of Experimental Botany, 63*(4), 1593–1608.

Lee, W. M., Kwak, J. I., & An, Y. J. (2012). Effect of silver nanoparticles in crop plants *Phaseolus radiatus* and *Sorghum bicolor*: Media effect on phytotoxicity. *Chemosphere, 86*(5), 491–499.

Lin, D., & Xing, B. (2007). Phytotoxicity of nanoparticles: Inhibition of seed germination and root growth. *Environmental Pollution, 150*(2), 243–250.

Lin, D., & Xing, B. (2008). Root uptake and phytotoxicity of ZnO nanoparticles. *Environmental Science and Technology, 42*(15), 5580–5585.

Linh, T. M., Mai, N. C., Hoe, P. T., Lien, L. Q., Ban, N. K., & Hien, L. T. T., et al. (2020). Metal-based nanoparticles enhance drought tolerance in soybean. *Journal of Nanomaterials*.

Ma, X., Sharifan, H., Dou, F., & Sun, W. (2020). Simultaneous reduction of arsenic (As) and cadmium (Cd) accumulation in rice by zinc oxide nanoparticles. *Chemical Engineering Journal, 384*, 123802.

Manoj, S. R., Karthik, C., Kadirvelu, K., Arulselvi, P. I., Shanmugasundaram, T., Bruno, B., & Rajkumar, M. (2020). Understanding the molecular mechanisms for the enhanced phytoremediation of heavy metals through plant growth promoting rhizobacteria: A review. *Journal of Environmental Management, 254*, 109779.

Mohamed, A. K. S., Qayyum, M. F., Abdel-Hadi, A. M., Rehman, R. A., Ali, S., & Rizwan, M. (2017). Interactive effect of salinity and silver nanoparticles on photosynthetic and biochemical parameters of wheat. *Archives of Agronomy and Soil Science, 63*(12), 1736–1747.

Navarro, A., Bañón, S., Conejero, W., & Sánchez-Blanco, M. J. (2008). Ornamental characters, ion accumulation, and water status in *Arbutus unedo* seedlings irrigated with saline water and subsequent relief and transplanting. *Environmental and Experimental Botany, 62*(3), 364–370.

Nejatzadeh, F. (2021). Effect of silver nanoparticles on salt tolerance of *Satureja hortensis* l. during in vitro and in vivo germination tests. *Heliyon, 7*(2), e05981.

Noohpisheh, Z., Amiri, H., Farhadi, S., & Mohammadi-Gholami, A. (2020). Green synthesis of Ag–ZnO nanocomposites using *Trigonella foenum-graecum* leaf extract and their antibacterial, antifungal, antioxidant and photocatalytic properties. *Spectrochimica Acta Part a: Molecular and Biomolecular Spectroscopy, 240*, 118595.

Noori, A., Donnelly, T., Colbert, J., Cai, W., Newman, L. A., & White, J. C. (2020). Exposure of tomato (*Lycopersicon esculentum*) to silver nanoparticles and silver nitrate: Physiological and molecular response. *International Journal of Phytoremediation, 22*(1), 40–51.

Prażak, R., Święciło, A., Krzepiłko, A., Michałek, S., & Arczewska, M. (2020). Impact of Ag nanoparticles on seed germination and seedling growth of green beans in normal and chill temperatures. *Agriculture, 10*(8), 312.

Rajput, V. D., Minkina, T. M., Behal, A., Sushkova, S. N., Mandzhieva, S., & Singh, R., et al. (2018). Effects of zinc-oxide nanoparticles on soil, plants, animals and soil organisms: A review. *Environmental Nanotechnology, Monitoring and Management*, *9*, 76–84.

Rajput, V., Minkina, T., Mazarji, M., Shende, S., Sushkova, S., & Mandzhieva, S., et al. (2020). Accumulation of nanoparticles in the soil-plant systems and their effects on human health. *Annals of Agricultural Sciences, 65*(2), 137–143.

Rezvani, N., Sorooshzadeh, A., & Farhadi, N. (2012). Effect of nano-silver on the growth of saffron in flooding stress. *International Journal of Agricultural and Biosystems Engineering, 6*(1), 11–16.

Salam, A., Khan, A. R., Liu, L., Yang, S., Azhar, W., & Ulhassan, Z., et al. (2022). Seed priming with zinc oxide nanoparticles downplayed ultrastructural damage and improved photosynthetic apparatus in maize under cobalt stress. *Journal of Hazardous Materials, 423*, 127021.

Sassykova, L. R., Aubakirov, Y. A., Akhmetkaliyeva, M. S., Sassykova, A. R., Sendilvelan, S., & Prabhakar, M., et al. (2020). Heavy metals accumulation in plants of the dry-steppe zone of the East Kazakhstan region. *Materials Today: Proceedings, 33*, 1187–1191.

Semida, W. M., Abdelkhalik, A., Mohamed, G. F., Abd El-Mageed, T. A., Abd El-Mageed, S. A., Rady, M. M., & Ali, E. F. (2021). Foliar application of zinc oxide nanoparticles promotes drought stress tolerance in eggplant (*Solanum melongena* L.). *Plants, 10*(2), 421.

Shafqat, W., Jaskani, M. J., Maqbool, R., Chattha, W. S., Ali, Z., & Naqvi, S. A., et al. (2021). Heat shock protein and aquaporin expression enhance water conserving behavior of citrus under water deficits and high-temperature conditions. *Environmental and Experimental Botany, 181*, 104270.

Singh, R. P., Handa, R., & Manchanda, G. (2021). Nanoparticles in sustainable agriculture: An emerging opportunity. *Journal of Controlled Release, 329*, 1234–1248.

Singh, S., & Husen, A. (2020) Behavior of agricultural crops in relation to nanomaterials under adverse environmental conditions. In A. Husen & M. Jawaid (Eds.), *Nanomaterials for agriculture and forestry applications* (pp. 219–256). Elsevier Inc. https://doi.org/10.1016/B978-0-12-817852-2.00009-3

Singh, S., & Husen, A. (2019) Role of nanomaterials in the mitigation of abiotic stress in plants. In A. Husen & M. Iqbal (Eds.), *Nanomaterials and plant potential* (pp. 441–471). Springer International Publishing AG. https://doi.org/10.1007/978-3-030-05569-1_18

Song, Y., Jiang, M., Zhang, H., & Li, R. (2021). Zinc oxide nanoparticles alleviate chilling stress in rice (*Oryza sativa* L.) by regulating antioxidative system and chilling response transcription factors. *Molecules, 26*(8), 2196.

Souza, L. R., Corrêa, T. Z., Bruni, A. T., & da Veiga, M. A. (2021). The effects of solubility of silver nanoparticles, accumulation, and toxicity to the aquatic plant *Lemna minor*. *Environmental Science and Pollution Research, 28*(13), 16720–16733.

Spielman-Sun, E., Avellan, A., Bland, G. D., Tappero, R. V., Acerbo, A. S., & Unrine, J. M., et al. (2019). Nanoparticle surface charge influences translocation and leaf distribution in vascular plants with contrasting anatomy. *Environmental Science: Nano, 6*(8), 2508–2519.

Stegemeier, J. P., Schwab, F., Colman, B. P., Webb, S. M., Newville, M., & Lanzirotti, A., et al. (2015). Speciation matters: Bioavailability of silver and silver sulfide nanoparticles to alfalfa (*Medicago sativa*). *Environmental Science and Technology, 49*(14), 8451–8460.

Su, Y., Ashworth, V., Kim, C., Adeleye, A. S., Rolshausen, P., & Roper, C., et al. (2019). Delivery, uptake, fate, and transport of engineered nanoparticles in plants: a critical review and data analysis. *Environmental Science: Nano, 6*(8), 2311–2331.

Sun, L., Song, F., Guo, J., Zhu, X., Liu, S., Liu, F., & Li, X. (2020). Nano-ZnO-induced drought tolerance is associated with melatonin synthesis and metabolism in maize. *International Journal of Molecular Sciences, 21*(3), 782.

Upadhyaya, G., Das, A., & Ray, S. (2021). A rice R2R3-MYB (OsC1) transcriptional regulator improves oxidative stress tolerance by modulating anthocyanin biosynthesis. *Physiologia Plantarum, 173*(4), 2334–2349.

Venkatachalam, P., Jayaraj, M., Manikandan, R., Geetha, N., Rene, E. R., Sharma, N. C., & Sahi, S. V. (2017). Zinc oxide nanoparticles (ZnONPs) alleviate heavy metal-induced toxicity in *Leucaena leucocephala* seedlings: A physiochemical analysis. *Plant Physiology and Biochemistry, 110*, 59–69.

Wahid, I., Kumari, S., Ahmad, R., Hussain, S. J., Alamri, S., Siddiqui, M. H., & Khan, M. I. R. (2020). Silver nanoparticle regulates salt tolerance in wheat through changes in ABA concentration, ion homeostasis, and defense systems. *Biomolecules, 10*(11), 1506.

Wang, S., Fu, Y., Zheng, S., Xu, Y., & Sun, Y. (2022). Phytotoxicity and accumulation of copper-based nanoparticles in brassica under cadmium stress. *Nanomaterials, 12*(9), 1497.

Wang, P., Menzies, N. W., Lombi, E., McKenna, B. A., Johannessen, B., & Glover, C. J., et al. (2013). Fate of ZnO nanoparticles in soils and cowpea (*Vigna unguiculata*). *Environmental Science and Technology, 47*(23), 13822–13830.

Wu, J., Yu, Q., Bosker, T., Vijver, M. G., & Peijnenburg, W. J. (2020). Quantifying the relative contribution of particulate versus dissolved silver to toxicity and uptake kinetics of silver nanowires in lettuce: Impact of size and coating. *Nanotoxicology, 14*(10), 1399–1414.

Wu, H., & Li, Z. (2022). Nano-enabled agriculture: how nanoparticles cross barriers in plants? *Plant Communications*, 100346.

Yadu, B., Chandrakar, V., Meena, R. K., & Keshavkant, S. (2017). Glycine betaine reduces oxidative injury and enhances fluoride stress tolerance via improving antioxidant enzymes, proline, and genomic template stability in *Cajanus cajan* L. *South African Journal of Botany, 111*, 68–75.

Yang, K. Y., Doxey, S., McLean, J. E., Britt, D., Watson, A., & Al Qassy, D., et al. (2018a). Remodeling of root morphology by CuO and ZnO nanoparticles: effects on drought tolerance for plants colonized by a beneficial pseudomonad. *Botany*, *96*(3), 175–186.

Yang, Q., Shan, W., Hu, L., Zhao, Y., Hou, Y., & Yin, Y., et al. (2018b). Uptake and transformation of silver nanoparticles and ions by rice plants revealed by dual stable isotope tracing. *Environmental Science and Technology*, *53*(2), 625–633.

Zafar, S., Hasnain, Z., Aslam, N., Mumtaz, S., Jaafar, H. Z., & Wahab, P. E. M., et al. (2021). Impact of Zn nanoparticles synthesized via green and chemical approach on okra (*Abelmoschus esculentus* L.) growth under salt stress. *Sustainability*, *13*(7), 3694.

Zhou, T., Prasher, S., Qi, Z., George, S., Mawof, A., & Nzediegwu, C., et al. (2021). Impact of silver nanoparticles in wastewater on heavy metal transport in soil and uptake by radish plants. *Water, Air, and Soil Pollution*, *232*(7), 1–1.

Zhu, J., Li, J., Shen, Y., Liu, S., Zeng, N., & Zhan, X., et al. (2020). Mechanism of zinc oxide nanoparticle entry into wheat seedling leaves. *Environmental Science: Nano*, *7*(12), 3901–3913.

Assessing the Impact of Silver and Zinc on Soil Microbial Structure and Functionality

Versha Dixit, Urvashi Goyal, William Joseph Kamal, Kundan Kumar Chaubey, B. T. Manjunath, Jobi Xavier, Anish Kumar Pal, Deepak Kumar Verma, and Ashwani Kumar Sanghi

Abstract This study demonstrates the complex interrelationships between soil microbial populations, silver, and zinc. Metals infiltrate the soil, water, and food chain due to anthropogenic, mining, and other activities. Microorganisms are crucial for the productivity of the soil, plants, and food chain. The role of soil microorganisms in nitrogen cycling and overall soil health is vital. However, certain soil metals may affect the functionality of microbial communities. The diversity, abundance, and enzyme activity of soil microorganisms are impacted by soil contamination with silver and zinc. Understanding how these metals affect the structure and functionality of microorganisms is crucial for forecasting and managing soil health and nutrient cycle activities. The effects of silver and zinc on the soil microbial population were investigated using in vivo and in vitro methods. The majority of current research examining their influence uses molecular-based methodologies. The influence of metals on soil and microbial communities, as well as at the molecular level, has been the subject of extensive marker gene study, and this research is still ongoing.

V. Dixit · W. J. Kamal · B. T. Manjunath · J. Xavier
Department of Life Sciences, CHRIST (Deemed to Be University), Bangalore 560029, Karnataka, India

U. Goyal
Sanjay Gandhi Postgraduate Institute of Medical Sciences, New PMSSY Rd, Raibareli Rd., Lucknow 226014, Uttar Pradesh, India

K. K. Chaubey (✉) · A. K. Pal
Division of Research and Innovation, School of Applied and Life Sciences, Uttaranchal University, P.O. Chandanwari, Arcadia Grant Premnagar, Dehradun 248007, Uttarakhand, India
e-mail: kundan2006chaubey@gmail.com

K. K. Chaubey
Department of Biotechnology, School of Basic and Applied Sciences, Sanskriti University, Mathura 281401, Uttar Pradesh, India

D. K. Verma
Centre for Health Research and Innovation, Apeejay Business Centre, Arunachal Building, 6th Floor Barakhamba Road, New Delhi 110001, India

A. K. Sanghi
School of Allied Health Sciences, MVN University, Palwal 121105, Haryana, India

R. K. Bachheti et al. (eds.), *Metal and Metal-Oxide Based Nanomaterials*, Smart Nanomaterials Technology, https://doi.org/10.1007/978-981-99-7673-7_9

More research is needed to comprehend the mechanisms underlying these effects and develop mitigation strategies for the possibly deleterious effects of soil pollution with Ag and Zn.

Keywords Silver · Zinc · Soil microbial structure · Microbial functionality · Nutrient cycling

1 Introduction

Estimates place the number of microorganisms per gram of soil at 10^9 cells and 10^4 species, respectively. These microorganisms play a significant role in various soil processes, including biogeochemical cycling, plant production, and climate regulation. They are also crucial to maintaining the health of terrestrial ecosystems. As a nonrenewable resource, the soil is increasingly stressed by the environment, most frequently due to the expansion of human activity. Given the critical significance of maintaining soil functions, a lot of work has been put into understanding how soil ecosystems react to disturbance or environmental change, as well as the resilience and resistance of soil microorganisms (Griffiths & Philippot, 2013). Poor soil quality caused by pollution can have wide-ranging economic repercussions due to the costs of rehabilitation and the reduction in primary productivity and effects on the global climate. So that soil microbial health may be preserved, it is crucial to comprehend and quantify the effects of xenobiotic contaminants on soil bacteria (Meier et al., 2020).

Effective methods for examining microorganisms' variety, distribution, and behavior in soil habitats must be developed for a deeper understanding of soil health. Because of the apparent connections between microbial diversity, soil and plant quality, and ecosystem sustainability, soils' microbial properties are increasingly being assessed as sensitive soil health markers. Although understanding microbial characteristics like biomass, activity, and diversity is crucial for scientists to advance their understanding of the elements influencing soil health, the results of such analyses may also be helpful to extension workers and farmers to develop usable soil quality indicators (Hill et al., 2000). A significant problem for soil ecology is the complexity and dynamic nature of the soil microbial community, whose composition differs between different compartments and layers (Trabelsi & Mhamdi, 2013).

Due to their toxicity, heavy metals in soil pose a severe hazard to human health and the ecosystem. Heavy metals (such as Pb and Zn) can contaminate soil due to anthropogenic activities such as mining, atmospheric deposition, sewage, mineral fertilizers, and pesticides (Fajardo et al., 2012). Metals are a xenobiotic contaminant in soils that warrants serious consideration. Remarkably, numerous metal-based nanomaterials have recently become valuable industrial materials. Due to their antibacterial properties, silver nanoparticles (AgNPs) have many uses, including water treatment, textile antimicrobials, food preservation, and medical sterilization (Gonfa et al., 2022, 2023; Husen & Siddiqi, 2023; Leyu et al., 2023). Including

AgNPs in consumer products has resulted in the accidental release of silver nanoparticles into ecosystems, even though these particles may have valuable applications due to their antibacterial capabilities (e.g., preventing disease or spoilage). Wastewater treatment facilities are the main entry point for AgNPs into the environment. Sewer sludge, frequently used as fertilizer for crops or dumped in landfills, retains over 90% of the AgNPs in wastewater (Meier et al., 2020).

It is also known that zinc oxide nanoparticles (ZnO NPs) reduce the soil's microbial population. As evidenced by reduced DNA at the same concentration of these nanoparticles, the makeup of the soil bacterial community was observed to change, with ZnO having a more significant impact than TiO_2NPs. These nanoparticles are also hazardous to higher trophic groups in the soil food cycle. The size and concentration of ZnO NPs in the soil affect how toxic they are. For example, ZnO NPs dramatically slowed the growth of the earthworm *Eisenia fetida* and the nematode species *C. elgans*. These effects of ZnO NPs on different trophic groups in the soil food web will have functional repercussions on activities in the soil ecosystem, such as nitrogen (N) and carbon (C) cycling (Rashid et al., 2017). The primary goal of this chapter is to provide a thorough explanation of silver and zinc, their introduction into the soil through diverse sources, and eventually to evaluate the effect of these metals on soil bacteria.

2 Overview of Silver and Zinc in the Environment

The crust of the Earth contains the naturally occurring element silver. The methods of mining and refining can be used to obtain it. Additionally, human activities, including industrial operations, trash incineration, and the use of silver-containing items, cause silver to be discharged into the environment (Briffa et al., 2020). Silver is used in a wide range of commercial and consumer goods. Its aesthetic appeal makes it frequently used in jewelry, silverware, and coinage. Additionally, it is used in batteries, solar panels, medical equipment, electronics, and photography supplies. The antibacterial qualities of silver also make it appropriate for use in textiles, water purification systems, and wound dressings (Godeto et al., 2023; Schwab et al., 2019). The crust of the Earth contains naturally occurring zinc, a vital element. The main methods of obtaining it are mining and extraction procedures. In addition, activities including industrial processes, burning fossil fuels, and using zinc-containing items can release zinc into the environment (Ferreira & Leite, 2015). Zinc is used in a wide range of products and services. It is frequently used to galvanize steel in order to prevent corrosion. Additionally, it plays a significant role in creating alloys such as bronze, brass, and others. Batteries, electrical appliances, rubber goods, cosmetics, paints, and fertilizers use zinc. Additionally, it is frequently given to animal feed and is a crucial micronutrient in agricultural practices (Sabnavis et al., 2018).

2.1 Routes of Entry into Soil

Atmospheric deposition: Through atmospheric deposition, silver can infiltrate the soil. Silver particles are released during industrial processes, trash incineration, and the burning of fossil fuels, and they land on the soil's surface (Purcell & Peters, 1998).

Wastewater and sewage sludge: Wastewater from homes and businesses may contain substances, including silver, that can be released into the environment. Silver can be introduced into the soil when sewage sludge is used as a fertilizer on agricultural land (Blaser et al., 2008).

Landfills and waste disposal: Silver can be released into the soil by leaching or direct contact if silver-containing objects, such as electronics or medical waste, are not disposed of properly (Chen et al., 2016).

Fertilizers and manures: Agricultural fertilizers and manures frequently contain zinc. Zinc may contaminate the soil when these substances are applied (Nicholson et al., 2003).

Industrial discharges: Wastewater containing zinc may be released into the environment directly by industrial activity or indirectly by inappropriate waste disposal, contaminating the ground and surrounding water sources (Chirila & Draghici, 2008).

Atmospheric deposition: Emissions from industrial operations, the burning of fossil fuels, and natural sources like dust and volcanic activity can all deposit zinc into the soil's surface (Sen & Peucker-Ehrenbrink, 2012).

2.2 Environmental Fate and Behavior of Silver and Zinc in Soil

Silver ions can attach to soil particles, decreasing their mobility. However, specific soil characteristics can increase silver's solubility and availability to plants and microorganisms, such as a low pH and a high quantity of organic matter (Lead, 2008). Silver degrades very slowly in soil. It can last long, although processes like sorption, precipitation, and transformation into less toxic forms may cause its availability and toxicity to decline over time (Olaniran et al., 2013). Silver ions can attach to soil particles, decreasing their mobility. However, specific soil characteristics can increase silver's solubility and availability to plants and microorganisms, such as a low pH and a high quantity of organic matter (Lead, 2008). Silver degrades very slowly in soil. It can last long, although processes like sorption, precipitation, and transformation into less toxic forms may cause its availability and toxicity to decline over time (Olaniran et al., 2013).

Zinc ions can be adsorbed onto soil particles, decreasing their mobility and bioavailability. However, under other circumstances, such as acidic soils, excessive zinc application, or low organic matter content, zinc can become more soluble and readily available for plant absorption or leaching into groundwater (Alloway, 2008;

Salem & Husen, 2023). Zinc is relatively stable in the soil. However, if it reacts with organic matter or produces insoluble compounds, such as precipitation, sorption, or complication, its availability and toxicity may decline with time. It is crucial to remember that many variables, such as soil characteristics, climate, and the specific forms and concentrations of these elements in the environment, might affect the fate and behavior of silver and zinc in soil (McLaughlin et al., 2011).

3 Soil Microbial Structure

The vast population of microorganisms living in soil are called "soil microbial communities." Bacteria, fungi, archaea, viruses, and other microbial life forms are examples of these microorganisms. They are essential to the activities of the soil ecosystem, including nutrient cycling, the breakdown of organic matter, soil formation, and interactions between plants and microbes (Islam et al., 2020). One of the most prevalent groups of soil microorganisms is bacteria. A single gram of soil contains hundreds of distinct species, demonstrating their enormous diversity. Numerous soil processes, including nutrient transformation, organic matter breakdown, and pathogen suppression, are aided by bacteria (Sokol et al., 2022).

Fungi: Another significant group of soil microorganisms is the fungi. They participate in soil breakdown processes that break down sophisticated organic molecules. Mycorrhizae, which fungi create with plant roots in a mutualistic relationship, help plants absorb nutrients (van der Heijden et al., 2008).

Archaea: Despite being less prevalent than bacteria and fungi, archaea are found in soil ecosystems. In specific situations, they are known to play a part in the cycling of nutrients, nitrogen fixation, and methane production and consumption (Aislabie et al., 2013).

Viruses: Viruses that infect bacteria, fungi, and other microbes are also present in soil. These viral interactions have an impact on nutrient cycling and the dynamics of microbial communities (Wilpiszeski et al., 2019).

3.1 Factors Influencing Soil Microbial Structure

Microbial community structure can be strongly influenced by soil properties such as texture, pH, moisture level, nutrient availability, and organic matter content. Various microorganisms prefer these soil characteristics, influencing their variety and abundance (Stibal et al., 2006). Agricultural techniques like crop rotation, tillage, fertilization, and pesticide use can impact the soil's microbial populations. For instance, excessive tillage can alter soil composition and microbial habitats, but other pesticides may target only a specific type of microbe. The kinds of plants in an ecosystem can affect the soil microbial communities through root exudation, which supplies nutrients and energy sources for microorganisms. The types and quantities of chemicals released by various plant species influence the composition of microbial communities. Environmental factors such as temperature, moisture, oxygen availability, and UV radiation can affect soil microbial populations. Different physiological responses of microorganisms to these elements result in variances in community organization across various settings (Wahid et al., 2020).

3.2 DNA Techniques for Assessing Soil Microbial Structure

DNA sequencing: Microbial DNA collected from soil samples can be analyzed using high-throughput DNA sequencing techniques like amplicon or metagenomic sequencing. This method details the functional potential, community makeup, and diversity of microorganisms (Corcoll et al., 2017).

Phospholipid fatty acid (PLFA) analysis: The PLFA method analyzes soil samples to determine the lipid composition of microbial membranes. It offers information about the diversity, biomass, and community structure of particular microbial taxa, like bacteria and fungi (Zhang et al., 2012).

Microscopic techniques: Direct microscopic inspection, fluorescent in situ hybridization (FISH), and immunofluorescence are three microscopy-based techniques that enable the visualization and identification of particular microbial groups or activity in soil samples (Moter & Göbel, 2000).

Stable isotope probing (SIP): In stable isotope probing (SIP), particular microbial groups are labeled in soil samples using isotopically labeled substrates, and their incorporation into microbial biomass is followed. This method aids in locating the active microbial populations engaged in particular nutrient cycling processes (Friedrich, 2006).

Functional gene analysis: Molecular approaches like quantitative PCR (qPCR) or microarray analysis can offer insights into functional microbial groups' presence and potential activity by focusing on specific functional genes engaged in important soil processes, such as nitrogen cycling or carbon decomposition (Yergeau et al., 2009). These approaches can be utilized singly or in combination to understand better soil microorganisms' composition, variety, and potential functional roles.

4 Impact of Silver on Soil Microbial Structure and Functionality

The pure form of metallic silver is inert, and mostly it will not react with any human tissue or microorganisms until it is completely ionized (Melaiye & Youngs, 2005). Silver is used in wound management and can be seen from the eighteenth century onwards; silver in silver nitrate can also treat ulcers (Klasen, 2000). Silver-based dressings have different release patterns for the amounts of silver ions based on different materials (Ford et al., 2022). It mainly happened to replace antibiotics for managing infected wounds. The problem with silver-based dressings is due to the low-level release of the ions, which sometimes can be the reason for developing resistance (Thomas & McCubbin, 2003). These Silver ions are disruptive and can cause bacterial cell disintegration and alter the biochemical function, which has minimal toxicity for mammalian cells (Möhler et al., 2018). Silver is also used as an efficient chemotherapeutic, anti-fungal, wound care product, medical devices, antibacterial, and dental fillings (Lansdown, 2010).

4.1 *Effects of Silver on Soil Microbial Composition and Diversity*

Soil is a source of many kinds of microbes, and silver has been used as an antimicrobial agent for a long time. Recent findings in research show that silver is being used to potentize and expand antibiotic activity (Vasileiadis et al., 2015). The silver can alter the soil microbial diversity and be based on the kind of plant species that are grown in that soil (Pallavi et al., 2016). Many microbes are resistant to silver, found in the environment and clinical isolates (Deshpande & Chopade, 1994). In the study done by Liu et al. (2017), the silver nanoparticles showed a significant impact on the structure of the bacterial community in the soil in the growth of wheat plants.

4.2 *Mechanisms of Silver-Induced Microbial Toxicity*

The bacteria and the other eukaryotic cells have the nature of getting toxic effects from silver ions, as they bind to the cell membrane and cause impairment in the permeability of the membrane based on the dosage (Terzioğlu et al., 2022). Some studies state that the bioavailable Ag released from the AgNP plays a significant role in Ag toxicity. In different environments like the aquatic environment, Ag is the key component in influencing toxicity (Barros et al., 2019). The small-sized particles, which are AgNPs, are more toxic to the prokaryotic cells other than the different forms of Ag, and even in minute concentrations, they can produce resistance in bacteria (Choi & Hu, 2009).

4.3 *Functional Implications of Silver Exposure on Soil Microbial Communities*

The functional diversity of the soil microbes had a huge impact because of the shapes of the silver nanoparticles; as the concentration of the silver in the soil increased,

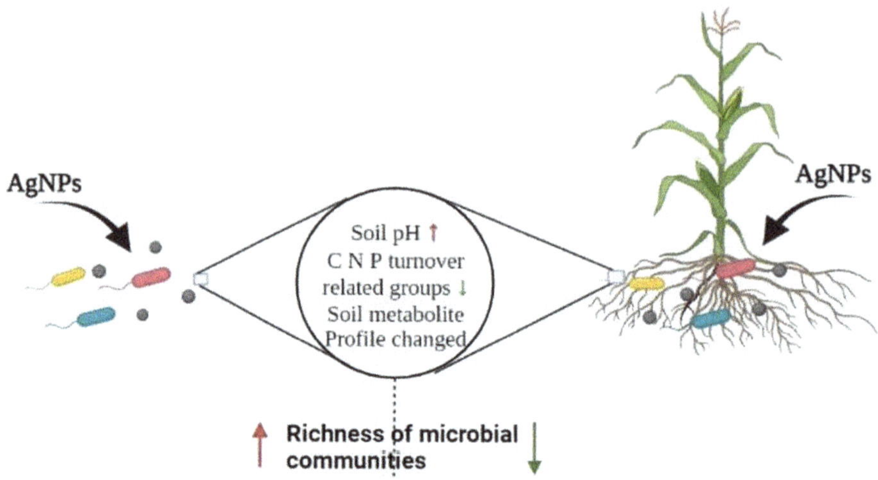

Fig. 1 Impact of silver on soil microbial structure and functionality

it showed a significant variation in the soil microbiota, and the exposure time also played a critical role in this (Zhai et al., 2016). Some studies state that silver can cause an inhibitory effect on some soil microorganisms, structure, and enzyme activities (Mishra et al., 2020). According to (Grün et al., 2019), many factors like concentration, soil texture, functionalization, and exposure time will significantly impact the soil microbes, mainly in long-term exposure (Fig. 1).

5 Impact of Zinc on Soil Microbial Structure and Functionality

Zinc is a mineral present in the soil, and it helps in plant growth, promoting the yield and development of the plant. The spatial availability of zinc in the soil varies based on its access to soil-zinc binding ability and bioavailable forms (Baran et al., 2018). Zinc in different forms is used in a wide range of commercial applications and thus is expected to have more expected to be found in soil (Rajput et al., 2022). Recently, zinc has accumulated in agricultural soils, so the soil function and properties have been sparsely reported (Wan et al., 2020).

5.1 *Effects of Zinc on Soil Microbial Composition and Diversity*

Zinc in the form of biosolids, when added to the soil, can improve the organic content but can cause the same amount of harm by adding heavy metals into the soil (Mossa et al., 2017; Wan et al., 2020). The high amount of heavy metals, including zinc, in the soil can cause adverse effects on plant growth, soil microbial activity, and diversity and has a huge impact on the genetic structure (Mossa et al., 2017; Wan et al., 2020; Xie et al., 2016). According to the study done by Bruce et al., heavy metal biosolid particles can permanently degrade the microbial decomposer communities present in agricultural soils (Moffett et al., 2003).

5.2 *Mechanisms of Zinc-Induced Microbial Toxicity*

The high usage of antibiotics in agriculture and medicine is a part of it, but the heavy metals contribute to antibiotic resistance even in the absence of antibiotics (Peltier et al., 2010). Zinc also acts as an alkylating agent, which is cytotoxic to many microbial colonies (Tobey et al., 1982). The microbes in the soil can change how the plants absorb the nutrients, and along with them, the particles like zinc and make plants undergo different pathways like ROS (Fan et al., 2023).

5.3 *Functional Implications of Zinc Exposure on Soil Microbial Communities*

According to Fan et al. (2023), García-Gómez et al. (2018), with the variation in the pH, there was a stimulating effect on the microbial activity of the soils, and the functional richness of the microbial community was drastically reduced. The increase in the heavy metals in the soil not only added contamination to the soil but showed a significant reduction in the microbial biomass and their enzyme activity (Agarwal et al., 2023). In the recent past, nanomaterials are also a threat to the environment, causing an alteration in the composition of bacterial communities (Chavan & Nadanathangam, 2020). The soils are getting exposed to man-made nanoparticles because of the application of sewage sludge as a fertilizer, and some consider it an organic soil improver (Read et al., 2016). in a study (Seneviratne & Zavahir, 2021), it was found that bacterial growth in the soils where it was treated or heavy metals is added in other ways to the microbial processes to stress, nature, and size (Fig. 2).

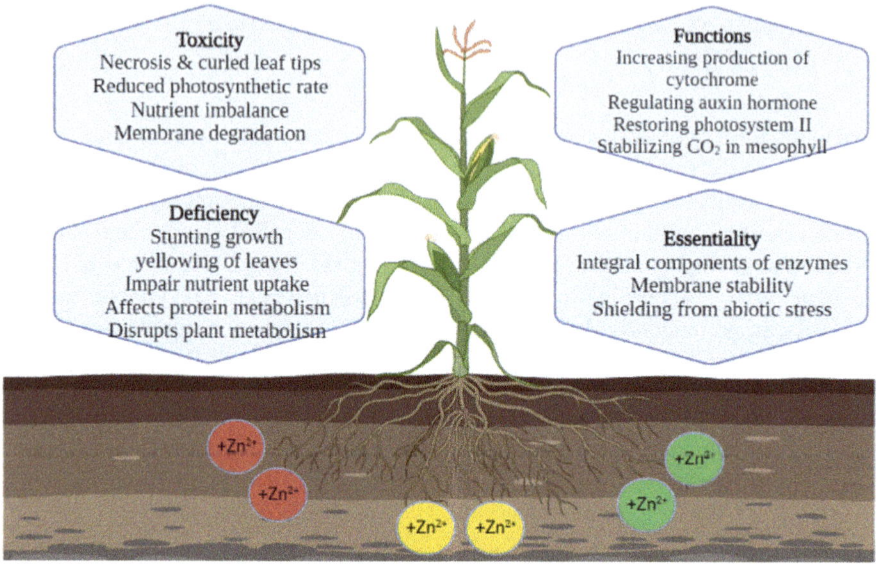

Fig. 2 Impact of zinc on soil microbial structure and functionality

6 Assessment Methods for Studying Microbial Responses to Silver and Zinc

6.1 Laboratory and Field-Based Approaches

The effects of silver and zinc oxide nanoparticles have only been studied in a few investigations using pure cultures of ecologically significant isolates (Bandyopadhyay et al., 2012). It has also been observed that the same nanoparticles generate comparable disruptions in soil bacterial communities (Sillen et al., 2015). These disturbances were tracked by following changes in microbial activity rates, microbial biomass, or microbial community makeup. Of these, the examination of the makeup of the microbial community offers more specific information on the phyla that may be impacted. One approach for assessing the complexity of microbial communities in various contexts is the sequencing and phylogenetic analysis of 16S rRNA genes using next-generation sequencing (NGS) technologies (Rastogi & Sani, 2011). A previous study published connection and binding interactions between metals and microbes. Proteins, lipids, and polysaccharides are only a few components that make up bacterial cells, which are crucial to the remediation process (Mukhopadhyay et al., 2002). When exposed to metal ions, the binding sites that absorb the metals develop and change. FT-IR is a powerful method for determining changes to bacterial cell walls' chemical composition and molecular elements (Anusha & Natarajan, 2020).

6.2 *Molecular Techniques for Analyzing Soil Microbial Communities*

Different stress response pathways have been developed to help cells eliminate silver ions and other metals. In addition, efflux pumps for other metal ions, like copper, can provide some silver resistance. Some genetic components, such as the sil genes on pMG101, have been characterized that confer silver resistance (Hobman & Crossman, 2015). Previous research has shown that soil microbial communities are hazardous to (Ag+) silver ions dissolved in AgNPs. Some researchers have employed soil microbiome sequencing to assess changes in response to environmental contaminants at the community structure or gene function level since next-generation sequencing (NGS) has become widespread (Ding et al., 2012; Yuan et al., 2019).

Since measuring gene functions is more expensive and technically difficult, metagenome-based methods are less biased than culture-based ones, yet much research has only looked at the present taxa. It is essential to understand the dynamic changes in the microbial communities' distribution of gene functions (Fondi et al., 2016). Functional redundancy across taxa may confuse bacterial responses to pollutants in marker gene studies that attempt to impute gene functions. It may also give an incomplete picture of community function due to poor database annotation and the challenge of obtaining species- or strain-level identifications (Meier et al., 2020).

7 Implications for Soil Health and Ecosystem Functioning

The soil microbiome is essential to maintain soil fertility, cycle nutrients, and sequester carbon. Furthermore, the health of plants and animals in various ecosystems is influenced by the soil microbiome both directly and indirectly. Research on the soil microbiome outlines the microorganisms that live there, their metabolic processes, and how fertile the soil is Islam et al. (2020). Trace elements in the context of soil remediation cannot be eliminated, like organic contaminants; they can only be moved. Different tactics have been employed to reduce the bioavailability and mobility of contaminants. Due to its function as an electron donor and its ability to act as a precipitant/sorbent substrate in its many crystalline forms, iron plays a significant part in the mobility, absorption, and degradation of environmental contaminants. Iron particles are a new generation of environmental remediation technologies that, at the nanoscale level, can offer economically viable solutions to some of the most challenging environmental cleaning issues (Fajardo et al., 2012).

It may be helpful to correlate the effects seen in laboratory-spun and field-contaminated soils using bioavailability parameters. The results of Zn on microbial processes were shown to be imperceptible in soil samples taken beneath galvanized pylons. However, microbial activity was decreased when uncontaminated soil was spiked with an equivalent amount of total Zn. According to similar thresholds calculated from the spiked soils, the Zn contents in the soil solution of the field-contaminated grounds were not high enough to anticipate significant adverse effects. That study's findings cannot be extrapolated to the association between soil characteristics and metal toxicity because it only considered three sites (Smolders et al., 2004).

7.1 *Effects of Silver and Zinc on Soil Nutrient Cycling*

It was hypothesized that community tolerance and acclimation processes to background Zn are more essential than metal solubility to explain variance in Zn toxicity among soils because Zn toxicity was a weaker function of soil pH (which affects Zn solubility) than background Zn. However, it was determined that there was still a significant amount of unexplained variation in the associations, necessitating experimental verification using a set of standardized toxicity endpoints (Smolders et al., 2004). According to reports, silver nanoparticles (AgNPs) increase soil nutrient availability by encouraging microbial activity and raising enzyme output. Additionally, leguminous plants' ability to fix nitrogen (N) has been reported to be improved by AgNPs, which promotes more significant plant growth and nutrient uptake. There have been reports of silver and zinc oxide nanoparticles having anti-phytopathogenic effects, improving seed germination and plant growth (Chavan & Nadanathangam, 2019). The agronomic productivity of the soil's microorganisms is crucial (Bhattacharyya & Jha, 2012). They facilitate plant growth, decompose organic waste, and recycle elements (Jacoby et al., 2017). These bacteria use any of the following techniques to directly promote plant growth: nitrogen fixation, phosphate solubilization, and the manufacture of phytohormones, siderophores, and antibiotics (Vejan et al., 2016).

This may have a negative impact on processes, including nitrification, organic carbon transformation, and chitin degradation (Grün et al., 2019). Various research has shown that AgNP's physicochemical and concurrent toxicological behavior varies depending on the soil type. AgNP may undergo various physicochemical transformations, including reduction, oxidation, aggregation, dissolution, complexation, and other secondary reactions (Masrahi et al., 2014). These transformations depend on pH, ionic strength, temperature, amount of dissolved ions and natural organic matter, oxygen concentration, grain size distribution, and others (Bundschuh et al., 2018). As a result, these changes impact the bioavailability and toxicity mechanism of AgNP. For instance, in comparison experiments with various soil types, AgNP toxicity was reduced in soils containing a higher percentage of clay because the AgNP was immobilized by hetero-aggregation with clay particles (Rahmatpour et al., 2017).

7.2 *Influence of Soil Microbial Changes on Plant Growth and Productivity*

Plant community structure can be used to anticipate compositional changes in the soil microbiome. As a result, different microbial communities should be present in soils that support a variety of plant species. Other plant species are sporadically linked to specific soil microbial taxa. Interestingly, mycorrhizal fungi, fungal diseases, and certain bacteria that fix nitrogen are often linked to particular plant species (Fitzpatrick et al., 2018). According to specific observations, it is possible to track changes in plant communities by looking at the general soil microbial community structures (Prober et al., 2015). In contrast, multiple studies have identified the plant varieties that have the most negligible impact on the makeup of the soil's microbial population (Tedersoo et al., 2016). A particular plant type may influence the makeup of the soil microbiome. A plant species is associated with specific microorganisms depending on the soil type. Similar to how many soil bacteria can be global and associated with various plant types (Islam et al., 2020).

A large amount of biologically accessible nitrogen is provided by nitrogen-fixing bacteria, also known as diazotrophic bacteria (DB), which are common rhizospheric communities that live inside and outside the roots of herbaceous plants and various other tree crops. Archaea are also essential for nitrogen cycling, significantly impacting soil quality. The oxidation of nitrite to nitrate by ammonia-oxidizing archaea (AOA) can cause soil to denitrify. According to metagenomic research, soils may differ substantially in terms of the relative abundance of AOA. Ammonia monooxygenase (AMO) enzymes are used by ammonia-oxidizing bacteria (AOB) and AOA to oxidize ammonia, producing nitrate as a byproduct (Islam et al., 2020).

7.3 *Ecological Implications of Altered Soil Microbial Communities*

Animals and plants growing in the same sediments or soils as microorganisms are far less vulnerable to the effects of heavy metal stress. Microorganisms sensitive to heavy metal pollution can be utilized as biological markers. With thorough consideration of microbial diversity and community structure, the microbial community function is ideal for use as an indicator to reflect heavy metal pollution (Li et al., 2020). Soil-living organisms, such as bacteria, fungi, and viruses, serve critical ecological functions and participate in the cycle of nutrients and minerals for plant growth (Lauber et al., 2013). Plants that live in the soil directly promote microbial growth while affecting the abiotic elements that subtly affect them (Santoyo et al., 2017). Geographic factors and soil characteristics significantly influence the architecture of soil microbial communities. However, according to Rillig et al. (2017), soil microorganisms significantly impact the development of soil aggregates. Mainly, soil moisture substantially influences microbial communities' composition more than nutrients (Banerjee et al., 2016).

According to Geyer et al. (2014), the abundance of bacterial species in soils from the polar desert was strongly associated with moisture content. To colonize the soil or rhizosphere regions close to root exudates or nutrients, bacteria must be mobile. For instance, in sandy soil with a moisture level of 16%, Azospirillum traveled 40–60 mm in 96 h; however, in soils with a moisture content of 10%, this distance was reduced to 20 mm during the same time (Ochoa et al., 2010). This emphasizes the connection between soil moisture and an organism's capacity to colonize the rhizosphere.

8 Mitigation Strategies and Regulatory Considerations

Because of inappropriate disposal, heavy metals adversely contaminate the soil through urbanization and industrialization (Anusha & Natarajan, 2020). Guidelines for the discharge of harmful materials into the environment have been strongly developed by numerous government organizations, including the World Health Organisation (WHO), the Environment Protection Agency (EPA), and the Ministry of Environment and Forests (MoEF) (Srivastava et al., 2007). However, many mining firms dump their industrial wastes into nearby landfill and water bodies. These mining soils unquestionably became a source of heavy metal pollution (Anusha & Natarajan, 2020). Humans, animals, plants, and microorganisms are all at risk due to heavy metal contamination (Garg et al., 2012).

The increased concentration of metals in the soil may also reduce its production. Therefore, creating innovative remediation techniques is crucial to cleaning up the polluted environment. The hazardous metals are typically detoxified using physical and chemical methods such as soil washing, soil replacement, vitrification, precipitation, solvent extraction, electrodialysis, and ion exchange (Pavel & Gavrilescu, 2008). However, these techniques have been found to have significant drawbacks, mainly expensive, time-consuming, complicated, and improper toxic removal. Due to the straightforward process, affordability, ease of accessibility, and environmentally friendly method, bioremediation has recently emerged as a feasible and alternative solution (Mishra, 2014).

Researchers have already discovered methods to overcome the exogenous. The microorganisms evolving their biological and metabolic activities have been highly suited to the specific contaminants in soil. Local microorganisms are, therefore, a superior option for any future bioremediation method (de Lorenzo & Loza-Tavera, 2014). A bacterium with metal resistance is a bacterium that can remove metals from the environment, which leads to the concept of bioremediation using bacteria (Nies, 1999). Bioremediation is a microbial-based method for removing pollutants from polluted soil, water, and sludge. Active and passive approaches to heavy metal bioremediation require microorganisms (Dixit et al., 2015).

The biosorption method of microbial remediation is the most efficient, practical, and appropriate for scale-up remediation operations. According to earlier studies,

Bacillus sp., Corynebacterium pseudotuberculosis, Pseudomonas, Brevibacterium linens, Enterobacter aerogenes, and Rhodococcus sp. were isolated as metal-tolerant and detoxifying native bacteria (Ali et al., 2012). Heavy metals (Co, Cu, and Zn) treated bacterial biomass showed alterations in the functional group in the spectroscopic investigation of the metal-resistant *rhizobacterium Azospirillus Brasiliense* (Kamnev et al., 2002).

9 Conclusion

The impact of silver (Ag) and zinc (Zn) on the structure and functionality of soil microbes showed that these metals significantly affect these communities' functional capabilities. The presence of Ag and Zn changed the variety, composition, and abundance of microorganisms. Furthermore, the presence of these metals affected the activity of enzymes in the cycling of nutrients. It is crucial to remember that not all of Ag and Zn's effects on soil microbial structure and functionality were detrimental. Other enzymes involved in metal detoxification and resistance exhibited increased activity, although higher concentrations of these metals often resulted in decreased microbial diversity and altered enzyme activity. Additionally, mitigation measures are essential to reduce the possible drawbacks of soil contamination with Ag and Zn. It is necessary to investigate how to manage soil to diminish metal inputs, such as by disposing of waste properly and using amendments that can immobilize or remove metals. Furthermore, strengthening the resistance of soil microbes by applying microbial inoculants or bioaugmentation techniques may help lessen the adverse impacts of metal pollution on soil microbe communities. For sustainable soil management techniques, a deeper comprehension of the effects of Ag and Zn on soil microbial structure and functionality is necessary. We can create plans to safeguard and improve soil health by considering the complex connections between metals and microbial communities. New strategies can be developed to protect and improve soil health, nutrient cycling, and ecological functioning.

References

Agarwal, P., Vibhandik, R., Agrahari, R., Daverey, A., & Rani, R. (2023). Role of root exudates on the soil microbial diversity and biogeochemistry of heavy metals. *Applied Biochemistry and Biotechnology*. https://doi.org/10.1007/s12010-023-04465-2

Aislabie, J., Deslippe, J. R., & Dymond, J. (2013). Soil microbes and their contribution to soil services. *Ecosystem Services in New Zealand--Conditions and Trends. Manaaki Whenua Press, Lincoln, New Zealand, 1*(12), 143–161.

Ali, N., Dashti, N., Al-Mailem, D., Eliyas, M., & Radwan, S. (2012). Indigenous soil bacteria with the combined potential for hydrocarbon consumption and heavy metal resistance. *Environmental Science and Pollution Research International, 19*(3), 812–820.

Alloway, B. J. (2008). *Zinc in soils and crop nutrition*. topsoils.co.nz. http://www.topsoils.co.nz/wp-content/uploads/2014/09/Zinc-in-Soils-and-Crop-Nutrition-Brian-J.-Alloway.pdf

Anusha, P., & Natarajan, D. (2020). Bioremediation potency of multi-metal tolerant native bacteria Bacillus cereus isolated from bauxite mines, kolli hills, Tamilnadu-A lab to land approach. *Biocatalysis and Agricultural Biotechnology, 25*, 101581.

Bandyopadhyay, S., Peralta-Videa, J. R., Plascencia-Villa, G., José-Yacamán, M., & Gardea-Torresdey, J. L. (2012). Comparative toxicity assessment of CeO_2 and ZnO nanoparticles towards Sinorhizobium meliloti, a symbiotic alfalfa associated bacterium: Use of advanced microscopic and spectroscopic techniques. *Journal of Hazardous Materials, 241–242*, 379–386.

Banerjee, S., Helgason, B., Wang, L., Winsley, T., Ferrari, B. C., & Siciliano, S. D. (2016). Legacy effects of soil moisture on microbial community structure and N_2O emissions. *Soil Biology and Biochemistry, 95*, 40–50.

Baran, A., Wieczorek, J., Mazurek, R., Urbański, K., & Klimkowicz-Pawlas, A. (2018). Potential ecological risk assessment and predicting zinc accumulation in soils. *Environmental Geochemistry and Health, 40*(1), 435–450.

Barros, D., Pradhan, A., Mendes, V. M., Manadas, B., Santos, P. M., Pascoal, C., & Cássio, F. (2019). Proteomics and antioxidant enzymes reveal different mechanisms of toxicity induced by ionic and nanoparticulate silver in bacteria. *Environmental Science: Nano, 6*(4), 1207–1218.

Bhattacharyya, P. N., & Jha, D. K. (2012). Plant growth-promoting rhizobacteria (PGPR): Emergence in agriculture. *World Journal of Microbiology and Biotechnology, 28*(4), 1327–1350.

Blaser, S. A., Scheringer, M., Macleod, M., & Hungerbühler, K. (2008). Estimation of cumulative aquatic exposure and risk due to silver: Contribution of nano-functionalized plastics and textiles. *The Science of the Total Environment, 390*(2–3), 396–409.

Briffa, J., Sinagra, E., & Blundell, R. (2020). Heavy metal pollution in the environment and their toxicological effects on humans. *Heliyon, 6*(9), e04691.

Bundschuh, M., Filser, J., Lüderwald, S., McKee, M. S., Metreveli, G., Schaumann, G. E., Schulz, R., & Wagner, S. (2018). Nanoparticles in the environment: where do we come from, where do we go to? *Environmental Sciences Europe, 30*(1). https://doi.org/10.1186/s12302-018-0132-6

Chavan, S., & Nadanathangam, V. (2019). Effects of nanoparticles on plant growth-promoting bacteria in indian agricultural soil. *Agronomy, 9*(3), 140.

Chavan, S., & Nadanathangam, V. (2020). Shifts in metabolic patterns of soil bacterial communities on exposure to metal-engineered nanomaterials. *Ecotoxicology and Environmental Safety, 189*, 110012.

Chen, J. P., Wang, L. K., Wang, M.-H.S., Hung, Y.-T., & Shammas, N. K. (2016). *Remediation of heavy metals in the environment*. CRC Press.

Chirila, E., & Draghici, C. (2008). Contamination of soils by waste deposits. *Soil Chemical Pollution, Risk Assessment, Remediation and Security*, 13–25.

Choi, O. K., & Hu, Z. Q. (2009). Nitrification inhibition by silver nanoparticles. *Water Science and Technology: A Journal of the International Association on Water Pollution Research, 59*(9), 1699–1702.

Corcoll, N., Österlund, T., Sinclair, L., Eiler, A., Kristiansson, E., Backhaus, T., & Eriksson, K. M. (2017). Comparison of four DNA extraction methods for comprehensive assessment of 16S rRNA bacterial diversity in marine biofilms using high-throughput sequencing. *FEMS Microbiology Letters, 364*(14). https://doi.org/10.1093/femsle/fnx139

de Lorenzo, V., & Loza-Tavera, H. (2014). Microbial bioremediation of chemical pollutants: How bacteria cope with multi-stress environmental scenarios. In *Bacterial stress responses* (pp. 481–492). ASM Press.

Deshpande, L. M., & Chopade, B. A. (1994). Plasmid-mediated silver resistance in Acinetobacter baumannii. *Biometals: An International Journal on the Role of Metal Ions in Biology, Biochemistry, and Medicine, 7*(1), 49–56.

Ding, G.-C., Heuer, H., & Smalla, K. (2012). Dynamics of bacterial communities in two unpolluted soils after spiking with phenanthrene: Soil type-specific and common responders. *Frontiers in Microbiology, 3*, 290.

Dixit, R., Wasiullah, Malaviya, D., Pandiyan, K., Singh, U. B., Sahu, A., Shukla, R., Singh, B. P., Rai, J. P., Sharma, P. K., Lade, H., & Paul, D. (2015). Bioremediation of heavy metals from soil and aquatic environment: An overview of principles and criteria of fundamental processes. *Sustainability: Science Practice and Policy*, *7*(2), 2189–2212.

Fajardo, C., Ortíz, L. T., Rodríguez-Membibre, M. L., Nande, M., Lobo, M. C., & Martin, M. (2012). Assessing the impact of zero-valent iron (ZVI) nanotechnology on soil microbial structure and functionality: A molecular approach. *Chemosphere, 86*(8), 802–808.

Fan, P., Wu, L., Wang, Q., Wang, Y., Luo, H., Song, J., Yang, M., Yao, H., & Chen, S. (2023). Physiological and molecular mechanisms of medicinal plants in response to cadmium stress: Current status and future perspective. *Journal of Hazardous Materials, 450*, 131008.

Ferreira, H., & Leite, M. G. P. (2015). A Life Cycle Assessment study of iron ore mining. *Journal of Cleaner Production, 108*, 1081–1091.

Fitzpatrick, C. R., Copeland, J., Wang, P. W., Guttman, D. S., Kotanen, P. M., & Johnson, M. T. J. (2018). Assembly and ecological function of the root microbiome across angiosperm plant species. *Proceedings of the National Academy of Sciences of the United States of America, 115*(6), E1157–E1165.

Fondi, M., Karkman, A., Tamminen, M. V., Bosi, E., Virta, M., Fani, R., Alm, E., & McInerney, J. O. (2016). "Every gene is everywhere but the environment selects": Global geolocalization of gene sharing in environmental samples through network analysis. *Genome Biology and Evolution, 8*(5), 1388–1400.

Ford, M. B., Mende, K., Kaiser, S. J., Beckius, M. L., Lu, D., Stam, J., Li, P., Stewart, L., Tribble, D. R., & Blyth, D. M. (2022). Clinical characteristics and resistance patterns of pseudomonas aeruginosa isolated from combat casualties. *Military Medicine, 187*(3–4), 426–434.

Friedrich, M. W. (2006). Stable-isotope probing of DNA: Insights into the function of uncultivated microorganisms from isotopically labeled metagenomes. *Current Opinion in Biotechnology, 17*(1), 59–66.

García-Gómez, C., Fernández, M. D., García, S., Obrador, A. F., Letón, M., & Babín, M. (2018). Soil pH effects on the toxicity of zinc oxide nanoparticles to soil microbial community. *Environmental Science and Pollution Research International, 25*(28), 28140–28152.

Garg, S. K., Tripathi, M., & Srinath, T. (2012). Strategies for chromium bioremediation of tannery effluent. *Reviews of Environmental Contamination and Toxicology, 217*, 75–140.

Geyer, K. M., Altrichter, A. E., Takacs-Vesbach, C. D., Van Horn, D. J., Gooseff, M. N., & Barrett, J. E. (2014). Bacterial community composition of divergent soil habitats in a polar desert. *FEMS Microbiology Ecology, 89*(2), 490–494.

Godeto, Y. G., Ayele, A., Ahmed, I. N., Husen, A., & Bachheti, R. K. (2023). Medicinal plant-based metabolites in nanoparticles synthesis and their cutting-edge applications: An overview. In *Secondary metabolites from medicinal plants* (pp. 1–34). CRC Press.

Gonfa, Y. H., Gelagle, A. A., Hailegnaw, B., Kabeto, S. A., Workeneh, G. A., Tessema, F. B., Tadesse, M. G., Wabaidur, S. M., Dahlous, K. A., Abou Fayssal, S., Kumar, P., Adelodun, B., Bachheti, A., & Bachheti, R. K. (2023). Optimization, characterization, and biological applications of silver nanoparticles synthesized using essential oil of aerial part of *Laggera tomentosa*. *Sustainability, 15*(1), 797.

Gonfa, Y. H., Tessema, F. B., Bachheti, A., Tadesse, M. G., Eid, E. M., Abou Fayssal, S., Adelodun, B., Choi, K. S., Širić, I., Kumar, P., & Bachheti, R. K. (2022). Essential oil composition of aerial part of *Pluchea ovalis* (Pers.) DC., Silver nanoparticles synthesis, and larvicidal activities against fall armyworm. *Sustainability*, *14*(23), 15785.

Griffiths, B. S., & Philippot, L. (2013). Insights into the resistance and resilience of the soil microbial community. *FEMS Microbiology Reviews*. https://academic.oup.com/femsre/article-abstract/37/2/112/622548

Grün, A.-L., Manz, W., Kohl, Y. L., Meier, F., Straskraba, S., Jost, C., Drexel, R., & Emmerling, C. (2019). Impact of silver nanoparticles (AgNP) on soil microbial community depending on functionalization, concentration, exposure time, and soil texture. *Environmental Sciences Europe, 31*(1), 15.

van der Heijden, M. G. A., Bardgett, R. D., & van Straalen, N. M. (2008). The unseen majority: Soil microbes as drivers of plant diversity and productivity in terrestrial ecosystems. *Ecology Letters, 11*(3), 296–310.

Hill, G. T., Mitkowski, N. A., Aldrich-Wolfe, L., & Emele, L. R. (2000). Methods for assessing the composition and diversity of soil microbial communities. *Applied Soil Ecology: A Section of Agriculture, Ecosystems and Environment*. https://www.sciencedirect.com/science/article/pii/S092913930000069X

Hobman, J. L., & Crossman, L. C. (2015). Bacterial antimicrobial metal ion resistance. *Journal of Medical Microbiology, 64*(Pt 5), 471–497.

Husen, A., & Siddiqi, K. S. (2023). *Advances in smart nanomaterials and their applications*. Elsevier Inc. https://doi.org/10.1016/C2021-0-02202-1

Islam, W., Noman, A., Naveed, H., Huang, Z., & Chen, H. Y. H. (2020). Role of environmental factors in shaping the soil microbiome. *Environmental Science and Pollution Research International, 27*(33), 41225–41247.

Jacoby, R., Peukert, M., Succurro, A., Koprivova, A., & Kopriva, S. (2017). The role of soil microorganisms in plant mineral nutrition—current knowledge and future directions. *Frontiers in Plant Science, 8*. https://doi.org/10.3389/fpls.2017.01617

Kamnev, A. A., Antonyuk, L. P., Tugarova, A. V., Tarantilis, P. A., Polissiou, M. G., & Gardiner, P. H. E. (2002). Fourier transform infrared spectroscopic characterization of heavy metal-induced metabolic changes in the plant-associated soil bacterium Azospirillum brasilense Sp7. *Journal of Molecular Structure, 610*(1), 127–131.

Klasen, H. J. (2000). Historical review of the use of silver in the treatment of burns. I. Early uses. *Burns: Journal of the International Society for Burn Injuries, 26*(2), 117–130.

Lansdown, A. B. G. (2010). A pharmacological and toxicological profile of silver as an antimicrobial agent in medical devices. *Advances in Pharmacological Sciences, 2010*, 910686.

Lauber, C. L., Ramirez, K. S., Aanderud, Z., Lennon, J., & Fierer, N. (2013). Temporal variability in soil microbial communities across land-use types. *The ISME Journal, 7*(8), 1641–1650.

Lead, J. R. (2008). Nanomaterials in the environment: Behavior, fate, bioavailability. *Environmental Toxicology and Chemistry / SETAC, 27*(9), 1825–1851.

Leyu, A. M., Debebe, S. E., Bachheti, A., Rawat, Y. S., & Bachheti, R. K. (2023). Green Synthesis of gold and silver nanoparticles using invasive alien plant *Parthenium hysterophorus* and their antimicrobial and antioxidant activities. *Sustainability, 15*(12), 9456.

Li, C., Quan, Q., Gan, Y., Dong, J., Fang, J., Wang, L., & Liu, J. (2020). Effects of heavy metals on microbial communities in sediments and establishment of bioindicators based on microbial taxa and function for environmental monitoring and management. *The Science of the Total Environment, 749*, 141555.

Liu, G., Zhang, M., Jin, Y., Fan, X., Xu, J., Zhu, Y., Fu, Z., Pan, X., & Qian, H. (2017). The effects of low concentrations of silver nanoparticles on wheat growth, seed quality, and soil microbial communities. *Water, Air, and Soil Pollution, 228*(9). https://doi.org/10.1007/s11270-017-3523-1

Masrahi, A., VandeVoort, A. R., & Arai, Y. (2014). Effects of silver nanoparticle on soil-nitrification processes. *Archives of Environmental Contamination and Toxicology, 66*(4), 504–513.

McLaughlin, M. J., Smolders, E., Degryse, F., & Rietra, R. (2011). Uptake of metals from soil into vegetables. In F. A. Swartjes (Ed.), *Dealing with contaminated sites: From theory towards practical application* (pp. 325–367). Springer.

Meier, M. J., Dodge, A. E., Samarajeewa, A. D., & Beaudette, L. A. (2020). Soil exposed to silver nanoparticles reveals significant changes in community structure and altered microbial transcriptional profiles. *Environmental Pollution, 258*, 113816.

Melaiye, A., & Youngs, W. J. (2005). Silver and its application as an antimicrobial agent. In *Expert opinion on therapeutic patents*. https://doi.org/10.1517/13543776.15.2.125

Mishra, V. (2014). Biosorption of zinc ion: A deep comprehension. *Applied Water Science, 4*(4), 311–332.

Mishra, S., Yang, X., & Singh, H. B. (2020). Evidence for positive response of soil bacterial community structure and functions to biosynthesized silver nanoparticles: An approach to conquer nanotoxicity? *Journal of Environmental Management, 253*, 109584.

Moffett, B. F., Nicholson, F. A., Uwakwe, N. C., Chambers, B. J., Harris, J. A., & Hill, T. C. J. (2003). Zinc contamination decreases the bacterial diversity of agricultural soil. *FEMS Microbiology Ecology, 43*(1), 13–19.

Mossa, A.-W., Dickinson, M. J., West, H. M., Young, S. D., & Crout, N. M. J. (2017). The response of soil microbial diversity and abundance to long-term application of biosolids. *Environmental Pollution, 224*, 16–25.

Moter, A., & Göbel, U. B. (2000). Fluorescence in situ hybridization (FISH) for direct visualization of microorganisms. *Journal of Microbiological Methods, 41*(2), 85–112.

Mukhopadhyay, R., Rosen, B. P., Phung, L. T., & Silver, S. (2002). Microbial arsenic: From geocycles to genes and enzymes. *FEMS Microbiology Reviews, 26*(3), 311–325.

Möhler, J. S., Sim, W., Blaskovich, M. A. T., Cooper, M. A., & Ziora, Z. M. (2018). Silver bullets: A new luster on an old antimicrobial agent. *Biotechnology Advances, 36*(5), 1391–1411.

Nicholson, F. A., Smith, S. R., Alloway, B. J., Carlton-Smith, C., & Chambers, B. J. (2003). An inventory of heavy metals inputs to agricultural soils in England and Wales. *The Science of the Total Environment, 311*(1–3), 205–219.

Nies, D. H. (1999). Microbial heavy-metal resistance. *Applied Microbiology and Biotechnology, 51*(6), 730–750.

Ochoa, M. S., Madrigal, R. P., Martínez, M. T., & Carreón, A. (2010). Plantas, hongos micorrízicos y bacterias: Su compleja red de interacciones. *Biológicas, 12*(1), 65–71.

Olaniran, A. O., Balgobind, A., & Pillay, B. (2013). Bioavailability of heavy metals in soil: Impact on microbial biodegradation of organic compounds and possible improvement strategies. *International Journal of Molecular Sciences, 14*(5), 10197–10228.

Pallavi, M. C. M., Srivastava, R., Arora, S., & Sharma, A. K. (2016). Impact assessment of silver nanoparticles on plant growth and soil bacterial diversity. *3 Biotech, 6*(2), 254.

Pavel, L. V., & Gavrilescu, M. (2008). Overview of ex-situ decontamination techniques for soil cleanup. *Environmental Engineering and Management*. https://www.academia.edu/download/8327674/eemj_vol7_nr6_2008.pdf#page=174

Peltier, E., Vincent, J., Finn, C., & Graham, D. W. (2010). Zinc-induced antibiotic resistance in activated sludge bioreactors. *Water Research, 44*(13), 3829–3836.

Prober, S. M., Leff, J. W., Bates, S. T., Borer, E. T., Firn, J., Harpole, W. S., Lind, E. M., Seabloom, E. W., Adler, P. B., Bakker, J. D., Cleland, E. E., DeCrappeo, N. M., DeLorenze, E., Hagenah, N., Hautier, Y., Hofmockel, K. S., Kirkman, K. P., Knops, J. M. H., & La Pierre, K. J., et al. (2015). Plant diversity predicts beta but not alpha diversity of soil microbes across grasslands worldwide. *Ecology Letters, 18*(1), 85–95.

Purcell, T. W., & Peters, J. J. (1998). Sources of silver in the environment. *Environmental Toxicology and Chemistry/SETAC, 17*(4), 539–546.

Rahmatpour, S., Shirvani, M., Mosaddeghi, M. R., Nourbakhsh, F., & Bazarganipour, M. (2017). Dose–response effects of silver nanoparticles and silver nitrate on microbial and enzyme activities in calcareous soils. *Geoderma, 285*, 313–322.

Rajput, V. D., Verma, K. K., Sharma, N., & Minkina, T. (2022). *The role of nanoparticles in plant nutrition under soil pollution: Nanoscience in nutrient use efficiency*. Springer Nature.

Rashid, M. I., Shahzad, T., Shahid, M., Ismail, I. M. I., Shah, G. M., & Almeelbi, T. (2017). Zinc oxide nanoparticles affect carbon and nitrogen mineralization of Phoenix dactylifera leaf litter in a sandy soil. *Journal of Hazardous Materials, 324*(Pt B), 298–305.

Rastogi, G., & Sani, R. K. (2011). Molecular techniques to assess microbial community structure, function, and dynamics in the environment. In I. Ahmad, F. Ahmad, & J. Pichtel (Eds.), *Microbes and microbial technology: Agricultural and environmental applications* (pp. 29–57). Springer.

Read, D. S., Matzke, M., Gweon, H. S., Newbold, L. K., Heggelund, L., Ortiz, M. D., Lahive, E., Spurgeon, D., & Svendsen, C. (2016). Soil pH effects on the interactions between dissolved zinc,

non-nano- and nano-ZnO with soil bacterial communities. *Environmental Science and Pollution Research International, 23*(5), 4120–4128.
Rillig, M. C., Muller, L. A., & Lehmann, A. (2017). Soil aggregates as massively concurrent evolutionary incubators. *The ISME Journal, 11*(9), 1943–1948.
Sabnavis, M., Jagasheth, U. H., Avachat, H., & Mishra, M. (2018). Zinc Industry: the unsung metal of the economy. In *CARE ratings: professional risk opinion, Report, 31st October, 2018.* careratings.com.
Salem, S. S., Husen, A. (2023). Zinc oxide nanoparticles: Plant response, interaction, phytotoxicity, and defence. In: A. Husen (Ed.), *Nanomaterials and nanocomposites exposures to plants. Smart nanomaterials technology* (pp. 247–261). Springer. https://doi.org/10.1007/978-981-99-2419-6_12
Santoyo, G., Hernández-Pacheco, C., Hernández-Salmerón, J., & Hernández-León, R. (2017). The role of abiotic factors modulating the plant-microbe-soil interactions: toward sustainable agriculture. A review. *Spanish Journal of Agricultural Research = Revista de Investigacion Agraria, 15*(1), e03R01.
Schwab, F., Jacques, J. J., & Oliveira, B. F. (2019). 925 Silver and 18k gold electroplated jewelry: A comparative analysis of environmental impacts. *International Journal of Advances in Engineering & Technology; Bareilly, 12*(4), 67–79.
Sen, I. S., & Peucker-Ehrenbrink, B. (2012). Anthropogenic disturbance of element cycles at the Earth's surface. *Environmental Science and Technology, 46*(16), 8601–8609.
Seneviratne, G., & Zavahir, J. S. (2021). *Role of Microbial Communities for Sustainability.* Springer Nature.
Sillen, W. M. A., Thijs, S., Abbamondi, G. R., Janssen, J., Weyens, N., White, J. C., & Vangronsveld, J. (2015). Effects of silver nanoparticles on soil microorganisms and maize biomass are linked in the rhizosphere. *Soil Biology and Biochemistry, 91*, 14–22.
Smolders, E., Buekers, J., Oliver, I., & McLaughlin, M. J. (2004). Soil properties affecting toxicity of zinc to soil microbial properties in laboratory-spiked and field-contaminated soils. *Environmental Toxicology and Chemistry/SETAC, 23*(11), 2633–2640.
Sokol, N. W., Slessarev, E., Marschmann, G. L., Nicolas, A., Blazewicz, S. J., Brodie, E. L., Firestone, M. K., Foley, M. M., Hestrin, R., Hungate, B. A., Koch, B. J., Stone, B. W., Sullivan, M. B., Zablocki, O., LLNL Soil Microbiome Consortium, & Pett-Ridge, J. (2022). Life and death in the soil microbiome: how ecological processes influence biogeochemistry. *Nature Reviews Microbiology, 20*(7), 415–430.
Srivastava, V. C., Mall, I. D., & Mishra, I. M. (2007). Adsorption thermodynamics and isosteric heat of adsorption of toxic metal ions onto bagasse fly ash (BFA) and rice husk ash (RHA). *Chemical Engineering Journal, 132*(1), 267–278.
Stibal, M., Sabacká, M., & Kastovská, K. (2006). Microbial communities on glacier surfaces in Svalbard: Impact of physical and chemical properties on abundance and structure of cyanobacteria and algae. *Microbial Ecology, 52*(4), 644–654.
Tedersoo, L., Bahram, M., Cajthaml, T., Põlme, S., Hiiesalu, I., Anslan, S., Harend, H., Buegger, F., Pritsch, K., Koricheva, J., & Abarenkov, K. (2016). Tree diversity and species identity effects on soil fungi, protists, and animals are context dependent. *The ISME Journal, 10*(2), 346–362.
Terzioğlu, E., Arslan, M., Balaban, B. G., & Çakar, Z. P. (2022). Microbial silver resistance mechanisms: Recent developments. *World Journal of Microbiology and Biotechnology, 38*(9), 158.
Thomas, S., & McCubbin, P. (2003). A comparison of the antimicrobial effects of four silver-containing dressings on three organisms. *Journal of Wound Care, 12*(3), 101–107.
Tobey, R. A., Enger, M. D., Griffith, J. K., & Hildebrand, C. E. (1982). Zinc-induced resistance to alkylating agent toxicity. *Cancer Research, 42*(8), 2980–2984.
Trabelsi, D., & Mhamdi, R. (2013). Microbial inoculants and their impact on soil microbial communities: A review. *BioMed Research International.* https://www.hindawi.com/journals/bmri/2013/863240/

Vasileiadis, S., Puglisi, E., Trevisan, M., Scheckel, K. G., Langdon, K. A., McLaughlin, M. J., Lombi, E., & Donner, E. (2015). Changes in soil bacterial communities and diversity in response to long-term silver exposure. *FEMS Microbiology Ecology, 91*(10). https://doi.org/10.1093/femsec/fiv114

Vejan, P., Abdullah, R., Khadiran, T., Ismail, S., & Nasrulhaq Boyce, A. (2016). Role of plant growth promoting rhizobacteria in agricultural sustainability—A review. *Molecules, 21*(5), 573.

Wahid, F., Sharif, M., Ali, A., Fahad, S., Adnan, M., Noor, M., Mian, I. A., Khan, I. A., Alam, M., Saeed, M., Ilyas, M., Ullah, R., Ilahi, H., & Azeem, M. (2020). Plant-microbes interactions and functions in changing climate. In S. Fahad, M. Hasanuzzaman, M. Alam, H. Ullah, M. Saeed, I. Ali Khan, & M. Adnan (Eds.), *Environment, climate, plant and vegetation growth* (pp. 397–419). Springer International Publishing.

Wan, Y., Jiang, B., Wei, D., & Ma, Y. (2020). Corrigendum to "Ecological criteria for zinc in Chinese soil as affected by soil properties" [Ecotoxicol. Environ. Saf. 194 (2020) 110418]. *Ecotoxicology and Environmental Safety, 196*, 110553.

Wilpiszeski, R. L., Aufrecht, J. A., Retterer, S. T., Sullivan, M. B., Graham, D. E., Pierce, E. M., Zablocki, O. D., Palumbo, A. V., & Elias, D. A. (2019). Soil aggregate microbial communities: towards understanding microbiome interactions at biologically relevant scales. *Applied and Environmental Microbiology, 85*(14). https://doi.org/10.1128/AEM.00324-19

Xie, Y., Fan, J., Zhu, W., Amombo, E., Lou, Y., Chen, L., & Fu, J. (2016). Effect of heavy metals pollution on soil microbial diversity and bermudagrass genetic variation. *Frontiers in Plant Science, 7*, 755.

Yergeau, E., Arbour, M., Brousseau, R., Juck, D., Lawrence, J. R., Masson, L., Whyte, L. G., & Greer, C. W. (2009). Microarray and real-time PCR analyses of the responses of high-arctic soil bacteria to hydrocarbon pollution and bioremediation treatments. *Applied and Environmental Microbiology, 75*(19), 6258–6267.

Yuan, L., Li, Z.-H., Zhang, M.-Q., Shao, W., Fan, Y.-Y., & Sheng, G.-P. (2019). Mercury/silver resistance genes and their association with antibiotic resistance genes and microbial community in a municipal wastewater treatment plant. *The Science of the Total Environment, 657*, 1014–1022.

Zhai, Y., Hunting, E. R., Wouters, M., Peijnenburg, W. J. G. M., & Vijver, M. G. (2016). Silver nanoparticles, ions, and shape governing soil microbial functional diversity: Nano shapes micro. *Frontiers in Microbiology, 7*, 1123.

Zhang, Q.-C., Shamsi, I. H., Xu, D.-T., Wang, G.-H., Lin, X.-Y., Jilani, G., Hussain, N., & Chaudhry, A. N. (2012). Chemical fertilizer and organic manure inputs in soil exhibit a vice-versa pattern of microbial community structure. *Applied Soil Ecology: A Section of Agriculture, Ecosystems and Environment, 57*, 1–8.

Role of Gold Nanoparticles in Plant Protection Against Pathogen

Urvashi Goyal, Kundan Kumar Chaubey, S. D. Pandey, Deepak Kumar Verma, Archana Bachheti, and Rakesh Kumar Bachheti

Abstract Numerous pathogens attack agricultural crops and cause enormous economic damage. To manage plant diseases, chemical pesticides, and fungicides have drawbacks, such as the development of resistance, environmental damage, and unfavorable impacts. To protect plants from infections, scientists and researchers are looking for more sustainable alternatives. Gold nanoparticles (AUNPs) are ideal for various applications, including medicine, electronics, and agriculture, due to their distinct physical and chemical properties. AuNPs have become a potentially effective substitute for plant pathogen resistance. In addition to possessing both direct and indirect antibacterial activity against plant diseases, it has been demonstrated that

Equal Contribution to Urvashi Goyal and Kundan Kumar Chaubey.

U. Goyal
Sanjay Gandhi Postgraduate Institute of Medical Sciences, New PMSSY Rd., Raibareli Rd., Lucknow 226014, Uttar Pradesh, India

K. K. Chaubey
Division of Research and Innovation, School of Applied and Life Sciences, Uttaranchal University, Arcadia Grant, P.O. Chandanwari, Premnagar, Dehradun 248007, Uttarakhand, India

School of Basic and Applied Sciences, Sanskriti University, Mathura, Uttar Pradesh-281401, India

S. D. Pandey
School of Applied and Life Sciences, Uttaranchal Institute of Technology, Uttaranchal University, Arcadia Grant, P.O. Chandanwari, Premnagar, Dehradun 248007, Uttarakhand, India

D. K. Verma
Centre for Health Research and Innovation, Apeejay Business Centre, 6th Floor, Arunachal Building, Barakhamba Road, New Delhi 110001, India

A. Bachheti
Department of Environmental Science, Graphic Era University, Dehradun 248002, Uttarakhand, India

R. K. Bachheti (✉)
Department of Allied Sciences, Graphic Era Hill University (G.E.H.U), Society Area, Clement Town, Dehradun 248002, India
e-mail: rkbachheti@gmail.com

Department of Industrial Chemistry, Addis Ababa Science and Technology University, P.O. Box 16417 Addis Ababa, Ethiopia

R. K. Bachheti et al. (eds.), *Metal and Metal-Oxide Based Nanomaterials*, Smart Nanomaterials Technology, https://doi.org/10.1007/978-981-99-7673-7_10

AuNPs can improve plant growth and stress resistance. The purpose of this chapter is to provide a recent update on how AuNPs help plants defend themselves against diseases.

Keywords Gold nanoparticles · Plant protection · Pathogen · Infection · Protection · Impacts · Role

1 Introduction

Agriculture may suffer significant economic losses due to plant illnesses from pathogens, including fungi, bacteria, and viruses that significantly lower crop output and quality (Albahri et al., 2023; Oerke, 2006). Using chemical pesticides and fungicides to manage plant infections has a variety of disadvantages as well, such as the emergence of resistance, environmental pollution, and adverse effects on organisms other than the target pest (Khan et al., 2023; Pietrzak et al., 2019). Alternative and environmentally acceptable techniques are therefore needed to shield plants against diseases.

Chemical pesticides and fungicides, cultural practices including crop rotation and cleanliness, breeding for genetic resistance, and biotechnology are all examples of traditional strategies for protecting plants against diseases (Pietrzak et al., 2019). However, due to pathogen resistance brought on by the misuse of chemical pesticides and fungicides, which can harm the environment and human health, infections increasingly call for higher doses and more frequent applications (Oerke, 2006). AuNPs have developed into a possibly successful replacement for plant pathogen resistance. AuNPs are beneficial for a number of applications, including medicine, electronics, and agriculture, thanks to their distinct physical and chemical properties (Husen, 2017, 2022, 2023; Husen & Iqbal, 2019; Husen & Siddiqi, 2023; Husen et al., 2019; Leyu et al., 2023; Pietrzak et al., 2019; Raja et al., 2023; Siddiqi & Husen, 2016, 2017). According to studies by Mishra et al. (2020), AuNPs have been demonstrated to have both direct and indirect antibacterial activity against plant diseases and the capacity to improve plant growth and stress resistance. AuNPs are also viable for low-cost and eco-friendly plant protection approaches because of their low toxicity and precise manufacturing (Dutta et al., 2021). This study topic is intriguing and promising for developing new and long-lasting plant protection strategies because of the importance of safeguarding plants against pathogens, the drawbacks of current techniques, and the potential advantages of AuNPs in plant protection.

2 Properties of Gold Nanoparticles

Due to their small size and high surface area to volume ratio, gold nanoparticles (AuNPs) have distinct physical and chemical characteristics. AuNPs can be manufactured in various geometries, such as spheres, rods, triangles, and cubes, and their diameter typically ranges from 1 to 100 nm (nm) (Dutta et al., 2021; Husen et al., 2019). The intense surface plasmon resonance that AuNPs display due to their small size gives them their distinctive color, which ranges from red to purple (Dutta et al., 2021). AuNPs are adaptable for various applications since their highly reactive surface may be functionalized with different chemical and inorganic substances (Bhusal et al., 2020) (Fig. 1).

AuNPs can be produced using chemical reduction, physical, or biological methods (Dutta et al., 2021). When using chemical reduction techniques, gold ions are reduced by a reducing agent like sodium borohydride or citrate. The vaporization of gold and the condensation of nanoparticles are involved in physical techniques like laser ablation and thermal evaporation. Plant extracts or microorganisms are used in biological processes like "green synthesis" to create AuNPs (Bhusal et al., 2020). Each approach has benefits and drawbacks, with green synthesis standing out due to its affordability, environmental friendliness, and possibility for mass manufacturing (Dutta et al., 2021).

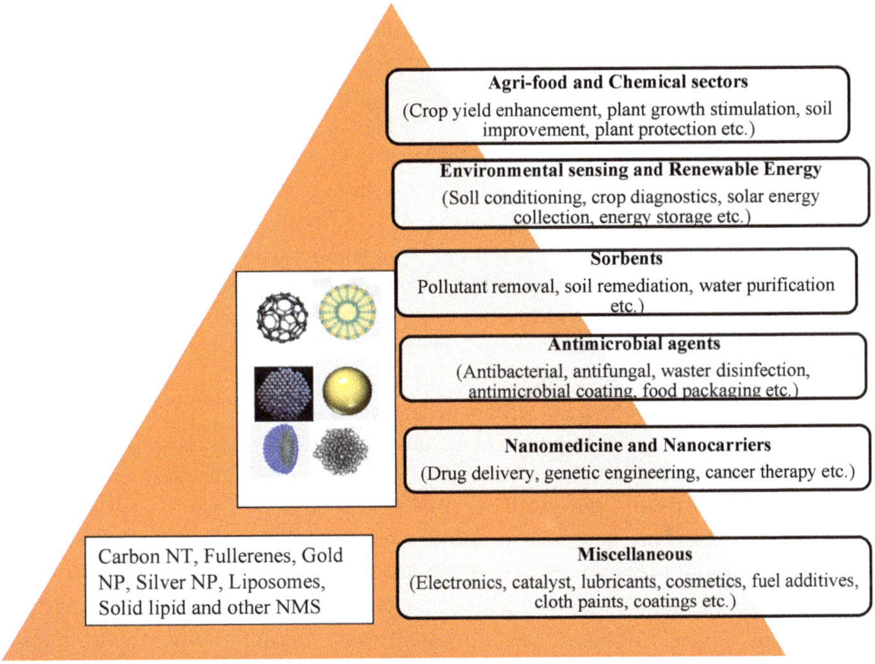

Fig. 1 Properties of Gold Nanoparticles

3 AuNPs in Plant Protection

AuNPs are superior to conventional plant protection techniques in many ways (Table 1), including biocompatibility, low toxicity, and the capacity to target certain pathogens (Pietrzak et al., 2019). Numerous techniques, such as foliar spraying, treating seeds, and soil application, can apply AuNPs to plants. By enhancing their distribution and boosting their efficacy, AuNPs can also improve the effectiveness of conventional plant protection techniques like fungicides and bactericides (Mishra et al., 2020). AuNPs can also increase plant growth and stress tolerance, making them more resistant to disease attacks (Pietrzak et al., 2019).

4 Mechanisms of Action of AuNPs in Plant Protection

It has been demonstrated that AuNPs have direct antibacterial activity against various plant diseases. The ability of AuNPs to penetrate pathogen cell walls and membranes because of their tiny size (usually less than 100 nm) causes membrane integrity to be disrupted and cellular contents to leak, which ultimately causes cell death (Kumar et al., 2021). Electrostatic, hydrophobic, hydrogen bonding, and van der Waals forces are only a few ways AuNPs and pathogen cells interact. These interactions may cause AuNPs to bind to the surfaces of the pathogen cells, which may ultimately cause the integrity of the cell membrane to be compromised (Kumar et al., 2021).

AuNPs can directly alter the structure of pathogen cell membranes and interact with pathogen cells. According to Mishra et al. (2020), AuNPs can pierce the lipid bilayer of a cell's membrane, causing the membrane's integrity to be compromised and its contents to flow out, which ultimately causes cell death. The high surface area to volume ratio of AuNPs, which enables them to interact with numerous lipid molecules and weaken the membrane structure, is responsible for this cell membrane rupture. Additionally, AuNPs can turn on the plant's pathogen defense mechanisms. In response to GNP exposure, plants release reactive oxygen species (ROS) and defense-related enzymes, including peroxidases and chitinases, which can destroy pathogen cell walls and prevent their growth Djiwanti and Kaushik (2019). In order to prevent pathogen invasion, the plants' ROS can also function as signaling molecules, activating a variety of defense mechanisms Djiwanti and Kaushik (2019). It has been demonstrated that AuNPs improve plant development and stress tolerance, which reduces the plant's susceptibility to disease attack. AuNPs can boost plant growth and health by increasing the production of photosynthetic pigments, nutrient and water uptake, and plant water consumption efficiency (Pietrzak et al., 2019).

By reducing the negative impacts of abiotic stresses like salt and drought, which can render plants more vulnerable to pathogen attack, AuNPs can also increase plant stress tolerance (Pietrzak et al., 2019). Direct and indirect antibacterial activity, plant growth stimulation, and stress tolerance are frequent mechanisms of action for

Table 1 Gold nanoparticles in diagnostic of pathogens, plant protection, and other applications

Serial number	Targets	Impact	References
1	Sclerotium rolfsii	Using 80 ppm concentration of gold nanoparticles, Sclerotium rolfsii growth was significantly inhibited	Takur et al. (2022)
2	Fusarium oxysporum	The 80 ppm concentration of gold nanoparticles changed the morphology and resulted in 70% growth inhibition after a week of incubation	Takur et al. (2022)
3	Alternaria alernata	Using gold nanoparticles, the 60%–70% growth inhibition and change in morphology were also observed	Takur et al. (2022)
4	Alternaria mali	Using gold nanoparticles, a 50%–60% of inhibition was observed at 60 and 80 ppm. A considerable change was observed in the morphology of mycelium after one week of incubation	Takur et al. (2022)
5	Magnaporthe grisea	150 ppm gold nanoparticles significantly inhibit the pathogens' ability to grow on their mycelia	Kaman et al. (2022)
6	Barley yellow mosaic virus	Poly-dispersed gold nanoparticles helped melt and dissolve the Barley yellow mosaic virus particles conferring resistance to the plant	Alkubaisi et al. (2015)
7	Barley yellow dwarf virus	Puffed virus-like particles decorated with gold NPs destroyed and vanished virus particles	Alkubaisi and Aref (2017)
8	Barley mosaic virus	AuNPs damaging as well as completely vanishing Virus–like particles	Kumar et al. (2022)
9	Karnal bunt disease	Gold nanoparticle-based optical immunosensors for detecting Karnal bunt disease in wheat	Shipway et al. (2000)
10	Tilletia indica	Nano-gold-based immunosensor to detect Karnal bunt-causing pathogen in wheat	Singh et al. (2010)
11	Bacterial infection	The Au NP-based probes have been successfully used to detect bacterial infections	Gill et al. (2008)
12	Gene editing	Cationic arginine gold nanoparticles (ArgNPs), assembled Cas9En (Etag)-RNP (ribonucleoproteins), used to deliver sgRNA with about 30% effective cytoplasmic/ nuclear gene editing efficiency in cultivated cell lines	Elamawi et al. (2018)
13	Helicobacter pylori	Gold nanoparticle probes were used for colorimetric detection of Helicobacter pylori DNA using isothermal helicase-dependent amplification	Gill et al. (2008)

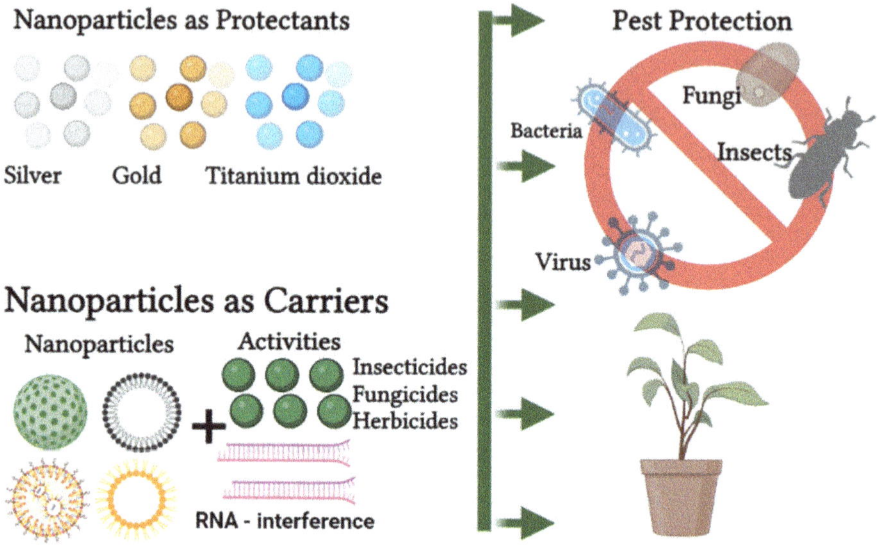

Fig. 2 Mechanisms of Action of AuNPs in Plant Protection

AuNPs in plant protection. These mechanisms make AuNPs suitable for developing modern, long-lasting plant protection methods (Fig. 2).

5 Application of AuNPs in Plant Protection

AuNPs can be used in various ways for plant protection, including foliar spray, seed treatment, and soil application. Foliar spray is a typical application technique in which a GNP solution is sprayed into plant leaves (Mishra et al., 2020). While soil application involves incorporating AuNPs into the soil before or after planting, seed treatment entails coating the seeds with a solution of AuNPs before sowing (Dutta et al., 2021). The target pathogen, the type of plant, and the desired outcome all influence the application strategy. The effectiveness of AuNPs in defending various plant species against pathogenic attacks has been the subject of numerous investigations. A foliar application of AuNPs, for instance, Sharma et al. (2019) observed, dramatically lessened the severity of the powdery mildew disease in wheat plants.

Similarly, Singh et al. (2018) discovered that seed treatment with AuNPs increased tomato plant growth and yield while decreasing the occurrence of the disease known as bacterial wilt. In another investigation, Patel et al. (2019) showed that soil application of AuNPs enhanced maize plant development and yield while lowering the prevalence of fungal infections. AuNPs may be superior to conventional chemical pesticides and fungicides because they effectively defend plants against infections. According to research by Mishra et al. (2020), AuNPs have been demonstrated to

Fig. 3 AuNPs in Plant Protection

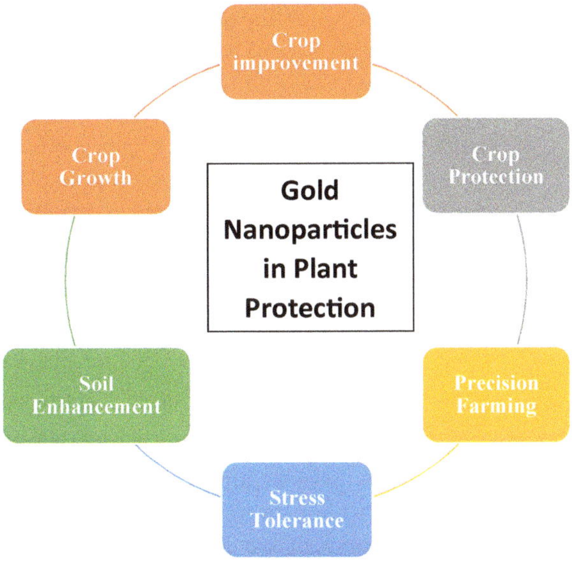

exhibit broad-spectrum activity against a variety of plant diseases, including fungi, bacteria, and viruses. Additionally, AuNPs have been shown to support plant growth and stress tolerance, which can improve plant health and yield (Dutta et al., 2021). Additionally, because they can be created using eco-friendly, cost-effective, and effective methods, AuNPs provide a sustainable alternative to chemical pesticides and fungicides Djiwanti and Kaushik (2019) (Fig. 3).

The use of gold nanoparticles (AuNPs) has shown promise in a number of areas, including plant protection. The following are some applications for AuNPs in plant protection:

1. **Disease control**: Specific proteins or chemical substances with antibacterial characteristics can be functionalized onto AuNPs. The growth of pathogenic microorganisms that cause plant illnesses, such as bacteria, fungi, and viruses, can be inhibited using these functionalized AuNPs. AuNPs' surface modifications enable targeted distribution to the diseased plant tissues while also enhancing their antibacterial action (Wang et al., 2017)
2. **Pesticide delivery**: AuNPs can be utilized as pesticide delivery vehicles for plants. Pesticides' stability, solubility, and effectiveness can be increased by attaching or encapsulating them on the surface of AuNPs. AuNPs can release pesticides under regulated conditions, which can increase their efficiency, decrease the amount of pesticide required, and reduce environmental pollution (Chaud et al., 2021; Guleria et al. 2022).
3. **Enhanced plant growth**: According to reports, AuNPs encourage the growth and development of plants. They can be applied to plants as nanofertilizers to deliver vital nutrients, boosting nutrient uptake and overall growth. Additionally, AuNPs have the capacity to serve as nanocarriers for bioactive substances, hormones, or

growth regulators, enabling their targeted transport to particular plant tissues and fostering the physiological responses that are sought (El-saadony et al., 2021).

4. **Stress tolerance**: Plants frequently experience a variety of environmental challenges, including salt, drought, and heavy metal toxicity. AuNPs have demonstrated potential for reducing the negative effects that these pressures have on plants. They have the ability to serve as antioxidants by scavenging reactive oxygen species (ROS) produced under stressful circumstances. By attaching to heavy metals and lowering their toxicity to plants, AuNPs can also aid in the detoxification of heavy metals (Nawaz et al., 2023).
5. **Sensor platforms**: AuNPs can be utilized as sensing platforms to find ambient contaminants or plant diseases. Target molecules can be detected in plant samples by using functionalized AuNPs that are created to bind exclusively to those compounds. This makes it possible to identify toxins or pathogens quickly and sensitively, which helps with early illness detection or environmental monitoring

Although AuNPs have a number of potential uses for plant protection, further study is necessary to improve their formulations, determine how they will affect plants and the environment over time, and determine whether or not they can be used on a wide scale practically.

6 Potential Risks and Concerns

There are concerns about the toxicity of AuNPs to humans and the environment, even though they may be effective for safeguarding plants. AuNPs have been shown to cause cytotoxicity, genotoxicity, and oxidative stress in various cell lines and animal models (Ahamed et al., 2011; Bai et al., 2010; Saini et al., 2021). Since AuNPs can build up in soil and water systems and impact microbial communities, there are worries regarding their environmental toxicity (Judy et al., 2016). The advantages and disadvantages of utilizing AuNPs for plant protection must be carefully weighed. To correctly design safety norms and guidelines, more research is required to completely comprehend the potential dangers associated with using AuNPs.

7 Conclusion and Future Prospectus

AuNPs have demonstrated tremendous promise as a novel and efficient strategy for protecting against plant diseases. They can interact directly with pathogen cells and enhance plant defense mechanisms due to their small size, wide surface area, and unique physical and chemical properties. In addition, AuNPs may enhance plant growth and stress resistance, making them an effective substitute for conventional plant protection techniques. Various concerns have also been raised by different researchers regarding the safety issues (such as toxicity for humans and the environment) associated with using AuNPs. More investigation is, therefore, necessary to know the Pros and cons of using AuNPs for plant protection. The use of AuNPs for plant protection has excellent potential benefits despite the present constraints and uncertainties, and more research will likely lead to further development and optimization of their application. AuNPs have the potential to make a significant contribution to sustainable agriculture and plant preservation with appropriate assessment and management.

References

Ahamed, M., Akhtar, M. J., Raja, M., Ahmad, I., & Siddiqui, M. K. J. (2011). Altered oxidative stress indices in organs of mice exposed to single and multiple doses of gold nanoparticles. *Oxidative Medicine and Cellular Longevity, 2011*, 1–11.

Albahri, G., Alyamani, A. A., Badran, A., Hijazi, A., Nasser, M., Maresca, M., & Baydoun, E. (2023). Enhancing essential grains yield for sustainable food security and bio-safe agriculture through latest innovative approaches. *Agronomy, 13*(7), 1709.

Alkubaisi, N., & Aref, N. (2017). Dispersed gold nanoparticles potentially ruin gold barley yellow dwarf virus and eliminate virus infectivity hazards. *Applied Nanoscience, 7*, 31–40. https://doi.org/10.1007/s13204-016-0540-0

Alkubaisi, N. A., Aref, N. M., & Hendi, A. A. (2015). Method of inhibiting plant virus using gold nanoparticles. US Patents US9198434B1. 2015 Dec 1;1.

Bai, W., Chen, Y., Wang, S., Gao, L., & Li, T. (2010). Comparative toxicity of several metal oxide nanoparticle aqueous suspensions to zebrafish (Danio rerio) early developmental stage. *Journal of Environmental Science and Health, Part A, 45*(6), 720–727.

Bhusal, S., Tiwari, M., & Singh, P. (2020). Gold nanoparticles: Preparation, functionalization, and applications in biosensors. *Journal of Nanoparticle Research, 22*(7), 183.

Chaud, M., Souto, E. B., Zielinska, A., Severino, P., Batain, F., Oliveira-Junior, J., & Alves, T. (2021). Nanopesticides in agriculture: benefits and challenge in agricultural productivity, toxicological risks to human health and environment. *Toxics, 9*(6), 131. https://doi.org/10.3390/toxics9060131

Djiwanti, S.R. & Kaushik, S. (2019) Nanopesticide: Future application of nanomaterials in plant protection. *Plant Nanobionics: Approaches in Nanoparticles, Biosynthesis, and Toxicity, 2*, 255–298.

Dutta, A. K., Kesharwani, A., Kumara, R., Dutta, P., & Iyer, P. K. (2021). Green synthesis of gold nanoparticles: A review. *Green Chemistry Letters and Reviews, 14*(1), 33–60.

El-Saadony, M. T., ALmoshadak, A. S., Shafi, M. E., Albaqami, N. M., Saad, A. M., & El-Tahan, A. M., et al. (2021). Vital roles of sustainable nano-fertilizers in improving plant quality and quantity-an updated review. *Saudi Journal of Biological Sciences, 28*(12), 7349–7359.

Elamawi, R. M., Al-Harbi, R. E., & Hendi, A. A. (2018). Biosynthesis and characterization of silver nanoparticles using Trichoderma longibrachiatum and their effect on phytopathogenic fungi. *Egyptian Journal of Biological Pest Control, 28*(1), 1–11.

Gill, P., Alvandi, A. H., Abdul-Tehrani, H., & Sadeghizadeh, M. (2008). Colorimetric detection of Helicobacter pylori DNA using isothermal helicase-dependent amplification and gold nanoparticle probes. *Diagnostic Microbiology and Infectious Disease, 62*, 119–124.

Guleria, G., Thakur, S., Shandilya, M., Sharma, S., Thakur, S., & Kalia, S. (2022). Nanotechnology for sustainable agro-food systems: The need and role of nanoparticles in protecting plants and improving crop productivity. *Plant Physiology and Biochemistry, 194*, 533–549.

Husen, A. (2017). Gold nanoparticles from plant system: synthesis, characterization and their application. In M. Ghorbanpourn, K. Manika & A. Varma (Eds.), Nanoscience and plant–soil systems

(Vol. 48, pp. 455–479). Springer International Publishing AG. https://doi.org/10.1007/978-3-319-46835-8_17

Husen, A. (2022). *Engineered nanomaterials for sustainable agricultural production, soil improvement and stress management*. Elsevier Inc. https://doi.org/10.1016/C2021-0-00054-7

Husen, A. (2023) *Nanomaterials and nanocomposites exposures to plants (Response, Interaction, Phytotoxicity and Defense Mechanisms)*. Springer Nature Singapore Pte Ltd. https://doi.org/10.1007/978-981-99-2419-6

Husen, A., & Iqbal, M. (2019). *Nanomaterials and plant potential*. Springer International Publishing AG. https://doi.org/10.1007/978-3-030-05569-1

Husen, A., Rahman, Q. I., Iqbal, M., & Yassin, M. O. (2019) Plant-mediated fabrication of gold nanoparticles and their applications. In A. Husen, & M. Iqbal (Eds.), *Nanomaterials and plant potential* (pp. 71–110). Springer International Publishing AG. https://doi.org/10.1007/978-3-030-05569-1_3

Husen, A., & Siddiqi, K. S. (2023) *Advances in smart nanomaterials and their applications*. Elsevier Inc. https://doi.org/10.1016/C2021-0-02202-1

Judy, J. D., Unrine, J. M., Rao, W., Wirick, S., & Bertsch, P. M. (2016). Bioavailability, toxicity, and fate of manufactured nanomaterials in terrestrial ecosystems. *Nanomaterials in the Environment*, 259–287.

Kadam, U. S., & Hong, J. C. (2022). Recent advances in aptameric biosensors designed to detect toxic contaminants from food, water, human fluids, and the environment. *Trends in Environmental Analytical Chemistry*, e00184.

Kaman, P., Dutta, P., & Bhattacharyya, A. (2022). *Synthesis of gold nanoparticles from Metarhizium anisopliae for management of blast disease of rice and its effect on soil biological index and physicochemical properties*. https://doi.org/10.21203/rs.3.rs-2080559/v1

Khan, B. A., Nadeem, M. A., Nawaz, H., Amin, M. M., Abbasi, G. H., & Nadeem, M., et al. (2023). Pesticides: impacts on agriculture productivity, environment, and management strategies. In *Emerging contaminants and plants: Interactions, adaptations and remediation technologies* (pp. 109–134). Springer International Publishing.

Kumar, A., Choudhary, A., Kaur, H., Guha, S., Mehta, S., & Husen, A. (2022). Potential applications of engineered nanoparticles in plant disease management: A critical update. *Chemosphere, 295*, 133798.

Kumar, V., Yadav, S. K., & Yadav, S. C. (2021). Mechanisms of antimicrobial activity of metal nanoparticles. In Antimicrobial nanotechnology (pp. 67–86). Springer.

Leyu, A. M., Debebe, S. E., Bachheti, A., Rawat, Y. S., & Bachheti, R. K. (2023). Green synthesis of gold and silver nanoparticles using invasive alien plant parthenium hysterophorus and their antimicrobial and antioxidant activities. *Sustainability, 15*(12), 9456.

Mishra, R., Das, J., & Das, S. (2020). Green synthesis of gold nanoparticles (GNPs) and its application in plant disease management. In *Nanotechnology in agriculture and food science* (pp. 245–264). Springer.

Nawaz, A., Rehman, H. U., Usman, M., Wakeel, A., Shahid, M. S., & Alam, S., et al. (2023). Nanobiotechnology in crop stress management: an overview of novel applications. *Discover Nano*, *18*(1), 74.

Oerke, E. C. (2006). Crop losses to pests. *The Journal of Agricultural Science, 144*(1), 31–43.

Patel, P., Parmar, R., Chudasama, C., & Jadeja, R. (2019). Antimicrobial potential of gold nanoparticles synthesized from Trigonella foenum-graecum seed extract against phytopathogenic fungi. *Applied Nanoscience, 9*(6), 1021–1032.

Pietrzak, K., Cwikła-Bundyra, D., Gadzała-Kopciuch, R., & Kapusta-Duch, J. (2019). The influence of gold nanoparticles on plants and their applications in agriculture. *Nanomaterials, 9*(11), 1570.

Raja, M. A., Ahmad, M. A., Daniyal, M., & Husen, A. (2023). Management of wastewater and other environmental issues using smart nanomaterials. In A. Husen, & K. S. Siddiqi (Eds.), *Advances in smart nanomaterials and their applications*. Elsevier Inc. https://doi.org/10.1016/B978-0-323-99546-7.00003-3

Sharma, G., Kumar, A., Naushad, M., & Das, R. (2019). Gold nanoparticles for disease control in agriculture: A review. *Journal of Nanoscience and Nanotechnology, 19*(11), 6831–6842.

Shipway, A. N., Katz, E., & Willner, I. (2000). Nanoparticle arrays on surfaces for electronic, optical, and sensor applications. *Chem Phys Chem, 1*, 18–52.

Siddiqi, K. S., & Husen, A. (2016). Engineered gold nanoparticles and plant adaptation potential. *Nanoscale Research Letters, 11*(400), 1–10. https://doi.org/10.1186/s11671-016-1607-2

Siddiqi, K. S., & Husen, A. (2017). Recent advances in plant-mediated engineered gold nanoparticles and their application in biological system. *Journal of Trace Elements in Medicine and Biology, 40*, 10–23. https://doi.org/10.1016/j.jtemb.2016.11.012

Singh, S., Singh, M., Agrawal, V. V., & Kumar, A. (2010). An attempt to develop surface plasmon resonance based immune sensor for Karnal bunt (Tilletia indica) diagnosis based on the experience of nano–gold-based lateral flow immune–dipstick test. *Thin Solid Films, 519*, 1156–1159.

Singh, R., Singh, D., Singh, V. P., Meena, K. K., & Dwivedi, S. (2018). Gold nanoparticles: Synthesis and applications in plant disease management. *Plant Pathology Journal, 34*(1), 1–10.

Wang, L., Hu, C., & Shao, L. (2017). The antimicrobial activity of nanoparticles: Present situation and prospects for the future. *International Journal of Nanomedicine, 12*, 1227–1249. https://doi.org/10.2147/IJN.S121956

Use of Smart Silver Nanoparticles in Drug Delivery System

Prakash Chandra Gupta, Nisha Sharma, Sweta Rai, and Priyanka Mishra

Abstract There are numerous applications for smart silver nanoparticles (AgNPs) in the scientific and technological domains. Such tiny particles are administered orally to enhance drug absorption, allow for dose reduction, and limit side effects. Oral delivery of AgNPs and herbal AgNPs are two options for a safer and more effective type of treatment. Due to their exceptional physical and chemical attributes, AgNPs have emerged among the most compelling nanomaterials in biomedicine. AgNPs' optical, electrical, and catalytic properties have led to rapid and widespread research. These traits, particularly the size, and form, are closely related to AgNPs characteristics. Several things should be considered when creating AgNPs with the desired size and shape. Further, regarding the various types of precursor salts, the production process must also account for additives such as reducing, capping, and stabilizing agents, besides considering the impact of reaction variables, namely reaction temperature, duration, pH, and additional energy. In addition to these approaches, biological synthesis utilizing microbes, fungi, and plant extract is a particularly simple, cost-effective, reliable, and sustainable option. AgNPs are primarily used as an antibacterial, anticancer, vaccine adjuvant, antidiabetic, and biosensor therapy, as well as to speed up bone and wound healing. The biological processes of AgNPs action are also covered, mainly the release of silver ions (Ag^+), the generation of reactive oxygen species (ROS), and membrane rupture.

Keywords Silver nanoparticles · Antimicrobial · Anticancer · Antidiabetic toxicity · Therapeutic

P. C. Gupta (✉) · N. Sharma · S. Rai · P. Mishra
School of Pharmaceutical Sciences, Chhatrapati Shahu Ji Maharaj University, Kalyanpur, Kanpur 208024, India
e-mail: prakashchandragupta@csjmu.ac.in

R. K. Bachheti et al. (eds.), *Metal and Metal-Oxide Based Nanomaterials*, Smart Nanomaterials Technology, https://doi.org/10.1007/978-981-99-7673-7_11

1 Introduction

The engineering and biological sciences have observed incredible advances in nanotechnology in the past few decades. Nanotechnology may be a rapidly expanding discipline that has likely been applied to a vast array of economically essential goods globally. Nanotechnology in the form of nanoparticles has influenced a diverse industry, namely electronics, information technology, manufactured goods, healthcare, and life sciences (Husen & Siddiqi, 2023). A nanoparticle is any material smaller than 100 nm in not less than one dimension (Ibrahim et al., 2019; Zebeaman et al., 2023). Unique optical, thermal, electrical, and magnetic properties are present in nanoscale materials. Because of their distinctive physicochemical characteristics, nanoparticles have drawn substantial consideration in biological applications (Bachheti & Bachheti, 2023; Husen, 2017; Mathur et al., 2018; Siddiqi & Husen, 2017, 2020; Siddiqi et al., 2018a, 2018b, 2018c). Medical research has demonstrated the therapeutic benefit of metal and metal oxide-supported nanomaterials. Some nanomaterials are fabricated by modifying other materials like copper, zinc, silver, and titanium. A few of them are also present in various areas where commercials are produced, such as the factories that make sunscreen and clothing that resists stains. They are also employed for pharmaceutical research, diagnostic kits, imaging, magnetic resonance imaging (MRI), drug delivery, and numerous other machinery and processes.

Metallic nanoparticles have long attracted scientists and are now widely used in biological sciences and engineering. Gold, silver, and iron oxide nanoparticles, chiefly magnetite (Fe_3O_4), are the most explored metal-based nanoparticles in the biomedical field. Silver nanoparticles (AgNPs) have emerged as a key product in nanotechnology. AgNPs have evolved into efficient tools for storing and delivering medications, making them smart. Nanoparticles are utilized for the regulated release of pharmaceuticals and albumin-drug complexes, in which nanoparticles serve as a carrier of drugs and selectively free them. Nanoparticles can be utilized as bullets of magic that target specific tissue cells.

Smart Silver Nanoparticles (AgNPs)

AgNPs are amongst the most valuable forms of metals in nanotechnology applications. AgNPs are tiny particles of silver engineered at the nanoscale level to exhibit unique properties and functionalities, making them act smartly. They usually are less than 100 nm in size and possess a large surface area-to-volume ratio, which gives them distinct characteristics compared to bulk silver. AgNPs' distinctive physicochemical and antibacterial attributes render them appropriate for a broad spectrum of applications. In 2009, the overall predicted output of AgNPs was around 500 tonnes per year, with a 900-tonne increase expected by 2025. Silver ions and metallic AgNPs are equally toxic to aquatic habitats and biological creatures. Silver consumption may provide health hazards due to metabolization and depositing in subcutaneous fat (Islam et al., 2021).

Silver components were frequently employed for curing wounds and burned prior to the discoveries of antibiotics. Because infections or diseases produced by bacteria, viruses, or fungi can result in dangerous situations for any patient, irrespective of age or gender, it is crucial to avoid any microbial activity in a wound. Smart AgNPs have been employed in various healthcare products like antimicrobial dressing/bandages, wound care products, breathing tubes, and catheters due to their unique antimicrobial characteristics to fight viruses, bacteria, and fungi. Smart AgNPs have a high potential for application as anticancer agents in treating human lung cancer cells. Furthermore, AgNPs have been shown to suppress HIV1 (Human Immunodeficiency Virus) cells. Recent research has also demonstrated that AgNPs synergistically impact when combined with specific antibiotics like penicillin G, amoxicillin, erythromycin, vancomycin, and clindamycin. Due to their remarkable antibacterial characteristics, low electrical and thermal resistance, and surface plasmon resonance, AgNPs particles are employed in various items, including consumer goods, healthcare, catalysis, electronics, and analytical equipment AgNPs are frequently coated or modified with diverse compounds to improve their stability, biocompatibility, and performance. They can be created with specific qualities in mind, such as antibacterial, antifungal, or antiviral activities. These nanoparticles have piqued researchers' curiosity in various sectors, including medical, electronics, and environmental applications.

1.1 Key Features and Applications of AgNPs

1. **Antibacterial Properties**

AgNPs have potent antibacterial activity due to their capability to release silver ions, which can disrupt the cell membranes and inhibit the growth of bacteria. They have been primarily studied for their potential use in wound healing, medical devices, and antimicrobial coatings (Cheng et al., 2016; Gonfa et al., 2022; Leyu et al., 2023).

2. **Drug Delivery Systems**

Smart AgNPs can be utilized as carriers for drug delivery. Their high surface area allows for efficient loading and controlled release of therapeutic agents. By functionalizing the nanoparticle surface, drugs can be targeted to specific cells or tissues, improving their effectiveness and minimizing side effects (Kim & Kim, 2018).

3. **Sensing and Imaging**

AgNPs can exhibit unique optical properties, like surface plasmon resonance, making them useful for sensing and imaging applications. They can be used as probes for detecting biological molecules, pollutants, or contaminants, and they have been explored for diagnostic purposes, including cancer detection (Mahmoudi et al., 2011).

4. **Electronics and Optics**

The conductive properties of AgNPs make them valuable in electronic applications. They can be incorporated into electronic circuits, transparent conductive films, or printed electronics. Additionally, their optical properties enable their use in optical devices, such as displays or sensors (Kang et al., 2016).

5. **Environmental Remediation**

AgNPs can be employed in environmental applications to remove pollutants from water or air. They can interact with various contaminants, including heavy metals and organic pollutants, through adsorption or catalytic processes, helping to mitigate environmental pollution. It is important to note that while smart silver nanoparticles offer promising benefits, their safety and potential environmental impacts are also subject to ongoing research. Proper regulation and risk assessment are essential to ensure their responsible use (Xu et al., 2010).

1.2 Basic Properties of AgNPs

1. **Size and Shape**

AgNPs typically have dimensions of 1–100 nm, with varying shapes, viz. spheres, rods, wires, or plates. Their small size and large surface area-to-volume ratio contribute to their distinctive properties and reactivity (Gao et al., 2018; Gonfa et al., 2023).

2. **Surface Plasmon Resonance**

AgNPs exhibit an optical event known as surface plasmon resonance (SPR), which is the collective oscillation of electrons on the nanoparticle's surface in response to incident light. SPR gives silver nanoparticles their distinct color, ranging from yellow to brown, depending on the particle size and shape (Rodriguez-Leon & Rendon-Romero, 2017).

3. **Chemical Reactivity**

The high reactivity of AgNPs is attributed to their large surface area, enabling strong interactions with other materials. AgNPs can undergo chemical reactions and interact with various substances, including gases, liquids, and solids (Bondarenko et al., 2013).

4. **Electrical Conductivity**

Silver is a highly conductive metal, and this property extends to AgNPs. Their conductive nature makes them useful in electronics, sensors, and energy device applications (Liao et al., 2019).

5. **Catalytic Activity**

AgNPs can exhibit catalytic behaviour, accelerating chemical reactions without being consumed. Their catalytic activity is influenced by size, shape, and surface chemistry. This property finds applications in chemical synthesis and environmental remediation (Le et al., 2020).

6. **Antibacterial Properties**

AgNPs possess inherent antimicrobial properties, primarily due to silver ions' release. These ions can disrupt the cell membranes and inhibit the growth of bacteria, making silver nanoparticles useful for various antibacterial applications (Ma et al., 2012a, 2012b).

7. **Stability and Aggregation**

AgNPs can aggregate or agglomerate over time, leading to changes in their properties. Surface modifications or coatings often enhance their stability and prevent aggregation. It is important to note that the properties of AgNPs can vary based on factors such as size, shape, surface chemistry, and synthesis methods. These properties can be further modified through functionalization or coating techniques to tailor their behaviour for specific applications (Yang et al., 2015) (Fig. 1).

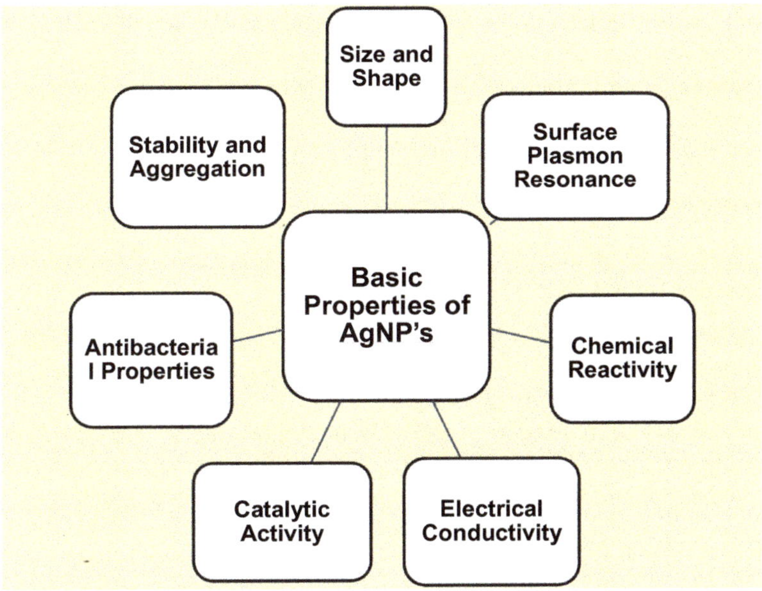

Fig. 1 Basic properties of AgNPs

1.3 Techniques for the Formulation of AgNPs in Drug Delivery Systems

Some commonly used techniques, along with examples, are discussed below:

1. **Chemical Reduction Method**

This approach comprises the reduction of silver ions in the presence of a stabilizing agent employing a reducing agent (Kimling et al., 2006; Mallikarjuna et al., 2017).

- **Citrate Reduction**: AgNPs can be synthesized using citrate as a reducing and stabilizing agent. This method, known as the Turkevich method, yields monodisperse nanoparticles suitable for drug delivery applications.
- **Polyol Reduction**: Polyols, such as ethylene glycol or glycerol, can be reduced and stabilizing agents. By controlling the reaction conditions, including temperature and reaction time, AgNPs with desired properties can be synthesized.

2. **Green Synthesis**

This approach utilizes natural extracts or biomolecules as reducing and stabilizing agents, offering a more environmentally friendly and biocompatible method (Prasad et al., 2012).

- **Plant Extracts**: Various plant extracts, such as aloe, green tea, or grape seed extract, can reduce silver ions and produce AgNPs. These extracts contain natural compounds that act as reducing and stabilizing agents.
- **Microorganisms**: Certain microorganisms, like bacteria or fungi, can also synthesize AgNPs. For example, *Escherichia coli* or *Fusarium oxysporum* can effectively reduce silver ions to form nanoparticles.

3. **Electrochemical Methods**

Electrochemical techniques involve the application of an electric current to reduce silver ions and form AgNPs (Xiao et al., 2013).

- **Electrodeposition**: In this method, a direct current is passed through an electrolyte solution containing silver ions and a suitable electrode. This leads to the deposition of AgNPs on the electrode surface.
- **Electrochemical Reduction**: AgNPs can be synthesized by applying a potential to an electrochemical cell containing a silver electrode and a counter electrode. This induces the reduction of silver ions and the formation of AgNPs.

4. **Microemulsion Method**

This method uses a microemulsion system consisting of water, oil, surfactant, and co-surfactant. AgNPs can be synthesized within the confined space of the microemulsion droplets. The composition and ratio of the components can be adjusted to control the size and properties of the nanoparticles (Shahiwala & Misra, 2004).

It is important to note that the choice of formulation technique depends on factors such as the desired particle size, shape, stability, and the compatibility of the method with the drug delivery system requirements. When working with silver nanoparticles or nanomaterials, it is crucial to follow appropriate safety guidelines and regulations to ensure materials' safe handling and disposal.

1.4 Polymers Used in the Formulation of AgNPs for Drug Delivery Systems

Numerous polymers have been successfully employed in developing smart AgNPs for drug delivery systems. Here are some commonly used polymers, along with examples of their application in research:

1. **Polyvinyl pyrrolidone (PVP)**

Polyvinyl pyrrolidone (PVP) is a commonly used polymer in the formulation of AgNPs for drug delivery systems. PVP acts as a stabilizing agent, preventing the agglomeration of nanoparticles and enhancing their stability. It also provides a biocompatible and protective coating, enabling efficient drug loading and controlled release. In a study by Sreekanth et al., PVP-coated AgNPs were synthesized and utilized to deliver doxorubicin, an anticancer drug. The PVP coating improved the stability of the nanoparticles and enabled efficient drug loading. The doxorubicin-loaded AgNPs demonstrated sustained drug release, enhancing their anticancer efficacy against triple-negative breast cancer cells (Ahmed et al., 2022).

2. **Polyethylene glycol (PEG)**

Polyethylene glycol (PEG) is commonly used to form AgNPs for drug delivery systems. PEG serves as a stabilizing agent, providing a protective coating around the nanoparticles, enhancing their stability and biocompatibility. Additionally, PEGylation can prolong the circulation time of nanoparticles in the body, improve their dispersibility, and reduce nonspecific interactions with biological components. In a study by Patra et al. PEGylated AgNPs were developed and utilized for the delivery of antibiotics. The PEG coating enhanced the stability of the AgNPs and facilitated their controlled release of the loaded antibiotics. The PEGylated AgNPs exhibited improved antibacterial activity against various bacterial strains (Patra & Baek, 2017).

3. **Chitosan**

Chitosan is a versatile polymer commonly used to form AgNPs for drug delivery systems. Chitosan possesses several advantageous properties, including biocompatibility, biodegradability, muco adhesiveness, and the ability to interact with biological membranes. When combined with AgNPs, chitosan can enhance the stability, biocompatibility, and controlled release of drugs. Khan et al. developed chitosan-coated AgNPs to deliver antibiotics to combat tuberculosis. The chitosan coating enhanced the stability and biocompatibility of the AgNPs. The loaded antibiotics exhibited sustained release, improved antimicrobial activity against Mycobacterium tuberculosis, and reduced cytotoxicity (Khan et al., 2018).

4. **Poly (lactic-co-glycolic acid) (PLGA)**

Poly (lactic-co-glycolic acid) (PLGA) is a commonly used biodegradable polymer in formulating AgNPs for drug delivery systems. PLGA offers several benefits, including biocompatibility, biodegradability, and tunable release kinetics. When combined with AgNPs, PLGA can provide a controlled release of drugs, improve stability, and enhance therapeutic efficacy. Singh et al. developed PLGA-based AgNPs for the treatment of chronic wounds. The PLGA nanospheres loaded with AgNPs showed sustained release of silver ions, providing antibacterial activity against wound pathogens. The formulation exhibited enhanced wound healing properties and demonstrated potential for treating chronic wounds (Naskar & Kim, 2020).

5. **Poly (N-isopropyl acrylamide) (PNIPAM)**

Poly (N-isopropyl acrylamide) (PNIPAM) is a temperature-responsive polymer utilized in the formulation of AgNPsfor drug delivery. PNIPAM exhibits a lower critical solution temperature around body temperature, allowing it to undergo a reversible phase transition from a hydrophilic to a hydrophobic state. This property can be harnessed to develop stimuli-responsive drug delivery systems. While PNIPAM may not directly interact with AgNPs, it can be combined with other polymers or coatings to provide additional functionalities. In research by Yallapu et al., silver nanoparticles were coated with a PNIPAM-based polymer for targeted drug delivery to cancer cells. The PNIPAM coating facilitated the encapsulation of an anticancer drug, resulting in controlled drug release triggered by temperature changes. The formulation exhibited enhanced drug uptake by cancer cells and improved therapeutic efficacy (Qasimet al., 2018).

6. **Polysaccharides**

Polysaccharides, natural polymers derived from carbohydrates, have gained significant attention in transforming silver nanoparticles for drug delivery systems. Their biocompatibility, biodegradability, abundance, and versatility make them suitable for various biomedical applications (Abate et al., 2022). Polysaccharides can serve as stabilizers, coatings, or matrix materials for silver nanoparticles, offering controlled release, enhanced stability, and improved biocompatibility (Wang et al. 2016).

Alginate, a naturally occurring polysaccharide, has been utilized to formulate AgNPs for drug delivery applications. Alginate-coated AgNPs demonstrated sustained release of loaded drugs, such as antibiotics, and exhibited enhanced antimicrobial activity against bacterial pathogens (Jayakumar et al., 2010).

Another example involves using hyaluronic acid, a polysaccharide widely found in the human body, to formulate AgNPs. Hyaluronic acid-coated smart AgNPs demonstrated improved biocompatibility, prolonged circulation time, and targeted delivery to specific cell types, making them suitable for drug delivery applications. Polysaccharides offer a promising platform for developing advanced drug delivery systems (Lee et al., 2010).

1.5 Biomedical Uses of Smart AgNPs

1. **Smart AgNPs for cancer therapy and treatment**

The formation of altered cells that divide uncontrollably makes cancer one of the most challenging diseases to treat. Its etiology could be brought on by genetic dysregulation or mutations brought on by acute or long-term exposure to xenobiotics or environmental toxins. Cancer cells can metastasize or spread to other organs. Chemotherapy, surgery, and radiation therapy are the standard treatment modalities for cancer. Chemotherapy medications frequently do not dissolve well in water. Hydrophobic medicines must be delivered at greater dosages to obtain therapeutic concentrations due to their decreased biocompatibility. Additionally, a drug's low hydrosolubility results in high systemic toxicity, reduction in the drug's bioavailability, is not very targeted, severely harms healthy tissues, and results in adverse side effects. In this regard, using controlled release with target-oriented drug delivery systems may be an alternative to get around some drawbacks of traditional chemotherapy (Azharuddin et al., 2019).

Nanomaterials of various kinds have been investigated to enhance cancer diagnosis (nano biosensors and bioimaging agents) and treatment (drug nanocarriers, photodynamic therapy, gene, and immunotherapy). One of the primary qualities that make nanomaterials so intriguing for treating cancer is their tuneable surface, which not only enables the nanoparticles to be synthesized with distinct physicochemical properties but also promotes the attachment of various molecules and medications. Therefore, these nanocarriers (NCs) might resolve long-standing problems like low drug solubility, drug metabolism, and systemic half-life. The Enhanced Permeability and Retention effect EPR is another way nanocarriers may increase drug deposition in tumours. The EPR effect is based on the fact that cancers have leaky blood arteries and reduced lymphatic function, which, in contrast to normal tissues, makes it easier for nanoparticles to penetrate and accumulate in tumours (Jain & Stylianopoulos, 2010).

AgNPs have antitumoral action, making them extremely promising for cancer treatment. AgNPs use multiple pathways to exert their broad-spectrum anticancer activity. AgNPs have been shown to reduce the viability and growth of cancer cells in several in vitro and in vivo experiments. By damaging the cancer cells' ultrastructure and creating ROS and DNA damage, AgNPs can lead to apoptosis and necrosis. By modulating critical signaling pathways, including the hypoxia-inducible factor (HIF) pathway, and up or downregulating the expression of critical genes like p53, AgNPs can trigger apoptosis. Cell cycle arrest may also be seen in cancer cells, subG1 arrest, and death of certain cancer cells treated with AgNPs. Also, AgNPs can lessen distant metastasis by preventing cancer cell migration and angiogenesis (Chen et al., 2018).

A dosage and time-dependent relationship exists between the ultra-structural alterations of cancer cells exposed to AgNPs. The severity of the damage to the cell ultrastructure is often inversely proportional to the AgNPs content and exposure duration, damage to cytoplasmic organelles, apoptosis, necrosis, and autophagy as various modes of cell death. AgNPs-treated cancer cells develop autophagosomes in their cytoplasm, which are linked to apoptosis and necrosis. AgNPs induce autophagy in cancer cells without impairing lysosomal function by activating the PtdIns3K pathway, which also helps to stimulate the production of autophagosomes. Cancer cells exposed to AgNPs may exhibit morphological degeneration, and this may be due to structural and functional disruption of the actin cytoskeleton, which may also prevent cancer cell invasion and migration. The cellular membranes are damaged due to the free Ag^+ produced by AgNPs. Damaged cells begin to leak cytoplasmic components due to Ag^+, which is generated by AgNPs, inducing glutathione oxidation and increasing lipid peroxidation in cellular membranes (Farah et al., 2016).

2. **Production of ROS**

Oxygen ions, peroxides, and oxygenated free radicals, collectively known as ROS, are by-products of cellular aerobic metabolism. AgNPs can cause autophagy and death in cancer cells by various ROS-mediated stress responses when disseminated in the cells through endocytosis. Additionally, ROS production by AgNPs may impact cellular signal transduction pathways, which could contribute to the activation of apoptosis. AgNPs, for instance, can reduce ATP synthesis by interfering with the mitochondrial respiratory chain, inhibiting the activity of the mitochondria. Additionally, AgNPs' production of ROS may ultimately result in DNA damage. The destruction of the electron transport chain, impairment of mitochondrial function, and enhancement of mitochondrial outer membrane permeabilization (MOMP) and Cytochrome C release are all possible effects of superoxide radicals aimed at mitochondria. Several variables influence agNPs-induced ROS production. AgNPs of varied sizes and concentrations induce ROS more strongly and have more potent cytotoxicity; when cancer cells are exposed to AgNPs, ROS levels sharply increase (Mytych et al., 2017; Yang et al., 2016).

AgNPs have more complex anticancer mechanisms. AgNPs can cause cancer cells to die by damaging the cell's ultrastructure, producing ROS, causing DNA damage, inactivating proteins, and controlling a variety of signaling pathways. Additionally, by stifling angiogenesis within the lesion, AgNPs may prevent cancer cells from migrating and invading the area. The cytotoxicity of AgNPs, however, might restrict the uses for which they are intended in medicine. AgNPs can be functionalized as their surface allows the coordination of many ligands (Panzarini et al., 2017).

3. **AgNPs for wound healing**

The largest organ in the body, the skin, is essential for sensory functioning, homeostasis, temperature regulation, and defense against viruses, toxins, and injury

(Percival et al., 2015). The creation of a wound is determined by the disruption of skin integrity, which can occur as a result of a disease or have an unintentional or deliberate etiology (Singh et al., 2017a, 2017b).

Due to potent bactericidal properties and broad-spectrum antibacterial activity, silver compounds and ions have been widely employed for hygienic and therapeutic purposes. Different silver-containing formulations have been used to treat chronic wounds by utilizing their antibacterial characteristics. Silver has been used to treat wounds for a very long time. Wounds were treated in Egypt circa 1850 BC with Silver. Silver nitrate was first used to heal ulcers in the seventeenth and eighteenth centuries and was first used to treat burns in 1960. Due to rising bacterial resistance to antibiotics and advancements in polymer technology, Silver has recently attracted renewed interest after seeing a decline in use following the 1940 introduction of antibiotics. Because of this, many dressings on the market include Silver. Burns are treated with Silver, either in the form of impregnated bandages or a cream that contains silver sulfadiazine as the active ingredient. A variety of fibers or polymeric scaffolds that have been impregnated or coated with an Ag salt or metallic Ag in the nanoparticulate form are now accessible as silver-based dressings. Gram-positive and negative bacteria show rapid and all-encompassing antibacterial activity (Jarbrink et al., 2017).

Silver-based lotions, ointments, and AgNPs-based biomedical formulations like wound dressings are now commercially available for various medical applications due to their broad-spectrum antibacterial properties. In addition to playing a key role in preventing infection, silver nanoparticles can stimulate myofibroblast differentiation in fibroblasts, encouraging wound contraction, speeding up healing, and stimulating keratinocyte proliferation and migration. A useful technique to keep macrophage activation in check and control the local inflammatory response could be silver nanoparticles with the right size and concentration (Gunasekaran et al., 2012).

Compared to other silver salts, compounds, and solutions, AgNPs have a substantially bigger surface area and far more potent biological effects, such as an antibacterial effect. Broad-spectrum antibacterial properties of AgNPs enable them to quickly kill more than 650 different types of microorganisms (Sim et al., 2018). Most notably, AgNPs greatly reduce the likelihood of bacterial resistance. AgNPs may inhibit the respiratory chain, harmful bacteria's DNA, and destroy their cell walls, all of which would cause the bacteria to die. AgNPs enhance the synthesis of an anti-inflammatory cytokine, IL-10, at the skin wound in the early phase after injury. Still, they reduce the production of early pro-inflammatory cytokines, such as IL-6. This suggests that AgNPs suppress early inflammation (Vijayakumar et al., 2019).

4. **Antibacterial characteristics of AgNPs**

Various antimicrobial activities of AgNPs have been reported, while the precise mechanism remains unclear. AgNPs can continuously emit silver ions, which might be thought of as the method by which microorganisms are killed (Bapat et al., 2018). Because of their affinity for sulphur proteins and electrostatic attraction, silver ions can stick to cell walls and the cytoplasmic membrane. The adhering ions could increase the cytoplasmic membrane's permeability, resulting in the rupture of the bacterial envelope. The sticking ions can make the cytoplasmic membrane more permeable and lead to bacterial membrane rupture (Khorrami et al., 2018). Following the entry of free silver ions into cells, respiratory enzymes may become inactive, resulting in the generation of reactive oxygen species but not adenosine triphosphate (Ramkumar et al., 2017).

Reactive oxygen species have the potential to act as a main trigger for DNA modification and cell membrane deterioration. The interaction of silver ions with these elements might interfere with DNA replication, hinder cell growth, or may result in the death of microorganisms due to sulphur and phosphorus in DNA. Moreover, silver ions can cease protein production by denaturing cytoplasm ribosomes (Duran et al., 2016).

In addition to having the ability to release silver ions, AgNPs have antibacterial properties. After adhering to the cell surface, AgNPs can build up in the pits on the cell wall. Cell membrane denaturation may result from the accumulating AgNPs. Due to their nanoscale size, AgNPs are competent in penetrating bacterial cell walls and altering the structure of the cell membrane (Liao et al., 2019).

The biofilm quickly forms in the oral environment and protects bacteria from silver ions and nanoparticles by obstructing their mobility. Scientists found that biofilm bacteria retained some vitality even after killing all planktonic bacteria with AgNPs at the same dose (Saravanan et al., 2018). Due to its intricate architecture, the biofilm is, therefore, tolerant of AgNPs. The mobility and bioavailability of silver nanoparticles in a biofilm are primarily governed by their diffusion coefficients, which are typically correlated with size and physicochemical properties. First, as molar mass increases, these factors fall, making it more difficult for bigger AgNPs to enter the biofilm (Yin et al., 2019). Larger particles than 50 nm can seriously impede transmission via the biofilm. Second, due to the chemical composition of the nanoparticles, AgNPs may adsorb and build up in the biofilm, reducing its diffusion. Third, electrostatic interactions between bacteria and AgNPs can influence the penetration of charged nanoparticles into the biofilm (Pugazhendhi et al., 2018).

5. **Antifungal characteristics of AgNPs**

Over the years, there have increased interest in AgNPs because of their potential use in human and animal medicine to treat skin infections, such as dermatomycosis. Additionally, these particles would serve as a therapeutic alternative against resistant microbes and difficulties linked to antifungals, whether used alone or in conjunction with other medications (Rai et al., 2009). Whatever form it takes, such as silver ions (Ag^+), silver complexes, and metallic Silver, including AgNPs, has fungicidal and virucidal effects.

The consensus is that silver ions and AgNPs are more biocidal to prokaryotes than eukaryotic cells, creating a therapeutic window when fungi are killed but mammalian tissue is left unharmed (Grelich et al. 2012). It is important to emphasize that AgNPs are chemically less reactive than silver ions, which interact with a range of biomolecules within a cell, namely nucleic acids, cell wall components, sulfhydryl groups of metabolic enzymes, and sulfur-containing cell components like glutathione (Ahamed et al., 2008). However, it is also essential to remember that AgNPs are a source of silver ions. Because AgNPs are prone to oxidative breakdown, silver ions are continuously released (Kittler et al., 2010). AgNPs' physicochemical characteristics and environmental factors affect how quickly silver ions are released (Chambers et al., 2013).

Smaller AgNPs are more harmful than larger ones (Baker et al., 2005). Two significant factors are linked to this reliance. First, AgNPs with smaller diameters infiltrate cells and other microorganisms more effectively (Bae et al., 2011). Additionally, smaller AgNPs produce more silver ions in less time than bigger AgNPs due to their greater oxidative dissolution sensitivity (Ma et al., 2012a, 2012b; Peretyazhko et al., 2014).

The second crucial element impacting AgNPs' biocidal capabilities is their form (Helmlinger et al., 2016). According to the study's findings, truncated triangular AgNPs had an elevated level of biocidal activity, followed by nanospheres and nanorods. The formation of anisotropic AgNPs is persuaded and directed by distinct, high-molar-mass molecules of surfactants or polymers (Pal et al., 2007). On the surfaces of AgNPs, shape-controlling molecules are frequently deposited, and like Silver, they have biological activity. Because of this, a key factor affecting the biocidal activity of the entire nanometric system is the surface chemistry of AgNPs (Tang & Zheng, 2018).

AgNPs, which have been used for a while as antibacterial agents and are very efficient against fungi, are one type of nanoparticle. AgNPs function via several processes, one of which is their affinity for the phosphate groups in DNA (Silva et al., 2017). Nonlinear interactions at the plasma membrane allow protons to diffuse and kill cells (Salleh et al., 2020). Additionally, they can interfere with the electron transport chain, interact with sulfhydryl groups in proteins and enzymes, and alter membrane permeability to protons and phosphate groups (Badmus et al., 2020).

5. AgNP's as vaccine adjuvant

Vaccination is one of the best ways to avoid infectious diseases and control healthcare expenses (Talbird et al., 2015). The fact that AgNPs raised both antigen-specific IgG and IgE production in the blood suggests that AgNPs induced Th2-biased immune responses. The mechanism of the adjuvant was further investigated, and it was found that AgNPs may attract and activate local macrophages and leukocytes. The load of AgNPs might dramatically boost immune responses by producing antibodies that neutralize the rabies virus, and even the smallest amount of virus-loaded AgNPs reduced cell viability. This restricted the use of AgNPs as a rabies virus adjuvant. After testing the green synthesis of AgNPs using *Eucalyptus procera* leaf extract and using AgNPs as an adjuvant in the rabies veterinary vaccine, they estimated the efficiency of the vaccination in mice and dogs. They verified that the vaccine was safe after loading an appropriate amount of AgNPs (Asgary et al., 2016).

Following recent vaccination developments, new antiviral treatments are compelling to tackle viral infection-related public health issues. Recently, there has been talk about using nanomaterials to increase the effectiveness of vaccines (Sano et al., 2017). Most likely due to a depot effect linked to prolonged antigen maintenance at entrance points (Okamoto et al., 2009). In addition, metallic nanoparticles like AgNPs have inherent antibacterial qualities (either through direct contact with microbial surfaces or because they disrupt their metabolism) (Lara et al., 2011). AgNPs have been demonstrated to increase the formation of ovalbumin (OVA) specific IgG antibodies in the serum of mice parenterally immunized with OVA, supporting this theory (Xu et al., 2013) and enhancing an intra-peritoneal rabies vaccine given to mice (Asgary et al., 2016) even if via unidentified processes. Nevertheless, despite these developments, it is becoming increasingly clear that systemic routes of adjuvant/antigen administration do not favour significant immune mucosal responses, in contrast to local administration (Sano et al., 2017).

Both the heat-killed IAV capsids and the model antigen KLH can be physically bound and sequestered by AgNPs. This increased antigen availability could improve antigen retention in the lung compartment when delivered intravenously (Anderson et al., 2015; Smulski & Eibel, 2018). This led to more effective local immune activation. In vivo, pulmonary injection of AgNPs caused a temporary localized inflammation, which upregulated IL-12. Although traditionally thought of as possible sources of IL-12, macrophages, and DCs (Wright et al., 2008), since AgNPs alone did not stimulate the APCs' IL-12 secretion in vitro, it is likely that the IL-12 secretion seen was caused by indirect activation of these cells by the favorable pro-inflammatory environment after AgNPs instillation. Another adjuvant, like MF59 or aluminum derivates, was, shown to activate DC in a manner similar to AgNPs indirectly. When administered intravenously, AgNPs did indeed produce more potent specific IgA antibodies.

1.6 Other Medical Applications

1. **Silver nanoparticles for antidiabetic potential**

Hyperglycaemia is a feature of the metabolic illness known as diabetes mellitus (DM). Recent studies have discovered that AgNPs made from plant extracts have antidiabetic potential (Sengottaiyan et al., 2016). *Solanum nigrum* leaf extract was used to create AgNPs, and the anti-hyperglycemic impact on rats with diabetes brought on by alloxan was assessed. When given AgNPs for 14 or 21 days, it was discovered that the blood glucose level dropped without experiencing any major acute toxicity. When compared to the usual antidiabetic medicine glibenclamide, AgNPs had a good hypoglycaemic impact (Saratale et al., 2017). AgNPs could effectively increase insulin secretion and sensitivity by activating the protein kinase C and PI3K pathways at the level of the insulin receptor substrate and inhibiting the protein kinase C isozymes. The effectiveness of AgNPs in lowering insulin resistance and DNA damage was emphasized.

2. **Biosensing and imaging surface**

Noble metals with Raman signals have drawn interest from increased Raman scattering (SERS) in various application tactics, including biological sensing, analytical chemistry, and materials research (Yamamoto et al., 2014). AgNPs, one of these nanomaterials, can be utilized as an affordable surface-enhanced Raman scattering substrate. AgNPs containing nanoparticles can be utilized as biosensors to identify infections, enzymes, tumor cell molecular markers, blood sugar, and other substances (Jiang et al., 2018).

AgNPs were used as a conductive additive to create a high-sensitivity nonenzymatic biosensor for glucose detection. AgNPs' porous nanostructures and the wide surface areas of their carriers improved the interaction sites between them and the electrode and glucose, which could hasten the transfer of electrons from AgNPs and boost the biosensor's sensitivity. For instance, hybrid Ag@NGO nanoparticles were created using a one-step reduction technique (Zeng et al., 2018). The nanosized graphene oxide (NGO) among these platforms served as inert protective layers and offered an incredibly thin protective layer for AgNPs. Ag@NGO demonstrated the benefits of SERS biosensing and drug delivery by using SERS biosensing to monitor biomolecule signals in tumor cells and interacting with the anticancer medication doxorubicin by forming covalent bonds. These findings demonstrate that AgNPs have many possible applications due to their SERS biosensing capabilities.

3. **Silver nanoparticles for dentistry**

Since the nineteenth century, Silver has been used in dentistry for various purposes, mainly because silver ions have antibacterial properties. Silver nanoparticles, which have an antibacterial effect primarily because of the slow release of silver ions, have changed how Silver is used in dentistry in the twenty-first century (Paiva et al., 2018; Venugopal et al., 2017). One of the factors contributing to oral problems is the production of plaque biofilm. Due to AgNPs' antibacterial properties, various dental biomaterials have been combined with them to decrease biofilm formation. Dentures are made from polymethyl methacrylate (PMMA), also known as acrylic resins and composite resins, but they are likely to harbor dangerous organisms due to the rough surface. The antibacterial impact of the dental material can be enhanced by AgNPs added into PMMA (Oei et al., 2012).

Studies in dentistry have shown that the fields of oral microbiology, preventive dentistry, prosthodontics, orthodontics, endodontics, and periodontics all utilize silver nanoparticles. In other research, the antibacterial activities of silver nanoparticles were tested against the most prevalent oral infections to further examine the possibilities of employing them (Zarco et al., 2012).

The oral cavity is home to about 700 different bacterial species, as well as fungi, viruses, and protozoa. The antibacterial effectiveness of silver nanoparticles is inversely correlated with their size. Smaller AgNPs effective against *Streptococcus orallis* and *S. mutans* biofilms have higher biofilm inhibition results than larger particles. Larger AgNPs, on the other hand, have been shown to have important antibacterial properties against a variety of dental plaque pathogens, exhibiting acceptable bacterial growth suppression even at lower doses (Espinosa-Cristobal et al., 2019).

Dental caries is a polymicrobial dysbiosis disease that is brought on by an imbalance between demineralization and remineralization. Silver ions can accumulate in carious lesions and precipitate, which causes the enamel to harden. Dental surgeons utilize sodium fluoride varnish in their daily clinical practice to remineralize developing lesions. However, when 5% of nanosilver is added to sodium fluoride varnish, caries lesions in the remaining teeth are 77% inhibited from progressing without leaving a metallic taste or uncomfortable ulcerations. Sodium fluoride or the conventional silver compound, Silver Diamine Fluoride (SDF), can be substituted for varnish with Silver Nano Fluoride (NSF), which is simple to administer, can only be used once a year, and has an acceptable cost–benefit ratio (Nozari et al., 2017).

5. Silver nanoparticles for catheter modification

The most frequent infections in healthcare facilities and among the general public are urinary tract infections (UTIs) (Das et al., 2006). Latex or silicone tubes are used to hold urinary catheters. Biofilms can easily grow on the inner or exterior surfaces of inserted catheters. *E. coli, Klebsiella pneumoniae, Enterococcus faecalis, Propionibacterium mirabilis, Staphylococcus aureus, Staphylococcus epidermidis, Pseudomonas aeruginosa, and Candida* spp. are the most frequent biofilm-forming bacteria that cause CAUTIs (Ullah et al., 2009). Due to the presence of divalent cations like calcium and magnesium in urine as well as the urine's relatively acidic pH, the attachment and colonization of bacteria on the surface of catheters are additionally facilitated. Antibiotic resistance is common among bacteria that create biofilms. Also, biofilms prevent antibiotics from penetrating, producing an unfavorable chemical microenvironment for antibiotic activity. To stop bacteria from colonizing catheters, numerous techniques have been devised. These include systemic antibiotics, applying antimicrobial ointments and lubricants, and including antimicrobials in collection bags. The majority of these tactics, meanwhile, have yet to be entirely successful. Ag^+ kills bacteria by interfering with their cell membranes, interacting with macromolecules like proteins and DNA, and inactivating crucial enzymes. This ability of silver ions (Ag^+) to kill bacteria has long been known. The catheters' internal and external surfaces are coated with an alloy made of Silver, gold, platinum, or nanoparticles (Stewart & Costerton, 2001; Tunney et al., 1999). Alternative techniques were introduced, such as coating or impregnating catheters with AgNPs both internally and externally. Recent research described AgNPs-modified catheters as harmless medical devices that limit infection-related problems by releasing AgNPs (Kumar & Sujitha, 2014).

1.7 Toxicities of AgNPs

The increasing observation of the potential harm that nanoparticles could cause to bodily organs and systems (Saha et al., 2014). There are many ways to provide AgNPs, but the most common ones are inhalation, ingestion, skin contact, and subcutaneous or intravenous injection. The absorbed AgNPs are dispersed throughout numerous systems (Lansdown, 2006; Singh et al., 2017a, 2017b), such as the dermal, respiratory, spleen, digestive, urinary, neurological, immunological, and reproductive systems, with the majority of AgNPs being dispersed in the spleen, liver, kidney, and lung, with little AgNPs deposition being shown in the teeth and bones.

1. **Eye toxicity**

The AgNPs agent may cause acute conjunctival irritation that is concentration dependant. However, there is yet no solid proof of toxicological effects. Early-stage people's eyes may experience developmental toxicity from AgNPs, which may ultimately cause a variety of eye abnormalities such as edema, visceral abnormalities, congenital cardiac problems, spinal abnormalities, and eye anomalies. Different eye abnormalities, including microphthalmia, exophthalmia, cyclopia, and anophthalmia, were present in the AgNPs-treated group (Maneewattanapinyo et al., 2011; Wu et al., 2010).

2. **Respiratory toxicity**

Acute lung toxicity caused by AgNPs can impair lung function, and the severity of the damage is correlated with particle accumulation and clearance (Shimada et al., 2006). Lung damage may be exacerbated by ultrafine particles' induction of oxidative stress and apoptosis. Additionally, nanoparticles demonstrated size-dependent pulmonary toxicity, meaning that smaller particles can cause lung inflammation and tissue damage more than bigger particles (Kaewamatawong et al., 2005; Maneewattanapinyo et al., 2011). However, AgNPs have the potential to cause dose-dependent pulmonary damage. The inflammatory lesions in mice may be caused by proinflammatory cytokines, including IL-1 and TNF, generated by alveolar macrophages and airway epithelial cells. Pulmonary fibrosis may result from the AgNPs' aggregation, which immediately impacts the basement membrane and upsets the balance between the synthesis and breakdown of the extracellular matrix. Furthermore, they assumed that AgNPs caused oxidative stress in the lung. They also understood that one of the potential lung protective mechanisms might be the expression of metallothionein (MT) triggered by AgNPs.

3. **Hepatobiliary system toxicity**

The liver is one of the most commonly assaulted organs. It tends to sequester, degrade, and collect some of the ingested nanoparticles, suggesting that the liver may be involved in nanoparticle metabolism. Conversely, the gallbladder gathers, stores, and excretes bile or biological waste into the gut. It is known that the liver uses this mechanism to export a variety of metal nanoparticles, including AgNPs. As a result, hepatocytes are extensively researched about AgNPs' liver damage (De Maglie et al., 2015). AgNPs are injected intravenously because of hepatobiliary damages such as portal vein injury, hepatocyte necrosis, and small bleeding surrounding the biliary tract. Second, they noted that AgNPs could cause mild splenomegaly, acute tubular necrosis, and apoptosis (Sadauskas et al., 2007; Tak et al., 2015).

4. **Central nervous system toxicity**

The brain and the spinal cord are the two components of the central nervous system. The interneuronal area in the central nervous system is largely occupied by neuroglial cells, which act as supportive non-nervous cells. Injected AgNPs pass through the blood–brain barrier (BBB) and enter the brain, whereas inhaled AgNPs pass through the olfactory and BBB to enter the central nervous system (Peters et al., 2006). Due to AgNPs' unique physicochemical characteristics, they can cause and exacerbate neurotoxicity and inflammation in nerve cells, astrocytes, and extravascular lymphocytes and enhance BBB permeability. Instead of Ag^+, the non-cytotoxic dose of AgNPs may cause neuro-inflammation by encouraging the release of several astrocyte cytokines, such as CINC-2a/b, CINC-3, IL-10, IP-10, L-selectin, and thymus chemokine (Xu et al., 2015).

5. **Kidney toxicity**

The kidney helps to regulate osmotic pressure, electrolyte concentration, drug metabolism, and toxic emissions, as well as the volume and pH of body fluids. Mammalian kidneys that have been exposed to AgNPs may exhibit abnormal renal function. In the majority of the tissues that were investigated, including the brain, lung, liver, dermis, blood, and testes, AgNPs show a dose-dependent accumulation. However, there is a gender-related variation in the kidney's accumulation of Silver.

6. **Immune system toxicity**

Our immune system, which serves as a natural host defense barrier, comprises immune cells, tissues, and organs. It may constantly interact with the internal environment, shielding us from infections in the outside world and giving us the innate ability to distinguish between friends and enemies inside our bodies. Davies et al. (2013), Shin et al. (2007) discovered that AgNPs caused cytotoxicity in peripheral blood mononuclear cells and, in a concentration-dependent manner, reduced the proliferation and production of cytokines, including IL-5, INF, and TNF. AgNPs may accumulate in immune organs and impact the number of immune cells and cytokine production (De Jong et al., 2013).

7. **Reproductive system toxicity**

The placenta, blood-testis, and epithelial barriers are examples of biological barriers that can shield the reproductive system from toxicity and infection. The testis, epididymis, ovary, and uterus are reproductive organs where AgNPs can deposit after overcoming biological obstacles. AgNPs may thereby harm germ cells and cells associated with them, including Sertoli cells, Leydig cells, primary and secondary follicles, and germline stem cells (Ong et al., 2016; Zhang et al., 2015). Additionally, AgNPs can alter sexual behavior by interfering with the hormone release from the glands and reproductive organs. Additional research verified that AgNPs cause reproductive toxicity by upping inflammation, altering DNA structure, suppressing gene expression, reducing mitochondrial function, generating ROS generation, and causing apoptosis.

1.8 Advantages and Disadvantages of Silver Nanoparticles in Drug Delivery

1. Enhanced Drug Solubility and Stability: Ravishankar and Jamuna (2011).

Advantages:

- Silver nanoparticles can improve the solubility of poorly soluble drugs, enhancing their bioavailability and therapeutic efficacy.
- They can stabilize labile drugs and prevent degradation, extending their shelf life.

Disadvantages:

- Silver nanoparticles' stability can be a concern, as they may undergo aggregation or oxidation, leading to changes in their physicochemical properties and potential loss of drug delivery functionality.
- The stability of drug-loaded silver nanoparticles during storage and transportation must be carefully addressed to maintain their therapeutic effectiveness.

2. Targeted Drug Delivery: Qureshi (2013)

Advantages:

- Functionalizing silver nanoparticles with ligands or antibodies can enable targeted drug delivery to specific cells or tissues.
- This targeted approach enhances the accumulation of drugs at the desired site, reducing off-target effects and improving treatment outcomes.
- The small size of nanoparticles facilitates their penetration into tissues and cells, enabling drug delivery to locations that may be otherwise inaccessible.

Disadvantages:

- Achieving precise targeting can be challenging, as it requires accurate identification of specific receptors or markers on the target cells.
- Surface modification of nanoparticles for targeting purposes may introduce additional complexities and costs to the drug delivery system.
- Unintended accumulation of silver nanoparticles in non-targeted tissues or organs could lead to potential toxicity concerns.

3. Controlled Drug Release: Li et al. (2018).

Advantages:

- Silver nanoparticles can be designed to encapsulate drugs and release them in a controlled manner.
- This controlled release mechanism allows for sustained drug concentrations over a prolonged period, reducing the frequency of administration and improving patient compliance.

Disadvantages:

- Achieving optimal control over drug release kinetics can be challenging due to variations in nanoparticle properties and drug loading techniques.
- Controlling release rates over extended periods may require complex nanoparticle designs or additional external stimuli, which can increase the complexity and cost of the drug delivery system.

4. Antimicrobial Properties: Lara et al. (2010).

Advantages:

- Silver nanoparticles possess inherent antimicrobial properties, which can be advantageous in drug delivery systems targeting infections.
- They can directly inhibit the growth of bacteria, fungi, and viruses, helping to prevent or treat associated infections.

Disadvantages:

- The antimicrobial properties of silver nanoparticles may also pose a risk of toxicity to normal cells and tissues.
- Extensive use of silver nanoparticles as antimicrobial agents can contribute to developing antimicrobial resistance, limiting their long-term effectiveness.

5. Biocompatibility and Toxicity: (Asharani et al., 2008)

Advantages:

- Silver nanoparticles can be engineered to enhance their biocompatibility and minimize toxicity.
- Proper surface modifications and coatings can reduce the potential adverse effects on cells and tissues.

Disadvantages:

- The toxicity of silver nanoparticles, especially at higher concentrations or with prolonged exposure, remains a concern.
- The precise mechanisms of nanoparticle toxicity and their long-term effects on human health are still under investigation.

6. Manufacturing and Cost: Iravani (2011).

Advantages:

- Silver nanoparticles can be produced using various scalable and cost-effective synthesis methods.
- Their relatively low cost and widespread availability make them attractive for drug delivery applications.

Disadvantages:

- Large-scale production of silver nanoparticles with consistent quality and reproducibility can be challenging.
- The overall manufacturing cost of Silver nanoparticle-based drug delivery systems, including surface modification and characterization, may impact their commercial viability.

1.9 Conclusion

Due to their distinctive physical, chemical, optical, electrical, and catalytic features, AgNPs have been investigated quickly and thoroughly over many years. These traits, particularly the size and form, are closely related to AgNPs' characteristics. Physical, chemical, and biological processes can all be used to create AgNPs with various properties. The synthesis process can use external energy sources such as light, heat, electricity, sound, and microwave. In order to create AgNPs with the predicted size and form, several criteria need to be taken into account. The importance of reaction parameters, such as reaction temperature, time, pH, and additional energy sources, should be understood in the production process in addition to the many types of precursor salts and additives like reducing agents, capping agents, and stabilizers. Among these techniques, biological synthesis employing bacteria, fungi, and plant extract has shown to be an easy, cost-effective, dependable, and ecologically benign process. The possible pathogens must be carefully examined when using a biological process instead of a physical or chemical one, which both demand high temperatures and poisonous or dangerous chemicals. To better understand how to produce nanoparticles with controlled size and shape, we examine the AgNPs synthesis techniques and weigh the benefits and drawbacks.

AgNPs offer a wide range of potential uses in medicine. Antimicrobial and anticancer qualities among them have drawn increased attention. The size, concentration/dose, exposure time, stabilizer, and surface charges of AgNPs affect their antibacterial and anticancer properties. The postulated mechanisms for AgNPs' antimicrobial activity include causing DNA damage, ROS generation, and cell wall destruction. AgNPs' anticancer mechanisms are more intricate. By damaging the cell's ultrastructure, causing the creation of ROS and DNA damage, inactivating proteins, and controlling numerous signaling pathways, AgNPs can cause cancer cells to undergo apoptosis and necrosis. Additionally, AgNPs may prevent cancer cell invasion and migration by stifling angiogenesis within the lesion.

However, AgNPs' potential cytotoxicity might restrict their medical uses. In order to increase AgNPs' compatibility, surface functionalization properly is a significant concern. Multiple ligands can be coordinated on the AgNPs' surface, allowing for functionalization. More antibacterial and anticancer medicines incorporating AgNPs can be developed since the surface functionalization of AgNPs can simultaneously increase their biological safety and complicate drug delivery. Additionally, AgNPs can be employed as an adjuvant or addition in vaccinations, dental materials, and bone scaffolds. AgNPs' antidiabetic activity is also investigated. The distinct optical features of AgNPs provide them tremendous therapeutic potential in biosensing and imaging and their excellent antibacterial and anticancer capabilities. Multiple ligands can be coordinated on the AgNPs' surface, allowing for functionalization. The potential toxicity of AgNPs in various systems, including the skin, eyes, kidneys, respiratory system, hepatobiliary system, and immunological and reproductive systems, has been examined, even though most studies concentrate on the therapeutic uses of AgNPs.

To create more secure and biocompatible AgNPs-based agents, additional in-depth research is required to assess the biocompatibility and potential cytotoxicity of AgNPs. In the pertinent portions of this review, we individually introduce the production process and anticancer characteristics of nanoscale silver particles. Comparatively, to the AgNPs stated in this review, we created pure, tiny, and fine silver particles. This extremely small size, where silver particles display broad-spectrum anticancer effects without overt cytotoxicity, may serve as a threshold for silver particles in medical applications. This fascinating finding motivates us to investigate more prospective uses of AgNPs in nanomedicine.

References

Abate, L., Bachheti, A., Bachheti, R. K., Husen, A., Pandey, D. P., & Megra, M. B. (2022). Coir fiber-based cellulose, nanocellulose, and their cutting-edge applications. In: M. Jawaid (Ed.), *Coir fiber and its composites* (pp. 309–325). Elsevier Inc.

Ahamed, M., Karns, M., Goodson, M., Rowe, J., Hussain, S. M., Schlager, J. J., & Hong, Y. (2008). DNA damage response to different surface chemistry of silver nanoparticles in mammalian cells. *Toxicology and Applied Pharmacology, 233*(3), 404–410.

Ahmed, A., Ahmed, E. A., Abu-Baker, M., Abdelatif, A. H., Ahmed, A. H., Abdel Rahman, S., & Tawfeek, H. (2022). Design and optimization of PEGylated silver nanoparticles for efficient delivery of doxorubicin to cancer cells. *Journal of Drug Delivery Science and Technology, 71*, 103347.

Anderson, D. S., Silva, R. M., Lee, D., Edwards, P. C., Sharmah, A., Guo, T., Pinkerton, K. E., & Van Winkle, L. S. (2015). Persistence of silver nanoparticles in the rat lung: Influence of dose, size, and chemical composition. *Nanotoxicology, 9*, 591–602.

Asgary, V., Shoari, A., Baghbani-Arani, F., Sadat Shandiz, S. A., Khosravy, M. S., Janani, A., Bigdeli, R., Bashar, R., & Cohan, R. A. (2016). Green synthesis and evaluation of silver nanoparticles as adjuvant in rabies veterinary vaccine. *International Journal of Nanomedicine, 11*, 3597–3605.

Asharani, P. V., Lianwu, Y., Gong, Z., Valiyaveettil, S., & Sing, Y. T. (2008). Toxicity of silver nanoparticles in zebrafish models. *Nanotechnology, 19*(25), 255102.

Azharuddin, M., Zhu, G. H., Das, D., Ozgur, E., Uzun, L., Turner, A. P. F., & Patra, H. K. (2019). A repertoire of biomedical applications of noble metal nanoparticles. *Chemical Communications, 55*, 6964–6996.

Bachheti, R. K., & Bachheti, A. (Eds.). (2023). *Secondary metabolites from medicinal plants: Nanoparticles synthesis and their applications*. CRC Press. https://doi.org/10.1201/9781003213727

Badmus, J. A., Oyemomi, S. A., Adedosu, O. T., Yekeen, T. A., Azeez, M. A., Adebayo, E. A., Lateef, A., Badeggi, U. M., Botha, S., Hussein, A. A., & Marnewick, J. L. (2020). Photo-assisted bio-fabrication of silver nanoparticles using Annona muricata leaf extract: Exploring the antioxidant, antidiabetic, antimicrobial, and cytotoxic activities. *Heliyon, 6*(11), e05413.

Bae, E., Park, H., Yoon, J., Kim, Y., Choi, K., & Yi, J. (2011). Bacterial uptake of silver nanoparticles in the presence of humic acid and $AgNO_3$. *Korean Journal of Chemical Engineering, 28*, 267–271.

Baker, C., Pradhan, A., Pakstis, L., Pochan, D. J., & Shah, S. I. (2005). Synthesis and antibacterial properties of silver nanoparticles. *Journal of Nanoscience and Nanotechnology, 5*, 244–249.

Bapat, R. A., Chaubal, T. V., Joshi, C. P., Bapat, P. R., Choudhury, H., Pandey, M., Gorain, B., & Kesharwani, P. (2018). An overview of the application of silver nanoparticles for biomaterials in dentistry. *Materials Science and Engineering, c: Materials for Biological Applications, 91*, 881–898.

Bondarenko, O., Ivask, A., & Kahru, A. (2013). Toxicity of Ag, CuO, and ZnO nanoparticles to selected environmentally relevant test organisms and mammalian cells in vitro: A critical review. *Archives of Toxicology, 87*(7), 1181–1200.

Chambers, B. A., Afrooz, A. R., Bae, S., Aich, N., Katz, L., Saleh, N. B., & Kirisits, M. J. (2013). Effects of chloride and ionic strength on physical morphology, dissolution, and bacterial toxicity of silver nanoparticles. *Environmental Science and Technology, 48*, 761–769.

Chen, B., Zhang, Y., Yang, Y., Chen, S., Xu, A., Wu, L., et al. (2018). Involvement of telomerase activity inhibition and telomere dysfunction in silver nanoparticles anticancer effects. *Nanomedicine, 13*, 2067–2082.

Cheng, L., Zhang, K., Weir, M. D., Wu, E. J., Xu, S. M., Zhou, X. D., & Xu, H. H. (2016). Antibacterial effect and mechanism of silver nanoparticles against Streptococcus mutans. *Nanomedicine, 12*(3), 601–611.

Das, R. N., Chandrashekhar, T. S., Joshi, H. S., Gurung, M., Shrestha, N., & Shivananda, P. G. (2006). Frequency and susceptibility profile of pathogens causing urinary tract infections at a tertiary care hospital in western Nepal. *Singapore Medical Journal, 47*, 281–285.

Davies, L. C., Jenkins, S. J., Allen, J. E., & Taylor, P. R. (2013). Tissue-resident macrophages. *Nature Immunology, 14*, 986–995.

Duran, N., Nakazato, G., & Seabra, A. B. (2016). Antimicrobial activity of biogenic silver nanoparticles, and silver chloride nanoparticles: An overview and comments. *Applied Microbiology and Biotechnology, 100*(15), 6555–6570.

Espinosa-Cristóbal, L. F., Holguín-Meráz, C., Zaragoza-Contreras, E. A., Martínez-Martínez, R. E., Donohue-Cornejo, A., Loyola-Rodríguez, J. P., Cuevas-González, J. C., Reyes-López, S. Y. (2019). Antimicrobial and substantivity properties of silver nanoparticles against oral microbiomes clinically isolated from young and young-adult patients. *Journal of Nanomaterials*, ID3205971.

Farah, M. A., Ali, M. A., Chen, S. M., Li, Y., Al-Hemaid, F. M., Abou-Tarboush, F. M., et al. (2016). Silver nanoparticles synthesized from Adenium obesum leaf extract induced DNA damage, apoptosis, and autophagy via the generation of reactive oxygen species. *Colloids and Surfaces b, Biointerfaces, 141*, 158–169.

Gao, Y., Li, Y., & Yue, T. (2018). Synthesis of silver nanoparticles and their applications. In *Handbook of ecomaterials* (pp. 1–18). Springer.

Gonfa, Y. H., Gelagle, A. A., Hailegnaw, B., Kabeto, S. A., Workeneh, G. A., Tessema, F. B., Tadesse, M. G., Wabaidur, S. M., Dahlous, K. A., Abou Fayssal, S., Kumar, P., Adelodun, B., Bachheti, A., & Bachheti, R. K. (2023). Optimization, characterization, and biological applications of silver nanoparticles synthesized using essential oil of aerial part of *Laggera tomentosa. Sustainability, 15*(1), 797.

Gonfa, Y. H., Tessema, F. B., Bachheti, A., Tadesse, M. G., Eid, E. M., Abou Fayssal, S., Adelodun, B., Choi, K. S., Širić, I., Kumar, P., & Bachheti, R. K. (2022). Essential oil composition of aerial part of *Pluchea ovalis* (Pers.) DC., silver nanoparticles synthesis, and larvicidal activities against fall armyworm. *Sustainability, 14*(23), 15785.

Greulich, C., Braun, D., Peetsch, A., Diendorf, J., Siebers, B., Epple, M., & Koller, M. (2012). The toxic effect of silver ions and silver nanoparticles on bacteria and human cells occurs in the same concentration range. *RSC Advances, 2*, 6981–6987.

Gunasekaran, T., Nigusse, T., & Dhanaraju, M. D. (2012). Silver nanoparticles as real topical bullets for wound healing. *Journal of the American College of Clinical Wound Specialists, 4*, 82–96.

Helmlinger, J., Sengstock, C., Gross-Heitfeld, C., Mayer, C., Schildhauer, T. A., Koller, M., & Epple, M. (2016). Silver nanoparticles with different sizes and shapes: Equal cytotoxicity, but different antibacterial effects. *RSC Advances, 6*, 18490–18501.

Husen, A. (2017). Gold nanoparticles from plant system: synthesis, characterization and their application. In M. Ghorbanpourn, K. Manika, A. Varma (Eds.), *Nanoscience and plant–soil systems* (Vol. 48, pp. 455–479). Springer International Publishing AG. https://doi.org/10.1007/978-3-319-46835-8_17

Husen. A., & Siddiqi, K. S. (2023). *Advances in smart nanomaterials and their applications.* Elsevier Inc. https://doi.org/10.1016/C2021-0-02202-1

Ibrahim, K., Khalid, S., & Irdees, K. (2019). Nanoparticles: Properties, applications, and toxicities. *Arabian Journal of Chemistry, 12*, 908–931.

Iravani, S. (2011). Green synthesis of metal nanoparticles using plants. *Green Chemistry., 13*(10), 2638–2650.

Islam, M. A., Jacob, M. V., & Antunes, E. (2021). A critical review on silver nanoparticles: From synthesis and applications to its mitigation through low-cost adsorption by biochar. *Journal of Environmental Management, 281*, 111918.

Jacob, J. M., John, M. S., & Jacob, A. (2019). Bactericidal coating of paper towels via sustainable biosynthesis of silver nanoparticles using Ocimum sanctum leaf extract. *Materials Research Express, 6*(4), 045401.

Jain, R. K., & Stylianopoulos, T. (2010). Delivering nanomedicine to solid tumors. *Nature Reviews Clinical Oncology, 7*(11), 653–664.

Jarbrink, K., Ni, G., Sönnergren, H., Schmidtchen, A., Pang, C., Bajpai, R., & Car, J. (2017). The humanistic and economic burden of chronic wounds: A protocol for a systematic review. *Systematic Reviews, 6*, 15.

Jayakumar, R., Prabaharan, M., Nair, S. V., Tokura, S., & Selvamurugan, N. (2010). Novel carboxymethyl derivatives of chitin and chitosan materials and their biomedical applications. *Progress in Materials Science., 55*(7), 675–709.

Jiang, Y., Xiao Ning, Z., LiJuan, P., Shu, Y., Li, M., LiYa, Z., & Jing, G. (2018). Silver nanoparticles modified two-dimensional transition metal carbides as nanocarriers to fabricate acetylcholinesterase-based electrochemical biosensors. *Chemical Engineering Journal, 339*, 547–556.

De Jong, W. H., Van Der Ven, L. T., Sleijffers, A., Park, M. V., Jansen, E. H., & Van Loveren, H. (2013). Systemic and immunotoxicity of silver nanoparticles in an intravenous 28 days repeated dose toxicity study in rats. *Biomaterials, 34*, 8333–8343.

Kaewamatawong, T., Kawamura, N., Okajima, M., Sawada, M., Morita, T., & Shimada, A. (2005). Acute pulmonary toxicity caused by exposure to colloidal silica: Particle size dependent pathological changes in mice. *Toxicolpathol, 33*, 745–751.

Kang, T. W., Hong, S. W., & Han, S. W. (2016). Advances in transparent conductive films based on metal nanowire networks. *Materials Today, 19*(3), 157–168.

Khan, M. (2018). Designing chitosan-coated silver nanoparticles for controlling tuberculosis. *International Journal of Biological Macromolecules., 111*, 1152–1164.

Khorrami, S., Zarrabi, A., Khaleghi, M., Danaei, M., & Mozafari, M. R. (2018). Selective cytotoxicity of green synthesized silver nanoparticles against the MCF-7 tumor cell line and their enhanced antioxidant and antimicrobial properties. *International Journal of Nanomedicine, 13*, 8013–8024.

Kim, T. H., & Kim, M. (2018). Silver nanoparticles in cancer therapy: Therapeutic efficacy and safety concerns. *Nanomaterials, 8*(12), 1004.

Kimling, J., Maier, M., Okenve, B., Kotaidis, V., Ballot, H., & Plech, A. (2006). Turkevich method for gold nanoparticle synthesis revisited. *The Journal of Physical Chemistry B, 110*(32), 15700–15707.

Kittler, S., Greulich, C., Diendorf, J., Köller, M., & Epple, M. (2010). The toxicity of silver nanoparticles increases during storage because of slow dissolution under the release of silver ions. *Chemistry of Materials, 22*, 4548–4554.

Kumar, C. G., & Sujitha, P. (2014). Green synthesis of Kocuran-functionalized silver glyconanoparticles for use as antibiofilm coatings on silicone urethral catheters. *Nanotechnology, 25*, 1–16.

Lansdown, A. B. (2006). Silver in health care: Antimicrobial effects and safety in use. *Current Problems in Dermatology, 33*, 17–34.

Lara, H. H., Ayala-Nunez, N. V., Ixtepan-Turrent, L., & Rodríguez-Padilla, C. (2010). Mode of antiviral action of silver nanoparticles against HIV-1. *Journal of Nanobiotechnology, 8*(1), 1–10.

Lara, H. H., Garza-Treviño, E. N., Ixtepan-Turrent, L., & Singh, D. K. (2011). Silver nanoparticles are broad-spectrum bactericidal and virucidal compounds. *Journal of Nanobiotechnology, 9*, 30.

Le, T. A., Nguyen, V. H., Nguyen, V. T., Nguyen, Q. H., Nguyen, V. C., & Dang, T. T. A. (2020). Silver nanoparticles with different shapes and sizes: Synthesis, characterization, catalytic activity and applications in environmental remediation. *Journal of Environmental Chemical Engineering, 8*(4), 104126.

Lee, Y., Chung, H. J., Yeo, S., Ahn, C. H., Lee, H., Messersmith, Phillip B. & Park, T. G. (2010). Thermo-sensitive injectable and tissue adhesive sol–gel transition hyaluronic acid/pluronic composite hydrogels prepared from bio-inspired catechol-thiol reaction. *Soft Matter, 6*, 977–983.

Leyu, A. M., Debebe, S. E., Bachheti, A., Rawat, Y. S., & Bachheti, R. K. (2023). Green Synthesis of gold and silver nanoparticles using invasive alien plant *Parthenium hysterophorus* and their antimicrobial and antioxidant activities. *Sustainability, 15*(12), 9456.

Li, J., Zou, L., Chen, Z., & Liang, X. J. (2018). Recent advances in controlled drug delivery using nanocarriers. *Nanoscale, 10*(8), 3983–4001.

Liao, C., Li, Y., & Tjong, S. C. (2019a). Bactericidal and cytotoxic properties of silver nanoparticles. *International Journal of Molecular Sciences, 20*(2), 449.

Ma, R., Levard, C., Marinakos, S. M., Cheng, Y., & Liu, J. (2012b). Size-controlled dissolution of organic-coated silver nanoparticles. *Environmental Science and Technology, 46*(2), 752–759.

Ma, R., Levard, C., Marinakos, S. M., Cheng, Y., Liu, J., Michel, F. M., Brown, G. E., & Lowry, G. V. (2012a). Size-controlled Dissolution of organic-coated silver nanoparticles. *Environmental Science and Technology, 46*, 752–759.

De Maglie, M., Cella, C., & Bianchessi, S. (2015). Dose and batch-dependent hepatobiliary toxicity of 10 nm silver nanoparticles after single intravenous administration in mice. *International Journal of Health Animal Science and Food Safety, 10*, 13130.

Mahmoudi, M., Serpooshan, V., & Laurent, S. (2011). Engineered nanoparticles for biomolecular imaging. *Nanoscale, 3*(8), 3007–3026.

Mallikarjuna, K., Narasimha, G., Dillip, G. R., Praveen, B., Varada Reddy, A., Sreedhar, B., & Raju, B. D. P. (2017). Citrate reduced silver nanoparticles: Synthesis, anticancer and antibacterial activity. *Journal of Molecular Liquids, 229*, 271–278.

Maneewattanapinyo, P., Banlunara, W., Thammacharoen, C., Ekgasit, S., & Kaewamatawong, T. (2011). An evaluation of acute toxicity of colloidal silver nanoparticles. *Journal of Veterinary Medical Science, 73*(11), 1417–1423.

Mathur, P., Jha, S., Ramteke, S., & Jain, N. K. (2018). Pharmaceutical aspects of silver nanoparticles. *Artificial Cells, Nanomedicine, and Biotechnology., 46*(S1), S115–S126.

Meikle, T. G., Dyett, B. P., Strachan, J. B., White, J., Drummond, C. J., & Conn, C. E. (2020). Preparation, characterization, and antimicrobial activity of ribosome encapsulated metal nanocrystals. *ACS Applied Materials and Interfaces, 12*(6), 6944–6954.

Mytych, J., Zebrowski, J., Lewinska, A., & Wnuk, M. (2017). Prolonged effects of silver nanoparticles on p53/p21 pathway-mediated proliferation, DNA damage response, and methylation parameters in HT22 hippocampal neuronal cells. *Molecular Neurobiology, 54*, 1285–1300.

Naskar, A., & Kim, K. S. (2020). Recent advances in nanomaterial-based wound-healing therapeutics. *Pharmaceutics, 12*(6), 499.

Noronha, V. T., Paula, A. J., Durán, G., Galembeck, A., Cogo-Muller, K., Franz-Montan, M., & Durán, N. (2017). Silver nanoparticles in dentistry. *Dental Materials, 33*(10), 1110–1126.

Nozari, A., Ajami, S., Rafiei, A., & Niazi, E. (2017). Impact of nano hydroxyapatite, nano silver fluoride and sodium fluoride varnish on primary teeth enamel remineralization: An in vitro study. *Journal of Clinical and Diagnostic Research, 11*, zc97–zc100.

Oei, J. D., Zhao, W. W., Chu, L., DeSilva, M. N., Ghimire, A., & Rawls, H. R. (2012). Antimicrobial acrylic materials with in situ generated silver nanoparticles. *Journal of Biomedical Materials Research Part b, Applied Biomaterials, 100*, 409–415.

Okamoto, S., Matsuura, M., Akagi, T., Akashi, M., Tanimoto, T., Ishikawa, T., Takahashi, M., Yamanishi, K., & Mori, Y. (2009). Poly (gamma-glutamic acid) nano-particles combined with mucosal influenza virus hemagglutinin vaccine protects against influenza virus infection in mice. *Vaccine, 27*(42), 5896–5905.

Ong, C., Lee, Q. Y., Cai, Y., Liu, X., Ding, J., & Yung, L.-Y.L. (2016). Silver nanoparticles disrupt germline stem cell maintenance in the Drosophila testis. *Science and Reports, 6*, 20632.

Paiva, L., Fidalgo, T. K. S., da Costa, L. P., Maia, L. C., Balan, L., Anselme, K., Ploux, L., & Thiré, R. M. (2018). Antibacterial properties and compressive strength of new one-step preparation silver nanoparticles in glass ionomer cement (NanoAg-GIC). *Journal of Dentistry, 69*, 102–109.

Pal, S., Tak, Y. K., & Song, J. M. (2007). Does the antibacterial activity of silver nanoparticles depend on the shape of the nanoparticle? A study of the Gram-negative bacterium Escherichia coli. *Applied and Environment Microbiology, 73*, 1712–1720.

Panzarini, E., Mariano, S., Vergallo, C., Carata, E., Fimia, G. M., Mura, F., Rossi, M., Vergaro, V., Ciccarella, G., Corazzari, M., & Dini, L. (2017). Glucose capped silver nanoparticles induce cell cycle arrest in HeLa cells. *Toxicology in Vitro, 41*, 64–74.

Patra, J. K., & Baek, K.-H. (2017). Antibacterial activity and synergistic antibacterial potential of biosynthesized silver nanoparticles against foodborne pathogenic bacteria along with its anticandidal and antioxidant effects. *Frontiers in Microbiology, 8*, 167.

Percival, S. L., Suleman, L., Vuotto, C., & Donelli, G. (2015). Healthcare-associated infections, medical devices and biofilms: Risk, tolerance and control. *Journal of Medical Microbiology, 64*, 323–334.

Peretyazhko, T. S., Zhang, Q., & Colvin, V. L. (2014). Size-controlled dissolution of silver nanoparticles at neutral and acidic pH conditions: Kinetics and size changes. *Environmental Science and Technology, 48*, 11954–11961.

Peters, A., Veronesi, B., Calderón-Garcidueñas, L., Gehr, P., Chen, L. C., & Geiser, M. (2006). Translocation and potential neurological effects of fine and ultrafine particles are a critical update. *Particle and Fibre Toxicology, 3*, 13.

Prasad, R. Y., Mc Ginnis, J. F., & Singh, R. (2012). Green synthesis of silver nanoparticles using cranberry powder aqueous extract and its antibacterial activity against Escherichia coli. *Journal of Environmental Science and Health, Part A Environmental Science, 47*(10), 1355–1360.

Pugazhendhi, A., Prabakar, D., Jacob, J. M., Karuppusamy, I., & Saratale, R. G. (2018). Synthesis and characterization of silver nanoparticles using Gelidiumamansii and its antimicrobial property against various pathogenic bacteria. *Microbial Pathogenesis, 114*, 41–45.

Qasim, M., Udomluck, N., Chang, J., Park, H., & Kim, K. (2018). Antimicrobial activity of silver nanoparticles encapsulated in poly-N-isopropylacrylamide-based polymeric nanoparticles. *International Journal of Nanomedicine, 13*, 235–249.

Qureshi, A. T. (2013). *Silver nanoparticles as drug delivery systems* (p. 1069). LSU Doctoral Dissertations.

Rai, M., Yadav, A., & Gade, A. (2009). Silver nanoparticles as a new generation of antimicrobials. *Biotechnology Advances, 27*(1), 76–83.

Ramkumar, V. S., Pugazhendhi, A., Gopalakrishnan, K., Sivagurunathan, P., Saratale, G. D., Dung, T. N. B., & Kannapiran, E. (2017). Biofabrication and characterization of silver nanoparticles using aqueous extract of seaweed Enteromorpha compressa and its biomedical properties. *Biotechnology Reports (Amst), 14*, 1–7.

Ravishankar, R. V., & Jamuna, B. A. (2011). Nanoparticles and their potential application as antimicrobials. Science against microbial pathogens: Communicating current research and technological advances. In: A. Méndez-Vilas (Ed.), Formatex, microbiology series (No. 3, Vol. 1, pp. 197–209).

Rodriguez-Leon, E., & Rendon-Romero, J. (2017). Silver nanoparticles: Optical properties and their influence in surface-enhanced Raman spectroscopy. In *Surface-enhanced raman scattering-active nanostructures and single-molecules detection* (pp. 3–21).

Sadauskas, E., Wallin, H., Stoltenberg, M., Vogel, U., Doering, P., & Larsen, A. (2007). Kupffer cells are central in the removal of nanoparticles from the organism. *Particle and Fibre Toxicology, 4*, 10.

Saha, S. K., Das, S., Chowdhury, P., & Saha, S. K. (2014). Biocompatibility of a sonochemically synthesized poly(N-isopropyl acrylamide)/silica nanocomposite. *RSC Advances, 4*, 14457.

Salleh, A., Naomi, R., Utami, N. D., Mohammad, A. W., Mahmoudi, E., Mustafa, N., & Fauzi, M. B. (2020). The potential of silver nanoparticles for antiviral and antibacterial applications: A mechanism of action. *Nanomaterials, 10*, 1566.

Sano, K., Ainai, A., Suzuki, T., & Hasegawa, H. (2017). The road to a more effective influenza vaccine: Up to date studies and future prospects. *Vaccine, 35*(40), 5388–5395.

Saratale, G. D., Saratale, R. G., Benelli, G., Kumar, G., Pugazhendhi, A., Kim, D. S., & Shin, H. S. (2017). The antidiabetic potential of silver nanoparticles synthesized with Argyreia nervosa leaf extracts high synergistic antibacterial activity with standard antibiotics against foodborne bacteria. *Journal of Cluster Science, 28*, 1709–1727.

Saravanan, M., Arokiyaraj, S., Lakshmi, T., & Pugazhendhi, A. (2018). Synthesis of silver nanoparticles from Phenerochaetechrysosporium (MTCC-787) and their antibacterial activity against human pathogenic bacteria. *Microbial Pathogenesis, 117*, 68–72.

Sengottaiyan, A., Aravinthan, A., Sudhakar, C., Selvam, K., Srinivasan, P., & Govarthanan, M., et al. (2016). Synthesis and characterization of Solanum nigrum-mediated silver nanoparticles and its protective effect on alloxan-induced diabetic rats. *Journal of Nanostructure in Chemistry, 6*, 41–48.

Shahiwala, A., & Misra, A. (2004). Studies in topical application of nanoparticles containing antihypertensive agent. *An Official Journal of the American Association of Pharmaceutical Scientists, 5*(1), e7.

Shanmuganathan, R., MubarakAli, D., Prabakar, D., Muthukumar, H., Thajuddin, N., Kumar, S. S., & Pugazhendhi, A. (2018). An enhancement of antimicrobial efficacy of biogenic and ceftriaxone-conjugated silver nanoparticles: A green approach. *Environmental Science and Pollution Research International, 25*(11), 10362–10370.

Shimada, A., Kawamura, N., Okajima, M., Kaewamatawong, T., Inoue, H., & Morita, T. (2006). Translocation pathway of the intratracheally instilled ultrafine particles from the lung into the blood circulation in the mouse. *Toxicologic Pathology, 34*, 949–957.

Shin, S.-H., Ye, M.-K., Kim, H.-S., & Kang, H.-S. (2007). The effects of nano-silver on the proliferation and cytokine expression by peripheral blood mononuclear cells. *International Immunopharmacology, 7*, 1813–1818.

Siddiqi, K. S., & Husen, A. (2017). Recent advances in plant-mediated engineered gold nanoparticles and their application in biological system. *Journal of Trace Elements in Medicine and Biology, 40*, 10–23. https://doi.org/10.1016/j.jtemb.2016.11.012

Siddiqi, K. S., & Husen, A. (2020). Current status of plant metabolite-based fabrication of copper/copper oxide nanoparticles and their applications: A review. *Biomaterials Research, 24*(11), 1–15. https://doi.org/10.1186/s40824-020-00188-1

Siddiqi, K. S., Husen, A., & Rao, R. A. K. (2018a). A review on biosynthesis of silver nanoparticles and their biocidal properties. *Journal of Nanobiotechnology, 16*(14), 1–28.

Siddiqi, K. S., Husen, A., Sohrab, S. S., & Osman, M. (2018b). Recent status of nanomaterials fabrication and their potential applications in neurological disease management. *Nanoscale Research Letters, 13*(231), 1–17.

Siddiqi, K. S., Rahman, A., & Tajuddin, H. A. (2018c). Properties of zinc oxide nanoparticles and their activity against microbes. *Nanoscale Research Letters, 13*(141), 1–13.

Silva, L. P., Silveira, A. P., Bonatto, C. C., Reis, I. G., & Milreu, P. V. (2017). Silver nanoparticles as antimicrobial agents: Past, present, and future. In *Nanostructures for antimicrobial therapy* (pp. 577–596). Elsevier.

Sim, W., Barnard, R. T., Blaskovich, M. A. T., & Ziora, Z. M. (2018). Antimicrobial Silver in medicinal and consumer applications: A patent review of the past decade (2007–2017). *Antibiotics, 26*, 93.

Singh, S. P., Bhargava, C., Dubey, V., Mishra, A., & Singh, Y. (2017b). Silver nanoparticles: Biomedical applications, toxicity, and safety issues. *International Journal of Research in Pharmaceutical Sciences, 4*, 1–10.

Singh, S., Young, A., & McNaught, C. E. (2017a). The physiology of wound healing. *Surface Science Reports, 35*, 473–477.

Smulski, C. R., & Eibel, H. (2018). BAFF and BAFF-receptor in B cell selection and survival. *Frontiers in Immunology, 9*, 2285.

Stewart, P. S., & Costerton, J. W. (2001). Antibiotic resistance of bacteria in biofilms. *Lancet, 358*, 135–138.

Tak, Y. K., Pal, S., Naoghare, P. K., Rangasamy, S., & Song, J. M. (2015). Shape-dependent skin penetration of silver nanoparticles: Does it really matter? *Science and Reports, 5*, 16908.

Talbird, S. E., Graham, J., Mauskopf, J., Masseria, C., & Krishnarajah, G. (2015). Impact of tetanus, diphtheria, and acellular pertussis (Tdap) vaccine use in wound management on health care costs and pertussis cases management on health care costs and pertussis cases. *Journal of Managed Care and Specialty Pharmacy, 21*, 88–99.

Tang, S., & Zheng, J. (2018). Antibacterial activity of silver nanoparticles: Structural effects. *Advanced Healthcare Materials, 7*(13), e1701503.

Tunney, M. M., Jones, D. S., & Gorman, S. P. (1999). Biofilm and biofilm-related encrustation of urinary tract devices. In *Methods in Enzymology* (Vol. 310, pp. 558–566). Elsevier.

Ullah, F., Malik, S., & Ahmed, J. (2009). Antibiotic susceptibility pattern and esbl prevalence in nosocomial Escherichia coli from urinary tract infections in Pakistan. *African Journal of Biotechnology, 8*, 3921–3926.

Venugopal, A., Muthuchamy, N., Tejani, H., Gopalan, A. I., Lee, K. P., Lee, H. J., & Kyung, H. M. (2017). Incorporation of silver nanoparticles on the surface of orthodontic micro implants to achieve antimicrobial properties. *The Korean Journal of Orthodontics, 47*, 3–10.

Vijayakumar, V., Samal, S. K., Mohanty, S., & Nayak, S. K. (2019). Recent advancements in biopolymer and metal nanoparticle-based materials in diabetic wound healing management. *International Journal of Biological Macromolecules, 122*, 137–148.

Wang, J., Ge, B., Li, Z., Guan, F., & Li, F. (2016). Structural analysis and immunoregulation activity comparison of five polysaccharides from Angelica sinensis. *Carbohydrate Polymers, 140*, 6–12.

Wright, A. K., Briles, D. E., Metzger, D. W., & Gordon, S. B. (2008). Prospects for the use of interleukin12 as a mucosal adjuvant for vaccination of humans to protect against pneumococcal respiratory infection. *Vaccine, 26*, 4893–4903.

Wu, Y., Zhou, Q., Li, H., Liu, W., Wang, T., & Jiang, G. (2010). Effects of silver nanoparticles on the development and histopathology biomarkers of Japanese medaka (Oryziaslatipes) using the partial-life test. *Aquatic Toxicology, 100*, 160–167.

Xiao, J., Liu, Y., Zu, Y., Tang, J., & Wang, S. (2013). Synthesis of silver nanoparticles in aqueous solutions by an electrodeposition method and their characterization. *Journal of Nanomaterials*, 596239.

Xu, X., Ray, P. C., & Yu, H. (2010). Silver nanoparticles and their potential environmental and human health impacts. *Journal of Environmental Science and Health Part C, 28*(3), 189–212.

Xu, L., Shao, A., Zhao, Y., Wang, Z., Zhang, C., & Sun, Y. (2015). Neurotoxicity of silver nanoparticles in rat brain after intragastric exposure. *Journal of Nanoscience and Nanotechnology, 15*, 4215–4223.

Xu, Y., Tang, H., Liu, J. H., Wang, H., & Liu, Y. (2013). Evaluation of the adjuvant effect of silver nanoparticles both in vitro and in vivo. *Toxicology Letters, 219*, 42–48.

Yamamoto, Y. S., Ishikawa, M., Ozaki, Y., & Itoh, T. (2014). Fundamental studies on enhancement and blinking mechanism of surface-enhanced Raman scattering (SERS) and basic applications of SERS biological sensing. *Frontiers of Physics, 9*, 31–46.

Yang, Y., Xie, J., & Lee, J. Y. (2015). Silver nanoparticles: Synthesis, characterization, properties, applications, and therapeutic approaches. *International Journal of Molecular Sciences, 16*(8), 20395–20423.

Yang, T., Yao, Q., Cao, F., Liu, Q., Liu, B., & Wang, X. H. (2016). Silver nanoparticles inhibit the function of hypoxia-inducible factor-1 and target genes: Insight into the cytotoxicity and antiangiogenesis. *International Journal of Nanomedicine, 11*, 6679–6692.

Yin, I. X., Yu, O. Y., Zhao, I. S., Mei, M. L., Li, Q. L., Tang, J., & Chu, C. H. (2019). Developing bio-compatible silver nanoparticles using epically catechin gallate for dental use. *Archives of Oral Biology, 102*, 106–112.

Zarco, M. F., Vess, T. J., & Ginsburg, G. S. (2012). The oral microbiome in health and disease and the potential impact on personalized dental medicine. *Oral Diseases, 18*, 109–120.

Zebeaman, M., Bachheti, R. K., Bachheti, A., Pandey, D. P., & Husen, A. (2023). Green and Cost-effective nanoparticles synthesis from some frequently used medicinal plants and their various applications. In *Secondary metabolites from medicinal plants* (pp. 287–304). CRC Press.

Zeng, F., Xu, D., Zhan, C., Liang, C., Zhao, W., Zhang, J., Feng, H., & Ma, X. (2018). Surfactant-free synthesis of graphene oxide coated silver nanoparticles for users biosensing and intracellular drug delivery. *ACS Applied Nano Materials, 1*, 2748–2753.

Zhang, X.-F., Choi, Y.-J., Han, J. W., Kim, E., Park, J. H., & Gurunathan, S. (2015). Differential nanoreprotoxicity of silver nanoparticles in male somatic cells and spermatogonial stem cells. *International Journal of Nanomedicine, 10*, 1335–1357.

Role of Gold Nanoparticles for Targeted Drug Delivery

Prakash Chandra Gupta, Nisha Sharma, Priyanka Mishra, Sweta Rai, and Tarun Verma

Abstract The possibility of using nanotechnology to diagnose or treat diseases is growing. New and improved nanomaterials are now being used in pharmaceutical and biomedical applications because of advancements in nanotechnology. Creating and targeting sure systems still requires work today, despite recent advances, to be therapeutically valuable. Due to their high thermal stability and low size-to-volume ratio, metal nanoparticles are widely used in biological sectors. Gold nanoparticles (AuNPs) are an appropriate choice for biomedical applications owing to ease of synthesis, stability, customization, low toxicity, and simplicity in identification. AuNPs can potentially be used in the early detection, diagnosis, and treatment of diseases for medical imaging, drug delivery, and cancer therapy. The synthesis of AuNPs has been carried out using various chemical techniques over the past few decades. However, more recently, emphasis has been drawn to more contemporary environmentally friendly green technologies. Numerous functionalizing moieties, such as ligands, medicinal substances, DNA, amino acids, proteins, peptides, and oligonucleotides, can be coupled to AuNPs. Recent research suggests that AuNPs can be valuable drug carriers since they penetrate blood arteries to reach the target and enter within organelles. The latest developments in various AuNP synthesis techniques are the main topic of this chapter. Additionally, AuNPs functionalization techniques and application mechanisms in transporting pharmaceuticals and macromolecules. The chapter aims to explain the relationship between these nanomaterials in the complex environment to reach the target site and how to formulate the efficient, targeted drug delivery for complicated settings and constantly observe the toxicity on the rationale for designing such delivery complexes. This chapter will provide information on the research, potential, and limitations of developing nanoparticles.

Keywords Gold nanoparticles · Drug delivery · Biological synthesis · Functionalization · TDDS

P. C. Gupta (✉) · N. Sharma · P. Mishra · S. Rai · T. Verma
School of Pharmaceutical Sciences, Chhatrapati Shahu Ji Maharaj University, Kalyanpur, Kanpur 208024, India
e-mail: prakashchandragupta@csjmu.ac.in

R. K. Bachheti et al. (eds.), *Metal and Metal-Oxide Based Nanomaterials*, Smart Nanomaterials Technology, https://doi.org/10.1007/978-981-99-7673-7_12

1 Introduction

Nanotechnology is rapidly rising as the most popular area of profound scientific investigation, drawing significant funding from investors around the globe to investigate and discover new uses. Targeted drug delivery system (TDDS)-based pharmaceutical medication design and development are among the most crucial fields. TDDS are created to precisely reach their intended sites and deliver the payload at the specific location for better curative benefits, enhancing the drug's bioavailability to the tissue. The physicochemical characteristics of nanoparticles have influenced the concept. TDDS attempts to strengthen the effectiveness of therapeutics, achieve regulated distribution, enhance medication localization, and lower the toxic effects of drugs (Kumar et al., 2013). In this respect, metallic nanoparticles provide a fresh perspective on achieving these goals in treating many diseases, and the scientific community has been increasingly interested in them due to their simplicity and ease of creation (Bachheti & Bachheti 2023; Mody et al., 2013). Due to their peculiar chemistry and numerous design factors, diseases, including cancer and eye conditions, share many commonalities with prospective nano-based therapeutic interventions (Alexis et al., 2008). The objective of TDDS is to create an approach capable of administering medications with high levels of specificity and effectiveness at rates precisely calibrated to the biological demand of the body (Cunliffe et al., 2005). The main goal is to create a system that safeguards the payload and raises the therapeutic index (Schroeder et al., 2012).

In this regard, AuNPs are attracting attention in specialty drugs. AuNPs are incredibly significant among the diverse variety of nanomaterials employed in anticancer therapy due to their unique capacity to instantly release payload in response to diverse stimulation, such as molecule binding or changes in ionic concentration. AuNPs may additionally be combined with targeting ligands to reach sub-cellular regions in a specific tissue (Dreaden et al., 2011).

Scientists have grown interested in the fabrication of AuNPs over the years due to their distinct physical and chemical properties, which have potential in various fields such as drug administration, nonlinear optical devices, biolabeling, and catalysis (Ealias & Saravanakumar, 2017). Targeting nanoparticles to deliver drugs to specific tissues has several advantages, including increased bioavailability and solubility, reduced toxicity, improved medication impact in the tissue, and protection from therapeutic agent deterioration. (Jurgons et al., 2006) (Fig. 1).

1.1 An Overview of AuNPs

Since ancient times, gold has been used as a medicine, mainly in India and China, where it was associated with fertility and prolonged life. It is still utilized in various ayurvedic treatments in India today. It was used in the Middle Ages and current times

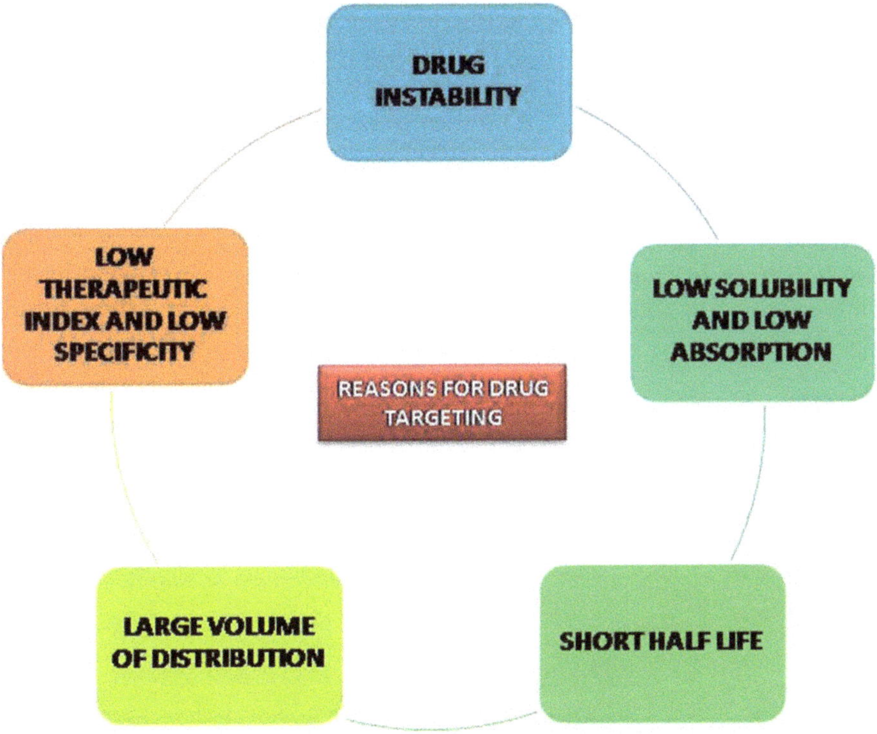

Fig. 1 Advantages of targeted drug delivery

for syphilis and neurological disorders. Gold has been used in the therapeutics of rheumatoid arthritis and tuberculosis throughout the twentieth century.

Francisci Antonii, a physician and philosopher, wrote a treatise in the seventeenth century that described the production of colloidal gold and provided insights into its therapeutic applications. He reported that AuNPs have curative characteristics. AuNPs consist of a gold core encased in an outer layer of organic ligands for protection. They are the safest and least hazardous metallic nanoparticles and exhibit remarkable optical, plasmonic, magnetic, and broad surface area (Leyu et al., 2023). They make the best vehicle for the delivery of medicinal compounds as of the simplicity of modification and the ability to be loaded with pharmacological therapies. In practice, multifunctional monolayer addition and controlled size particle creation have been achievable for a while now. Because of ligand place-exchange processes, these monolayers can also contain several medications or targeting agents, as well as minute molecules, peptides, proteins, antibodies, or oligonucleotides, because of ligand place-exchange processes (Yafout et al., 2021).

The biological application of metallic nanoparticles, particularly AuNPs, has attracted the most interest among the various nanomaterials because of their substantial advantages (Husen, 2017, 2022, 2023; Husen & Iqbal, 2019; Husen & Siddiqi,

2023; Husen et al., 2019; Siddiqi & Husen, 2016, 2017). Spherical, rod-like, cage-like, and other forms of AuNPs, as well as diameters ranging from 1 nm to more than 100 nm, are all easily synthesized. The shape and size of AuNPs significantly impact their optical and electrical characteristics. AuNPs have a negative charge, which makes it simple for various biomolecules, including medicines, to functionalize them. AuNPs are non-toxic and biocompatible. Surface plasmon resonance (SPR) bands are present in AuNPs and have an ultra-small, macroscopic quantum tunneling effect, and a special surface effect. Due to all these unique characteristics, AuNPs are trending as the utmost promising material for biomedical relevance, namely drug delivery, molecular imaging, and biosensing (Shah et al., 2014a, 2014b) (Fig. 2).

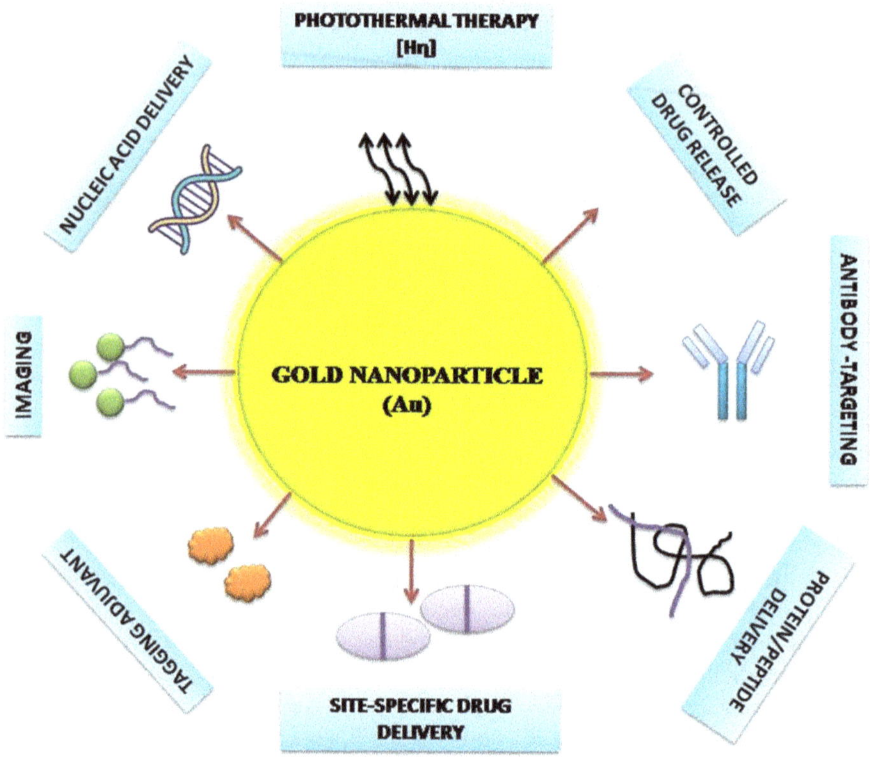

Fig. 2 Biomedical applications of gold nanoparticles

2 Synthesis of Gold Nanoparticles

The "bottom-up" and "top-down" approaches are two categories into which schemes for developing various AuNPs can be classed. The bottom-up approach uses thermal reduction, chemical, photochemical, electrochemical, sonochemical, and nanosphere lithography tools. In this process, atoms are assembled into the intended nanostructures (produced by ion reduction). The matter must be removed from the bulk material to obtain a particular nanostructure using top-down methods like photolithography and electron beam lithography. AuNPs can be produced by either of the approaches with the required size and shape. However, each has disadvantages, such as top-down approaches' significant material waste and bottom-up approaches' poor monodispersity. The following is a discussion of some of the more often utilized "bottom-up" methods for creating AuNPs (Shah et al., 2014a, 2014b).

2.1 *Turkevich Method*

The Turkevich approach was first 1stproposed in 1951 and has since become one of the most widely adopted processes to develop AuNPs in the 10 nm to 20 nm size range with a spherical shape. The conversion of gold ions (Au^{3+}) into gold atoms (Au^0) is the core concept of this technique. This transition happens when reducing substances like citrate, amino acids, ascorbic acid, or UV light are present. By utilizing a variety of capping and stabilizing chemicals, the size of the AuNPs can be substantially stabilized. Initially, the Turkevich method was restricted due to the limited array of AuNPs that could be produced using this method. Researchers have been able to manufacture particles of a wider range of sizes using this technology because of many improvements made to the technique since it was first developed. In 1973, Frens discovered that obtaining AuNPs of a certain extent, between 16 and 147 nm, can be accomplished by adjusting the proportion of reducing agents to stabilizing agents in the reaction. Later on, a greater understanding of the roles played by pH, temperature, and sodium citrate concentration made it possible to develop a model for forming particles (Sivaraman et al., 2011).

2.2 *Brust Method*

In 1994, the Brust technique was first explained. Using organic solvents and changing the thiol-to-gold ratio, this approach is a two-phase procedure that produces AuNPs between the size 1.5 and 5.2 nm (Burst et al., 1994). Faraday's two-phase system turned out to be an inspiration for the Brust approach. To complete the procedure, gold salt must be moved from an aqueous solution to an organic solvent such as toluene with the assistance of a phase transfer agent. Tetra octyl ammonium bromide is one example of such an agent. With the aid of an alkanethiol, sodium borohydride is subsequently added to reduce the gold. The reaction becomes orange to brown due to alkanethiols stabilizing AuNPs. Schriffin reported the purification of AuNPs stabilized with dodecanethiol from Tetra octyl ammonium bromide (Waters et al., 2003).

2.3 Seeded Growth Method

Although spherical AuNPs can be produced using the Turkevich and Brust techniques, AuNPs also exist in many nanostructures, including rods, cubes, and tubes. The method most frequently chosen to produce AuNPs in different shapes is seed-mediated. The main idea behind this process is to initially create seed particles by reducing gold salts with a potent reducing agent, such as sodium borohydride. The seed particles are then introduced to a metal salt solution with a mild reducing agent (ascorbic acid) and a structure-directing agent to stop additional nucleation and speed up the production of AuNPs in an anisotropic state. One can change the geometry of gold nanostructures by adjusting the concentration of seeds, reducing agents, and structure-directing agents (Jana et al., 2001).

2.4 Miscellaneous Methods

By applying excessive ligands (digestive ripening agents), digestive ripening is a practical way to create monodisperse gold nanoparticles from polydisperse nanoparticles. Employing high temperatures (around 138 °C) for 2 min, subsequently 110 °C for 5 h while alkanethiols are present, the colloidal suspension is heated. The temperature has a significant impact on the size of gold colloids. Amines, Silanes, Phosphines, many other ligands, and Thiols are used for gastric ripening (Prasad et al., 2002). In addition, various alternative techniques have been investigated for producing AuNPs, including those employing ultrasonic waves (Lee et al., 2012), microwaves (Kundu et al., 2008), laser ablation, the solvothermal approach, electrochemical reduction, and photochemical reduction (Ma et al., 2004).

3 Biofunctionalization of Gold Nanoparticles

Amongst the ultimate attributes of AuNPs in the biomedical field, surface biofunctionalization may be altered with various biomolecules, including antibodies, peptides, and DNA. Functionalization interactions are of two types, Covalent, and noncovalent interactions. van der Walls forces, electrostatic interactions, and hydrophobic trapping are responsible for noncovalent modifications (Han et al., 2006a, 2006b).

Particularly in biological research, the coupling is not sufficiently strong to form permanent surfaces that accept the required incubation conditions and cleaning processes. It is crucial to consider the effects of the ionic strength and pH of the surrounding media. Conversely, the covalent modifications provide excellent reproducibility and stability because they involve linker molecules or direct chemical attachment. High salt concentrations are tolerated through covalent contact, which is also very stable in heat conditions. Covalent interaction, on the other hand, is more challenging and occasionally requires intensive ligand preparation. This surface modification with noncovalent and covalent alterations can be used precisely for biological targeting, biosensing, and bio-diagnostics (Joshi et al., 2012).

AuNP's ability to transport drugs can be improved in two ways: lengthening their plasma half-life and causing accumulation within cells and enhancing cellular uptake and drug release inside desired cells. When AuNPs are conjugated with polyethylene glycol, macrophage and reticuloendothelial system clearance of the AuNPs is reduced.

To boost the uniqueness of their role as drug carriers in cells and increase their attraction to biological molecules, strategies have been developed for synthesizing AuNPs with functional moieties. Single or a blend of the following groups, including oligo or polypeptides, polyethylene glycol, oligonucleotides, bovine serum albumin, antisense or sense RNA, surface receptors of cells, antibodies, and else comparable moieties, is the basis of frequently utilized functionalization approaches (Kalimuthu et al., 2018).

3.1 PEGylation

AuNPs are frequently functionalized by PEGylation. AuNPs are functionalized with polyethylene glycol or biotin, peptides, or oligonucleotides to help target cells internalise them. Functionalized AuNPs (fAuNPs) can carry drugs because they attach to the cell membrane. Synthetic biomolecule-functionalized PEGylated AuNPs include lectin, lactose, and biotin. PEGylated AuNPs are essential in the cellular and intracellular targeting of biological molecules. Synthesis of Hetero-bifunctional PEGylated AuNPs with a thiol group and coumarin, a fluorescent dye, was done. Functionalized AuNPs (fAuNPs) with dye could enter cells and be tracked. The stability and functional virtue of PEGylated AuNPs are impacted by PEG molecular weight, functional groups, ligands, and AuNPs size (Scott et al., 2017).

Thioctic acid-anchored PEGylated AuNPs were tested in mice for tumor ablation. Fauna's internalization depended on nanoparticle size, PEG molecular weight, and PEGylation ligands. Physicochemical variables also affected AuNP's distribution into cells. AuNPs with functional moieties are synthesized to enhance their affinity to biological molecules and use them as cell-specific drug carriers. AuNPs can transport drugs by boosting cellular uptake and drug release inside specific cells or prolonging their plasma half-life and cell accumulation. Polyethylene glycol conjugation with AuNPs reduces macrophage and reticuloendothelial system clearance (Zhang et al., 2009).

3.2 Peptide/Amino Acid Conjugation

The specificity and efficiency of nanoparticle-based delivery systems can be enhanced by functionalaizing nanoparticles with amino acids and peptides, which has proven an efficient method in recent years (Zareet al., 2014). AuNPs functionalized with amino acids, viz. lysine, polylysine, and glycine, can attach DNA more effectively for gene delivery while avoiding toxicity. Primary ammonium groups of these amino acids share a more remarkable ability to bind to the cationic groups on DNA. In addition, it was discovered that lysine dendrons are superior to polylysine when it comes to expressing the reporter galactosidase gene (Satnami et al., 2015). Similarly, it was demonstrated that using amine-functionalized AuNPs bearing siRNA-PEG conjugates against human prostatic carcinoma cells helped suppress certain cancer genes (Lee et al., 2008). Due to the carboxyl group of amino acids facilitating the adhesion of proteins via their amine groups, it was found that carboxylated AuNPs that were synthesized employing glutamic acid were superior at synthesizing fAuNPs with proteins (Ghosh et al., 2008a, 2008b). This was discovered when researchers compared the two methods of synthesizing faunas. However, after the protein was attached to the AuNPs, there were observed to be conformational changes in the protein.

Peptide-functionalized AuNPs have been shown to stimulate macrophages, which suggests that they have the potential to be used as an adjuvant in the administration of vaccines. This is achievable since they can bind to different proteins and show that the immune system recognizes smaller molecules that the macrophages would not likely recognize differently. It was discovered that AuNPs that had been functionalized with an amyloid growth inhibitory peptide and were linked to Alzheimer's disease were beneficial for the transport of drugs inside cells. They can preferentially target the beta-amyloid fibers and the sweet arrow peptide, proteins that bone marrow-derived macrophages may recognize.

In addition, peptide-conjugated AuNPs are being utilized in developing a protein kinase test that uses imaging with secondary ions obtained from mass spectrometry. Compared to more traditional approaches, which use radioactive or fluorescent markers, this method, which assesses the variation in mass within the peptide substrate before and after kinase activation, is significantly less complicated. Similarly, bioconjugated gold nanorods have been used as imaging probes. A mouse monoclonal antibody was distinct to human epidermal growth factor receptor 2, over-expressed in SKBR3 breast cancer cells. It was coupled to AuNPs or nanorods and can be employed for biomedical imaging of carcinoma cells. Because Bombesin has a more specificity for gastrin-releasing peptides, which are over-expressed in cancer cells, AuNPs functionalized with Bombesin peptides can image cancer cells. Cancer cells over-express these peptides (Sun et al., 2008).

3.3 *Oligonucleotide Functionalized Nanoparticles*

Several distinctive scientists have developed strategies to functionalize nanoparticles made of gold and other materials by employing oligonucleotides on their own or with certain alterations. It is possible to synthesize DNA-conjugated nanostructures in a controlled manner. This can be accomplished either by attaching a predetermined number of single-stranded DNA molecules to the surface of the AuNPs using thiol caps or by saturating the exterior of the AuNPs with single-stranded DNA molecules. In yet another innovative piece of research, DNA-functionalized AuNPs were utilized to develop a chip-based DNA bio-bar code sensor to locate specific sequences of target DNA (Rosi et al., 2006).

4 Various Types of Gold Nanoparticles

4.1 *Au Nanospheres*

The term “Au nanospheres” is also used to describe the Au colloid. The diameters could be between 2 and 100 nm, and they could be prepared by decreasing the solution of aqueous HauCl while adding diverse reducing agents and doing so with various variables and circumstances. The reducing agent, where citrate was most frequently utilized, created monodisperse Au nanospheres (Hiramatsu & Osterloh, 2004). Nanosphere production would increase if citrate were used less. Altering the ratio of citrate and Au could likewise be used to regulate nanosphere size. The preceding technique’s main drawbacks were its diminished yield of Au nanospheres and its limited ability to use water as a solvent. To create air and thermally resistant Au nanospheres with decreased dispersity, Faraday employed a dual-phase technique in 1857, using tetraethyl ammonium bromide as the phase transfer reagent. If the reactants were introduced to the cooled solution rapidly, monodispersed Au nanospheres would result.

As a result, various strategies for synthesizing nanospheres were tested utilizing different reducing agents or ligands. Dendrimers could stabilize or act as templates for fabricating Au nanospheres. Regulating the synthesis parameters, like reactant, HAuCl, and blocked copolymer concentrations, would affect the size and form of Au nanospheres with absorption peaks at 510 and 550 nm. The absorption peak shifts to a longer wavelength when the particle size increases. Interestingly, various researchers have attempted to generate gold nanospheres in human cells (Anshup et al., 2005).

4.2 Au Nanorods

Several techniques have been researched for the development of Au nanorods. The template approach was used to synthesize Au nanorods, which relies on the electrochemical accumulation of Au within the pores of nanoporous polycarbonate or alumina template membranes. The size of the holes in the template membrane may serve as a predetermined guide for the diameter of the Au nanorod. The quantity of accumulated Au inside the membrane's pores may be used to regulate the length of an Au nanorod. The limitation of the process is that it yields limited Au nanorods, as only one layer of nanorods can be manufactured. The synthesis of Au nanorods using a synthetic electrochemical approach was also described; this approach allowed for the measurement of nanorod length, which affected the aspect ratio of longer diameter to shorter diameter (Huang et al., 2009).

The "Seed-mediated synthesis" is the most accepted and widely used approach to creating Au nanorods because it produces materials with more excellent aspect ratios than other techniques. To reduce the Au chloride, Au seed solution is typically produced by a potent reducing agent, such as $NaBH_4$. These seeds would serve as the location where nanorods would form and continue to grow in the presence of an $AuCl_2$ solution, a mild reducing agent, such as ascorbic acid, and hexadecyl trimethyl ammonium bromide. The Au nanorods' dimension could be altered by manipulating the Au seed solution concerning the Au precursor. Additionally, if $AgNO_3$ were included in the mixture, the yield of nanorods would rise. In addition to the seed-mediated process, several other methods and approaches have been documented. These include bio-reduction, the growth of gold nanorods on the surface of mica, and synthesis through a photochemical method (Nikoobakht & El-Sayed, 2003).

4.3 Au Nanoshells

A spherical nanoparticle known as a nanoshell has a thin metallic shell, often made of Au, and a dielectric center. These nanoshells used a plasmon, a quasi-particle created by collective excitation or quantum plasma oscillation, in which all ions and electrons may vibrate concurrently. Plasmon hybridization, which was related to simultaneous oscillation and produced higher or lower energy levels, was another name for this process. While the more significant energy levels could only combine feebly with the incident light, the lower energy level would strongly combine. Since thin shell layers would have stronger plasmon hybridization interactions, the wavelength of the light emitted could be controlled by coupling the thickness of the shell and the overall particle radius (Murphy et al., 2008). The nanoshells' major applications were in biomedical optical imaging, fluorescence intensification of feeble molecular emitters, therapeutic uses, surface amplification Raman spectroscopy, and surface amplification infrared absorption spectroscopy because of their highly reflective optical and chemical properties. Optical imaging may conclude the sample's structure, texture, anatomical makeup, and chemical composition as revealed by the deviation of light discharged from laser or infrared sources. All bimolecular absorbance levels may be at their lowest in the 700–900 nm near-infrared range, creating a clear window for optical imaging. Altering the makeup and dimensions of layers, which could be made via Surface Plasmon Resonance (SPR) with peaks in the visible and near-infrared spectrum, is another way to create Au nanoshells. The SPR peak of an Au nanoshell might be modified for a certain composition by adjusting the core size ratio to its shell thickness. Au nanoshells could be created with SPR peaks in the NIR region by covering them with silica or polymer beads. By employing the Stober technique to reduce Tetra-ethyl Ortho Silicate in Ethanol, silica cores were grown. Using the seed-mediated technique, Au solution was coated onto silica NPs.

Other methods demonstrated the attachment process of tiny Au nanospheres with a 2–4 nm diameter to the silica core. Amino-terminated silane was utilized as a linear molecule as long as the seed particle amalgamated and was incorporated into a single layer of the shell via decreasing excess Au (Brinson et al., 2008). The size of the Au nanoshell would depend on the silica core's diameter. The amount of deposited Au on the core surface may be used to influence the shell thickness. They were creating Au nanoparticles in situ using thermosensitive core–shell particles as a template, making it possible to create gold nanoshells. Using microgel as a core could lessen particle aggregation and aid in regulating the thickness of Au nanoshells produced by plating. Cores with smaller diameters than silica, roughly 80 nm, and minute size dispersion were created (Loo et al., 2005).

4.4 Au Nanocages

The galvanic replacement reaction of truncated silver nanocubes and aqueous HAuCl was used in 2006 to create gold nanocages, which consisted of controlled pores on the surface (Sun & Xia, 2002). In addition, it was seen that the reduction of pyrol could regulate the created morphologies of the silver nanostructures. In this instance, ethylene glycol was responsible for reducing AgNO, generating silver atoms. Further reduction resulted in the generation of nanocrystals or seeds (Chen et al., 2006a, 2006b). Adding additional silver atoms and the simultaneous management of the silver seed crystalline structures by adding polyvinylpyrrolidone resulted in the production of the desired nanostructures. This binding of polyvinylpyrrolidone to the surface presented the possibility of selective binding. Through the use of galvanic replacement, the silver nanostructures have the potential to be transformed into an internal hollow space within the gold nanostructures. This transformation can take place if the silver nanostructures are used as a sacrificial template. It was possible to precisely manage the size of the produced gold nanocages by modifying the molar ratio of silver to HAuCl. This allowed for the nanocages' wall thickness to be altered as well (Chenet al., 2005). The Au nanocages may offer significant benefits, including the following:

(i) By manipulating the proportion of Ag nanocubes and $HauCl_4$, the surface Plasmon resonance peaks of the Au nanocages may be modified, covering the spectral range of 500–1200 nm.
(ii) By adjusting the quantity of truncated angles and void diameters, Au nanocagees' absorption coefficient may be changed.
(iii) Despite their minute size (<50 nm), Au nanocages may exhibit resonance peaks in the near-infrared region.
(iv) With the surface modifications, Au nanocages can be applied to a variety of biomedical applications.

5 Targeted AuNPs for Drug Delivery

One of the most promising and rapidly developing applications for AuNPs in medicine is the targeted administration of medications. Antibiotic and antitumor drugs are the most often used target delivery agents (Elbayoumi, 2010). It was suggested that AuNPs could be ptrahemcytabin, 6 mercaptopurine, dodecyl-cysteine, sulfonamide, 5-fluorouracil, platinum complexes, tamoxifen, Herceptin, doxorubicin, prospidin, etc. Patra et al. (2008), Paciotti et al. (2004).

The fusion process was brought out via the immediate physical adsorption of the drugs over AuNPs or using alkanethiol linkers. Both (mostly) in vitro models utilizing tumor cell cultures and in vivo in mice with generated tumors of various types and sizes (Lewis lung carcinoma, pancreatic adenocarcinoma) were used to evaluate the effects of conjugates. The delivery system was created using the active ingredient as well as target molecules (such as cetuximab), which allow for more excellent complex anchoring and penetration into the target cells.

The use of multifunctional delivery systems was also suggested. In these systems, a gold nanoparticle is loaded with various therapeutic drugs (both hydrophilic and hydrophobic) and auxiliary agents, such as target molecules and dyes, for photo-dynamic therapy (Burygin et al., 2009). The majority of researchers emphasize the effectiveness of anticancer drugs combined with AuNPs. AuNPs are also considered items that can deliver antibiotics and other antibacterial agents. It has been shown that it is possible to create a stable combination of vancomycin with colloidal gold and that this complex is effective against a variety of enteropathogenic bacteria of *Escherichia coli, Enterococcus faecium*, and *Enterococcus faecalis*, including strains that are vancomycin-resistant. Similar outcomes were attained in a ciprofloxacin complex with gold nanoshells, which exhibited strong antibacterial action against *E. coli. Micrococcus luteus, Staphylococcus aureus, Pseudomonas aeruginosa, Aspergillus fumigatus*, and *A. niger* are significantly affected by the anti-leukemia medicine 5-fluorouracil when it is conjugated with colloidal gold. It should be highlighted that the optical spectra of the conjugates provided evidence that the drug complexes with AuNPs in all of the situations mentioned were stable. Contrarily, it was impossible to create stable complexes with gold nanoparticles for antibiotics that are effective against *E. coli, M. luteus, S. aureus*, and *P. aeruginosa*, such as ampicillin, streptomycin, kanamycin, gentamycin, neomycin, ciprofloxacin, gatifloxacin, and norfloxacin. However, depending on the antibiotic, their activity when combined with colloidal gold was higher by 12–40% than that of the antibiotic when taken alone (Rai et al., 2010).

These facts led to the conclusion that AuNPs increased the antibacterial activity of antibiotics. However, the processes underlying any potential enhancement of a drug's antibacterial action remain a mystery. Research has shown that in experiments on dense and liquid nutrient media, free gentamicin and a combination of it with AuNPs do not significantly vary in antibacterial activity. It is hypothesized that to increase the antibacterial activity, stable conjugates of nanoparticles coated with antibiotic molecules are needed (Beik et al., 2019). Thus, it was suggested to synthesize AuNPs using the antibiotic cefaclor directly. A stable conjugate was produced as a result. It was distinguished by its antibacterial activity against *S. aureus* and *E. coli*. Less information is available on other medications that have been conjugated with AuNPs. However, it is important to take note of the tocopherol complex with AuNPs' excellent antioxidant activity and the several prospective applications that have been suggested (Shao et al., 2013).

According to published data, AuNPs conjugated with the medication TAK799 displayed more significant efficacy against the human immunodeficiency virus than the drug alone because of the high local concentration. On rat models of diabetes mellitus, the process of administering insulin conjugated with colloidal gold orally and intravenously was developed. Reliable demonstration of a drop in blood sugar levels is comparable to that caused by the injection of insulin subcutaneously (Nam et al., 2013). The antirheumatic medication etanercept conjugated with gold nanorods has finally been characterized for its therapeutic impact. Including gene therapy in this section is important because it is the best course of action for inherited and acquired disorders. A strategy based on introducing genetic structures into cells and the organism for therapeutic reasons is implied by the term "gene therapy." Either the inserted gene was expressed, or the damaged or overexpressed gene's function was wholly or partially suppressed to produce the desired outcome. Recently, efforts were made to modify the structure and operation of the defective (affected) gene. In this situation, the delivery of the genetic material into the cytoplasm and cell nucleus can be effectively aided by using AuNPs (Putnam, 2006).

6 Targeted AuNPs for Gene Delivery

A promising approach for treating inherited and acquired disorders brought on by aberrant gene expression in gene delivery. As the delivery of genetic materials on their own suffers from unforeseen deterioration in the physiological environment, it combines the delivery of foreign genetic material to target cells utilizing particular vectors. Both viral and nonviral vectors are currently used in research and therapeutic applications. Nonviral vectors have been employed extensively in various gene delivery applications despite their low efficacy due to their adaptable and simple chemistry, cost-effectiveness, and improved safety profiles. Nonviral vectors, such as polymeric systems (such as dendrimers, micelles, and nanoparticles), liposomes, ceramic particles, carbon nanotubes, and metal nanoparticles (such as nanorods and nanoparticles), have been widely used as carrier systems (Morille et al., 2008; Packet al., 2005; Putnam, 2006).

Gold nanoparticles (AuNPs) are the top metal nanoparticles for gene delivery applications and other biomedical applications, such as diagnostic and therapeutic delivery vehicles, among the extensive range of carrier systems. Because they are stable, homogeneous, and biocompatible metal nanoparticles with distinctive electronic structures, size-related intensity display, and highly controllable electronic, magnetic, and optoelectronic capabilities, AuNPs have recently been used in these applications. Unlike polymeric nanoparticles, which reflect an electronic band recognized as "surface plasmon resonance (SPR)," the optoelectronic properties of AuNPs are greatly influenced by the particle's size, shape, interparticle distance, and protective. Additionally, a small change in the nanoparticles' particle size can significantly alter their characteristics, which can be detected through SPR (Li et al., 2019; Rietwyk & Peer, 2017).

Recent research suggests that double-stranded DNA (dsDNA), single-stranded DNA (ssDNA), and single-stranded RNA (ssRNA) could have exceptional ameliorative benefits since they can deliver all types of oligonucleotides, including plasmids. AuNPs are promising options for delivering DNA and RNA. Nucleic acids are shielded from nuclease degradation by AuNPs. They can aid in targeting and nucleic acid transfection of cells. Currently, AuNPs, namely nanospheres and nanorods, are employed for this objective. AuNP conjugates with oligonucleotide (and short interfering RNA) modifications have particular characteristics such that they could act as intracellular gene regulators. In the human body's peripheral blood mononuclear cells, immune-related genes and pathways can be activated by firmly functionalized gold nanoparticles (AuNPs), but not a long-lasting, lineage-restricted cell line. The use of oligonucleotide-modified AuNPs conjugates in translational research, the development of therapies, and the advancement of gene delivery technologies will all greatly benefit from these discoveries (Kim et al., 2012).

A review of Au NPs that had been electrostatically integrated into plasmid DNA and functionalized with cationic quaternary ammonium groups was conducted as early as 2001. The outcomes showed that the composite particle could control DNA transcription of T7 RNA polymerase and shield DNA from enzymatic destruction (Han et al., 2006a, 2006b; McIntosh et al., 2001). To silence genes, Mirkin et al. coated citrate-stabilized spherical Au NPs with a thick layer of ssDNA molecules functionalized with either one or more thiol groups (Rosi et al., 2006). The structures of this compound could withstand protease degradation and a large amount of endogenous glutathione, and it was efficiently internalized by cells. The efficiency of the medicines in gene silencing applications was improved by the better integration of complementary DNA with the grafted ssDNA. Au NRs can also deliver siRNA to specific cells or tissues because of their size- and shape-dependent optoelectronic characteristics. Recently, Prasad's team coupled AuNRs and cetyl trimethyl ammonium bromide (CTAB) to siRNA (against DARPP-32 gene in dopaminergic neuronal (DAN) cells), and they investigated the conjugates' uptake into the DAN cells (Bonoiuet al., 2009). Dark-field imaging and confocal microscopy were used to demonstrate that the Au NR-siRNA conjugates successfully delivered 98% of DAN cells siRNA. The study's findings have confirmed that AuNPs can be utilized as cutting-edge carriers to introduce genes into neuronal cells. To stimulate the development of human mesenchymal stem cells (MSCs), Zhao et al. created Au NR-based nanocarriers with poly-sodium 4-styrene sulfonate (PSS) and poly-allylamine hydrochloride (PAH) that can carry small interfering RNA (siRNA) contrary to LSD1 (Zhao et al., 2015).

The findings of their investigation demonstrated that Au NR-PSS-PAH-siRNA (Au NRssiRNA) nanocarriers were successfully internalized, preventing LSD1 from manifesting and triggering the in vitro differentiation of human MSCs into a hepatocyte lineage in response to hepatocyte growth factor (HGF). Their scientific discoveries might aid in creating more practical nano-platforms for the delivery of siRNA for tissue regeneration therapy. The findings of their investigation demonstrated that Au NR-PSS-PAH-siRNA (Au NRssiRNA) nanocarriers were successfully internalized, preventing LSD1 from manifesting and triggering the in vitro differentiation of human MSCs into a hepatocyte lineage in response to hepatocyte growth factor (HGF). Their findings could aid in creating more practical nanoplatforms for the delivery of siRNA for tissue regeneration therapy.

7 Targeted AuNPs for Protein Delivery

The interaction between proteins and gold nanoparticles (AuNP) has been steadily increasing in several possible biomedical applications. Physiological and therapeutic responses have been severely impacted by unpredictable deleterious protein effects, despite improvements in adjustable size, shape, and great biocompatibility. The development of clinical uses for AuNP, such as imaging agents or drug administration systems based on AuNP, was constrained by the intricacy and unpredictability of AuNP generation. As a result, extensive efforts have been made to characterize AuNP in vitro and extrapolate these results to in vivo protein. Response time still needs to be shorter. However, with the growing understanding of protein production and the special characteristics of AuNP, we are now motivated to pursue fruitful exploitations. Recent studies on the interaction between proteins and AuNP are listed for a thorough knowledge of this interaction's subsequent biological ramifications. It's important to note the new developments in utilizing prospects and future applications based on protein-AuNP interaction (Khandelia et al., 2013).

Additionally, AuNPs can be utilized as protein-delivery nanocarriers. The interfacial synergy between protein and AuNPs profoundly implies applications of AuNPs in biology and biomedicine. An important chemical probe for examining the structure, morphology, and stability of proteins on AuNPs is organothiol. The likelihood of organothiol incorporation into protein-shielded AuNPs and its possible repercussions on the functionality and toxicity of the protein-shielded AuNPs should be addressed in biological and biomedical sectors due to its large analogous abundance in biofluids carrying serum plasma (Liu et al., 2012). In earlier experiments, chitosan functionalized AuNPs to transport insulin (Bhumkar et al., 2007). A non-toxic biopolymer that can stabilize AuNPs is chitosan. Insulin is successfully administered transmucosal due to substantial insulin adsorption on the surface of chitosan-coated particles. According to Verma and his co- worker cationic tetra alkyl ammonium-functionalized Au NPs recognize the surface of an anionic protein by complementary electrostatic contact and inhibit its activity (Verma et al., 2004). By manipulating the protein-particle complex with SH, the activity was increased due to the release of free protein, demonstrating AuNPs as potential protein transporters. Before intravenous injection, Krol et al. employed AuNPs to conjugate with either human serum albumin (alb-AuNP) or apolipoprotein E (apoE-AuNP). Comparing protein conjugation to citrate-stabilized AuNPs, the results demonstrated that protein conjugation drastically reduced liver retention. Their research unequivocally supports the idea that sustained AuNP conjugation with albumin and apoE before intravenous injection boosts the specificity and efficacy of NPs in sick target organs, pointing to a possible application in nanomedicine and nano pharmacy (Schäffler et al., 2014).

8 Targeted AuNPs for Neurodegenerative Disorders

The term "neurodegenerative diseases" refers to a group of ailments that damage nerve cells and the nervous system due to the death of neurons and the networks that connect them. Due to the progressive death of neurons in the peripheral and central nervous systems, many illnesses cause disability (Khan et al., 2020). While many of these disorders are caused by unknown factors, there are times when they are brought on by illnesses like alcoholism, tumors, or strokes, as well as other factors, including genetic mutations, poisons, chemicals, and viruses (Polak & Shefi, 2015; Yang et al., 2016). Numerous neurotoxic events, including excessive inflammation, reactive oxygen species (ROS) triggering, and mitochondrial dysfunctions, can cause neurodegenerative disorders (Jyoti et al., 2011). Ataxia, dementia, or all of these symptoms are the primary symptoms linked to these conditions. Because they increase morbidity and mortality, these symptoms have social and economic repercussions (Xu et al., 2012).

The sole symptomatic alleviation offered by the treatments for these conditions is an increase in lifespan of a few years. The intricacy of the pathophysiology of neurodegenerative illnesses and the underlying cell interactions need to be better understood, restricting the development of therapeutic approaches. Therefore, much research is still being done to uncover therapeutic markers for such diseases (Wang et al., 2011). These physical limitations and the lack of specificity in current pharmacological treatments account for the ineffectiveness of most medications and neurosurgical techniques in non-treating illnesses of the central nervous system. Numerous characteristics of AuNPs have been used in biology. They are harmless, stable, and biocompatible. AuNPs may help treat neurodegenerative illnesses because of other inherent features. Among these innate qualities, the electrical conductivity of these particles helps restore the functionality of injured neurons. The antioxidant and anti-inflammatory properties of AuNPs can also be therapeutic in treating disorders characterized by inflammation and neuronal damage (Zhan et al., 2009).

AuNPs have been shown to promote neuronal excitability in vitro by boosting APs, lowering AP thresholds and durations, and decreasing hyperpolarization (Boisselier & Astruc, 2009). However, in pathological settings, AuNPs may have some harmful effects on neuronal cells. The differentiation, neuronal outgrowth, and maturation of neuronal cells are positively influenced by AuNPs adorned on electrospun nanofiber, and they may one day serve as an implanted cellular instrument for mending injured neural tissues (Pokharkar et al., 2011).

9 Targeted AuNPs for Infectious Diseases

Fungi, viruses, bacteria, and parasites are the main contributors to the spread of infectious diseases, accounting for many mortalities worldwide. There are two types of infectious diseases: emerging and re-emerging infectious diseases. Emerging infectious illnesses are recently discovered, while re-emerging infectious diseases are long-established but have developed drug resistance, which renders them challenging to address or manage (Morens & Fauci, 2013; Parretet al., 2008). While the human body's immune system is capable of protecting the body from illnesses, some infections are more easily spread than others and are more ferocious (NIH, 2007). However, not all illnesses spread like others. In general, diseases can spread when bacteria that cause infections enter the host body through natural orifices, developing at the site of the entrance before multiplying in the host cells and causing tissue damage. Drug resistance hinders the treatment of infectious diseases, highlighting the urgent need for novel therapies to overcome drug resistance. For the treatment of infectious disorders, therapies based on antibodies, metal-based nanoparticles, etc., have been created and are characterized by modest diameters between 10 and 100 nm, which explains why they interact well with biomolecules both within and outside of cells. Their large surface area encourages increased cell permeability (Casadevall, 1996; Saylor et al., 2009).

They can also be made more specialized by conjugating particular ligands, proteins, antibodies, medicines, and enzymes with special target cells with a specific binding activity. This increases their ability to deliver pharmaceuticals precisely to the diseased region and increases their therapeutic efficacy. Drugs, antibodies, proteins, and other substances can be conjugated onto metal nanoparticles to shield them from the body's immune system and increase the time that their blood circulates. Good physicochemical characteristics and surface charges can be found in metal nanoparticles (Mody et al., 2010).

Additionally, it has been claimed that gold nanoparticles function well as antibacterial agents. The capacity of AuNPs to penetrate the cell wall gives them their antibacterial properties. The integrity of bacterial membranes can be effectively damaged by cationic hydrophobic AuNPs, which also render bacterial cells poisonous. They had a negligible harmful effect on mammalian cells (Mohamed et al., 2017). Because of the differences in the structure of gram-positive and gram-negative bacteria's membranes, gold nanoparticles' antibacterial action differs against each kind of bacteria, necessitating a larger nanoparticle dose. Its antibacterial effectiveness results from decreased adenosine triphosphate synthase activity, which disrupts the metabolic process, and from a drop in the ribosomal subunit that binds tRNA, which causes the biological mechanism to break down (Li et al., 2014).

The bacteria and nanoparticles could contact directly thanks to the increased surface area. The nanoparticle disrupted the bacterial cytoplasm and protein, leading to cell death (Lima et al., 2013). Antibiotics and AuNPs together have increased antibacterial action. AuNPs may be adequate for both preventative and therapeutic purposes, reduce co-viral infection, and increase the effectiveness of antiviral medications. By interacting with the surface of the virion, they demonstrate antiviral activity by preventing the virus from attaching to receptors and preventing the first stages of viral reproduction. Though there aren't many reports, they suggest that more study is needed to develop metal nanoparticles that can treat viral infections (Mubarak Ali et al., 2011). It has also been recommended that AuNPs be used to treat parasitic illnesses such as malaria, leishmaniasis, and helminths. Additionally, it has changed the parasite's enzyme activity, killing it (Burygin et al., 2009). It works best when AuNPs are present in high concentrations because they cause an increase in intracellular ROS that permeates the cell membrane, interferes with the electron transport system, and prevents the parasite from producing ATP (Zharov et al., 2006).

10 Targeted AuNPs for Cardiovascular

The most common cause of death worldwide and a significant factor in healthcare expenses are cardiovascular diseases (CVDs). Even though the diagnosis of CVDs has advanced significantly, there is still a critical need for more precise early diagnosis and the development of novel diagnostic techniques (Flora & Nayak, 2019; Zamani et al., 2019). The word "CVDs" refers to a broad range of heart-related conditions, such as ischemic heart disease, cerebrovascular illnesses, arrhythmia, Marfan syndrome, thrombosis, pericardial and vascular diseases, heart failure, stroke, and cardiomyopathies (Bachheti et al., 2022; Braunwald, 2017). Drug delivery based on heart-targeted nanocarriers is an innovative, efficient, and successful method of treating many cardiac-related illnesses. Drug delivery systems based on nanocarriers avoid issues with conventional drug administration methods, such as their non-specificity, harmful side effects, and harm to healthy cells. It is possible to significantly change nanocarriers' in-vivo pharmacokinetic and pharmacodynamic data by altering their physicochemical features, such as size, shape, and surface changes, which would result in a better treatment plan.

Gold nanoparticles, for example, are unique sources of nanomaterials that can be used to target CVDs. The cellular response in CVDs is actively enhanced by the unique characteristics of nanomaterials, including surface energy and topographies. Due to their characteristics, including minimal cytotoxicity, stability, and biocompatibility, AuNPs make excellent carrier systems, particularly for treating CVDs. With a high degree of stability in the circulatory system and the capacity to break down in a suitable environment, once they have reached the target place, they offer a wide range of applications in biomedical science, such as molecular tagging and medication administration. As ligands, biological entities, namely medicines, proteins, antibodies, and genes, can all be functionalized with AuNPs due to their enormous surface area. The use of AuNPs in imaging CVDs with applications in atherosclerotic lesions and inflammation is still under development (Chen et al., 2006a, 2006b).

The characteristics of Au have also been used to prepare vascular drug-encapsulated nanostructures employing photothermal therapy methods. The following reaction can support NPs' thermal instability and help the medications they are encapsulating to release at the targeted spot. The function of macrophages in vascular lesions and their capacity to take up AuNPs to enable contrast-enhanced intravascular photoimaging of cardiovascular abnormalities. In addition, AuNPs exhibit powerful antioxidant characteristics that could be helpful in the management of CVDs. When delivering such NPs to specific plaques to diagnose photothermal revascularization of blocked arteries, the photothermal property of AuNPs was used. The ligand exchange reaction on citrate-coated AuNPs led to the development of AuNP conjugated with pure lisinopril, reduced thioctic lisinopril, or thioctic lisinopril. The angiotensin-converting enzyme (ACE) targeting was assessed using X-ray CT, employing the enhanced stability characteristics of thioctic lisinopril AuNPs (Chithrani & Chan, 2007). The photos revealed that the targeting of ACE and overexpression of ACE were associated with the development of pulmonary and cardiac fibrosis since the region of the heart and lungs had strong contrast. Using CT imaging, this novel approach might be an effective tool for tracking cardiovascular disease. For the treatment and diagnosis of CVDs, AuNPs may be utilized singly or in combination (Ghann et al., 2012).

11 Limitations of Gold Nanoparticles

While AuNPs have shown huge assurance in drug delivery applications, they have certain limitations that must be considered. Here are some of the key limitations associated with the use of AuNPs in drug delivery:

11.1 Limited Payload Capacity

One of the primary limitations of AuNPs is their relatively small size, which limits the amount of drug that can be loaded onto or encapsulated within the nanoparticles. This can be a challenge when delivering therapeutically relevant concentrations of drugs, especially for large drug molecules or high drug doses (Putzu et al., 2018).

11.2 Lack of Control Over Drug Release

AuNPs often exhibit burst release kinetics, where a significant amount of the loaded drug is rapidly released upon reaching the target site (Dreaden et al., 2012). This lack of controlled release can lead to suboptimal therapeutic outcomes and potential toxicity issues (Dykman & Khlebtsov, 2012).

Inefficient cellular uptake

AuNPs, particularly when coated with non-specific ligands or functional groups, may face challenges in efficiently entering target cells. This can hinder their ability to deliver drugs to the desired intracellular compartments (Ghosh et al., 2008a, 2008b).

Clearance by the reticuloendothelial system (RES)

Unmodified AuNPs can be rapidly recognized and cleared by the RES, primarily through uptake by macrophages in organs such as the liver and spleen. This rapid clearance limits the circulation time of AuNPs, reducing their potential for prolonged drug delivery (Huang & El-Sayed, 2010).

Lack of target specificity

While biofunctionalization of AuNPs with targeting ligands can improve selectivity, achieving precise targeting and uptake by specific cells or tissues can be challenging (Pissuwan et al., 2006). Factors such as ligand density, ligand affinity, and heterogeneity of target receptors can influence the effectiveness of targeting.

Potential toxicity concerns

While gold is generally considered biocompatible, the surface properties and modifications of AuNPs can influence their toxicity (Saha et al., 2012). Factors like nanoparticle size, shape, charge, and surface coating can impact cellular interactions and potentially induce cytotoxic effects. Additionally, the long-term fate and biodegradability of AuNPs in the body are still areas of thriving research and investigation (Zhang et al., 2015).

Addressing these limitations is a thriving field of research, and scientists are exploring various strategies to overcome these challenges. This includes developing advanced drug loading methods, designing stimuli-responsive drug release systems, optimizing surface modifications for improved cellular uptake and reduced clearance, and enhancing target specificity through rational ligand design.

It's worth noting that the limitations of AuNPs in drug delivery are not exclusive to gold nanoparticles and are common challenges encountered with nanomedicine in general. Researchers are continuously working on innovative solutions to maximize the potential of AuNPs and overcome these limitations for improved drug delivery efficacy and safety.

12 Conclusion

This chapter has covered the methods for producing gold nanoparticles, their functionalization, and their uses in pharmaceuticals. Because of their biocompatibility, ease of functionalization with different ligands, easily adjustable size and shape, and improved absorption properties due to their surface plasmon absorption band, AuNPs present themselves as suitable drug carriers. While AuNPs exhibit passive targeting towards cancers through the EPR effect, active targeting of AuNPs towards tumours is becoming increasingly popular due to the current trend in generating tailored therapies. The ability of a wide range of biological systems to behave as reducing agents during synthesis reactions has been investigated. Different moieties, such as PEG, amino acids and peptides, oligonucleotides, and antibodies, can modify the surface of AuNPs in various ways, making it easier for drugs and biomolecules to be loaded. PEGlyation of AuNPs is the most suitable functionalization method for therapeutic agent delivery in vivo since it is biocompatible and helps nano-drug carriers to subvert the immune system. By adding target-specific ligands to the surface of AuNPs, PEGlyation restrictions, such as AuNPs' diminished capacity to bind with the target receptor, can be overcome.

Using AuNPs as a delivery vehicle for large biomolecules is an inventive and intriguing field that has received much attention in recent years. However, more research is still needed to develop structures that can deliver conjugates intracellularly and intranuclearly with the fewest adverse effects. Effective and efficient payload release at the target site is another barrier to using AuNPs for medication delivery. The performance of this function has been attributed to various external and internal inputs. However, antibodies are more expensive than most other targeting moieties, and because of their enormous size, antibodies have poor tumour penetration. Much research is currently being done on antibody fragments and nanobodies to overcome this limited penetration, but this shrinkage shortens the lifetime in circulation. For example, how to create a better and more efficient diagnostic and therapeutic product for biomedical applications, AuNPs have opened new doors for the pharmaceutical business. The pharmaceutical industry has made particularly significant strides in tumour targeting. AuNPs may target and deliver anticancer medications directly into malignant cells with the benefit of being quick and affordable. Future studies will focus on creating a revolutionary gold nano drug delivery system capable of flawlessly delivering and discharging payload at the target place.

References

Alexis, F., Basto, P., Levy-Nissenbaum, E., Radovic-Moreno, A. F., Zhang, L., Pridgen, E., Wang, A. Z., Marein, S. L., Westerhof, K., Molnar, L. K., & Farokhzad, O. C. (2008). HER-2-Targeted nanoparticle–affibody bioconjugates for cancer therapy. *Chem Med Chem: Chemistry Enabling Drug Discovery, 3*(12), 1839–1843.

Anshup, V. J. S., Subramaniam, C., Kumar, R. R., Priya, S., Kumar, T. S., Omkumar, R. V., John, A., & Pradeep, T. (2005). Growth of gold nanoparticles in human cells. *Langmuir, 21*(25), 11562–11567.

Bachheti, R. K., & Bachheti, A. (Eds.). (2023). *Secondary metabolites from medicinal plants: Nanoparticles synthesis and their applications*. CRC Press. https://doi.org/10.1201/9781003213727

Bachheti, R. K., Worku, L. A., Gonfa, Y. H., Zebeaman, M., Pandey, D. P., & Bachheti, A. (2022). Prevention and treatment of cardiovascular diseases with plant phytochemicals: A review. *Evidence-Based Complementary and Alternative Medicine*. https://doi.org/10.1155/2022/5741198

Beik, J., Khateri, M., Khosravi, Z., Kamrava, S. K., Kooranifar, S., Ghaznavi, H., & Shakeri-Zadeh, A. (2019). Gold nanoparticles in combinatorial cancer therapy strategies. *Coordination Chemistry Reviews, 387*, 299–324.

Bhumkar, D. R., Joshi, H. M., Sastry, M., & Pokharkar, V. B. (2007). Chitosan reduced gold nanoparticles as novel carriers for transmucosal delivery of insulin. *Pharmaceutical Research, 24*, 1415–1426.

Boisselier, E., & Astruc, D. (2009). Gold nanoparticles in nanomedicine: Preparations, imaging, diagnostics, therapies and toxicity. *Chemical Society Reviews, 38*(6), 1759–1782.

Bonoiu, A. C., Mahajan, S. D., Ding, H., Roy, I., Yong, K. T., Kumar, R., Hu, R., Bergey, E. J., Schwartz, S. A., & Prasad, P. N. (2009). Nanotechnology approach for drug addiction therapy: Gene silencing using delivery of gold nanorod-siRNA nanoplex in dopaminergic neurons. *Proceedings of the National Academy of Sciences, 106*(14), 5546–5550.

Braunwald, E. (2017). Cardiomyopathies: An overview. *Circulation Research, 121*(7), 711–721.

Brinson, B. E., Lassiter, J. B., Levin, C. S., Bardhan, R., Mirin, N., & Halas, N. J. (2008). Nanoshells made easy: Improving Au layer growth on nanoparticle surfaces. *Langmuir, 24*, 14166–14171.

Brust, M., Walker, M., Bethell, D., Schiffrin, D. J., & Whyman, R. (1994). Synthesis of thiol-derivatized gold nanoparticles in a two-phase liquid–liquid system. *Journal of the Chemical Society, Chemical Communications., 7*, 801–802.

Burygin, G. L., Khlebtsov, B. N., Shantrokha, A. N., Dykman, L. A., Bogatyrev, V. A., & Khlebtsov, N. G. (2009). On the enhanced antibacterial activity of antibiotics mixed with gold nanoparticles. *Nanoscale Research Letters, 4*, 794–801.

Casadevall, A. (1996). Antibody-based therapies for emerging infectious diseases. *Emerging Infectious Diseases, 2*, 200.

Chen, C. C., Lin, Y. P., Wang, C. W., Tzeng, H. C., Wu, C. H., Chen, Y. C., Chen, C. P., Chen, L. C., & Wu, Y. C. (2006a). DNA−gold nanorod conjugates for remote control of localized gene expression by near-infrared irradiation. *Journal of the American Chemical Society, 128*(11), 3709–3715.

Chen, J., McLellan, J. M., Siekkinen, A., Xiong, Y., Li, Z. Y., & Xia, Y. (2006b). Facile synthesis of gold-silver nanocages with controllable pores on the surface. *Journal of the American Chemical Society, 128*(46), 14776–14777.

Chen, J., Saeki, F., Wiley, B. J., Cang, H., Cobb, M. J., Li, Z. Y., Au, L., Zhang, H., Kimmey, M. B., Li, X., & Xia, Y. (2005). Gold nanocages: Bioconjugation and their potential use as optical imaging contrast agents. *Nano Letters, 5*(3), 473–477.

Chithrani, B. D., & Chan, W. C. (2007). Elucidating the mechanism of cellular uptake and removal of protein-coated gold nanoparticles of different sizes and shapes. *Nano Letters, 7*(6), 1542–1550.

Cunliffe, D., Kirby, A., & Alexander, C. (2005). Molecularly imprinted drug delivery systems. *Advanced Drug Delivery Reviews, 57*(12), 1836–1853.

Dreaden, E. C., Alkilany, A. M., Huang, X., Murphy, C. J., & El-Sayed, M. A. (2012). The golden age: Gold nanoparticles for biomedicine. *Chemical Society Reviews, 41*(7), 2740–2779.

Dreaden, E. C., Mackey, M. A., Huang, X., Kang, B., & El-Sayed, M. A. (2011). Beating cancer in multiple ways using nanogold. *Chemical Society Reviews, 40*(7), 3391–3404.

Dykman, L., & Khlebtsov, N. (2012). Gold nanoparticles in biomedical applications: Recent advances and perspectives. *Chemical Society Reviews, 41*(6), 2256–2282.

Ealia, S. A., & Saravanakumar, M. P. (2017). A review of the classification, characterization, synthesis of nanoparticles and their application. *In IOP Conference Series: Materials Science and Engineering, 263*, 032019.

Elbayoumi, T. A. (2010). Nano drug-delivery systems in cancer therapy: Gains, pitfalls, and considerations in DMPK and PD. *Therapeutic Delivery, 1*(2), 215–219.

Flora, G. D., & Nayak, M. K. (2019). A brief review of cardiovascular diseases, associated risk factors, and current treatment regimes. *Current Pharmaceutical Design, 25*(38), 4063–4084.

Ghann, W. E., Aras, O., Fleiter, T., & Daniel, M. C. (2012). Syntheses and characterization of lisinopril-coated gold nanoparticles as highly stable targeted CT contrast agents in cardiovascular diseases. *Langmuir, 28*(28), 10398–10408.

Ghosh, P., Han, G., De, M., Kim, C. K., & Rotello, V. M. (2008a). Gold nanoparticles in delivery applications. *Advanced Drug Delivery Reviews, 60*(11), 1307–1315.

Ghosh, P. S., Kim, C. K., Han, G., Forbes, N. S., & Rotello, V. M. (2008b). Efficient gene delivery vectors by tuning the surface charge density of amino acid-functionalized gold nanoparticles. *ACS Nano, 2*, 2213–2218.

Han, G., Martin, C. T., & Rotello, V. M. (2006a). Stability of gold nanoparticle-bound DNA toward biological, physical, and chemical agents. *Chemical Biology and Drug Design, 67*(1), 78–82.

Han, G., You, C. C., Kim, B. J., Turingan, R. S., Forbes, N. S., Martin, C. T., & Rotello, V. M. (2006b). Light-regulated release of DNA and its delivery to nuclei by means of photolabile gold nanoparticles. *Angewandte Chemie, 118*(19), 3237–3241.

Hiramatsu, H., & Osterloh, F. E. (2004). A simple large-scale synthesis of nearly monodisperse gold and silver nanoparticles with adjustable sizes and with exchangeable surfactants. *Chemistry of Materials, 16*(13), 2509–2511.

Huang, X., & El-Sayed, M. A. (2010). Gold nanoparticles: Optical properties and implementations in cancer diagnosis and photothermal therapy. *Journal of Advanced Research, 1*(1), 13–28.

Huang, X., Neretina, S., & El-Sayed, M. A. (2009). Gold nanorods: From synthesis and properties to biological and biomedical applications. *Advanced Materials, 21*(48), 4880–4910.

Husen, A. (2017). Gold nanoparticles from plant system: synthesis, characterization and their application. In M. Ghorbanpourn, K. Manika, A. Varma (Eds.), *Nanoscience and plant–soil systems* (Vol. 48, pp. 455–479). Springer International Publishing AG. https://doi.org/10.1007/978-3-319-46835-8_17

Husen, A. (2022). *Engineered nanomaterials for sustainable agricultural production, soil improvement and stress management*. Elsevier Inc. https://doi.org/10.1016/C2021-0-00054-7

Husen, A. (2023). *Nanomaterials and nanocomposites exposures to plants (Response, Interaction, Phytotoxicity and Defense Mechanisms)*. Springer Nature Singapore Pte Ltd. https://doi.org/10.1007/978-981-99-2419-6

Husen, A., & Iqbal, M. (2019). *Nanomaterials and plant potential*. Springer International Publishing AG. https://doi.org/10.1007/978-3-030-05569-1

Husen, A., Rahman, Q. I., Iqbal, M., & Yassin, M. O. (2019). Plant-mediated fabrication of gold nanoparticles and their applications. In A. Husen, M. Iqbal (Eds.), Nanomaterials and plant potential (pp. 71–110). Springer International Publishing AG. https://doi.org/10.1007/978-3-030-05569-1_3

Husen, A., & Siddiqi, K. S. (2023) *Advances in smart nanomaterials and their applications*. Elsevier Inc. https://doi.org/10.1016/C2021-0-02202-1

Jana, N. R., Gearheart, L., & Murphy, C. J. (2001). Seed-mediated growth approach for shape-controlled synthesis of spheroidal and rod-like gold nanoparticles using a surfactant template. *Advanced Materials, 13*(18), 1389–1393.

Joshi, P., Chakraborti, S., Ramirez-Vick, J. E., Ansari, Z. A., Shanker, V., Chakrabarti, P., & Singh, S. P. (2012). The anticancer activity of chloroquine-gold nanoparticles against MCF-7 breast cancer cells. *Colloids and Surfaces b: Biointerfaces, 95*, 195–200.

Jurgons, R., Seliger, C., Hilpert, A., Trahms, L., Odenbach, S., & Alexiou, C. (2006). Drug-loaded magnetic nanoparticles for cancer therapy. *Journal of Physics: Condensed Matter., 18*(38), S2893.

Jyoti, A., Singh, S. P., Yashpal, M., Dwivedi, P. D., & Shanker, R. (2011). Rapid detection of enterotoxigenic Escherichia coli gene using bio-conjugated gold nanoparticles. *Journal of Biomedical Nanotechnology, 7*(1), 170–171.

Kalimuthu, K., Lubin, B. C., Bazylevich, A., Gellerman, G., Shpilberg, O., Luboshits, G., & Firer, M. A. (2018). Gold nanoparticles stabilize peptide-drug-conjugates for sustained targeted drug delivery to cancer cells. *Journal of Nanobiotechnology, 16*(1), 1–3.

Khan, A., Jahan, S., Imtiyaz, Z., Alshahrani, S., AntarMakeen, H., Mohammed Alshehri, B., Kumar, A., Arafah, A., & Rehman, M. U. (2020). Neuroprotection: Targeting multiple pathways by naturally occurring phytochemicals. *Biomedicines, 8*(8), 284.

Khandelia, R., Jaiswal, A., Ghosh, S. S., & Chattopadhyay, A. (2013). Gold nanoparticle–protein agglomerates as versatile nanocarriers for drug delivery. *Small (weinheim an Der Bergstrasse, Germany), 9*(20), 3494–3505.

Kim, E. Y., Schulz, R., Swantek, P., Kunstman, K., Malim, M. H., & Wolinsky, S. M. (2012). Gold nanoparticle-mediated gene delivery induces widespread changes in the expression of innate immunity genes. *Gene Therapy, 19*(3), 347–353.

Kumar, A., Zhang, X., & Liang, X. J. (2013). Gold nanoparticles: An emerging paradigm for targeted drug delivery system. *Biotechnology Advances, 31*(5), 593–606.

Kundu, S., Peng, L., & Liang, H. (2008). A new route to obtain high-yield multiple-shaped gold nanoparticles in aqueous solution using microwave irradiation. *Inorganic Chemistry, 47*(14), 6344–6352.

Lee, S. H., Bae, K. H., Kim, S. H., Lee, K. R., & Park, T. G. (2008). Amine-functionalized gold nanoparticles as non-cytotoxic and efficient intracellular siRNA delivery carriers. *International Journal of Pharmaceutics, 364*(1), 94–101.

Lee, J. H., Choi, S. U., Jang, S. P., & Lee, S. Y. (2012). Production of aqueous spherical gold nanoparticles using a conventional ultrasonic bath. *Nanoscale Research Letters, 7*, 1–7.

Leyu, A. M., Debebe, S. E., Bachheti, A., Rawat, Y. S., & Bachheti, R. K. (2023). Green Synthesis of gold and silver nanoparticles using invasive alien plant *Parthenium hysterophorus* and their antimicrobial and antioxidant activities. *Sustainability, 15*(12), 9456.

Li, W., Cao, Z., Liu, R., Liu, L., Li, H., Li, X., Chen, Y., Lu, C., & Liu, Y. (2019). AuNPs as an important inorganic nanoparticle applied in drug carrier systems. *Artificial Cells, Nanomedicine, and Biotechnology, 47*(1), 4222–4233.

Li, X., Robinson, S. M., Gupta, A., Saha, K., Jiang, Z., Moyano, D. F., Sahar, A., Riley, M. A., & Rotello, V. M. (2014). Functional gold nanoparticles as potent antimicrobial agents against multi-drug-resistant bacteria. *ACS Nano, 8*, 10682–10686.

Lima, E., Guerra, R., Lara, V., & Guzmán, A. (2013). Gold nanoparticles as efficient antimicrobial agents for Escherichia coli and Salmonella typhi. *Chemistry Central Journal, 7*, 1–7.

Liu, L., Li, S., Liu, L., Deng, D., & Xia, N. (2012). Simple, sensitive, and selective detection of dopamine using dithiobis (succinimidyl propionate)-modified gold nanoparticles as colorimetric probes. *The Analyst, 137*(16), 3794–3799.

Loo, C., Lowery, A., Halas, N., West, J., & Drezek, R. (2005). Immunotargetednanoshells for integrated cancer imaging and therapy. *Nano Letters, 5*(4), 709–711.

Ma, H., Yin, B., Wang, S., Jiao, Y., Pan, W., Huang, S., Chen, S., & Meng, F. (2004). Synthesis of silver and gold nanoparticles by a novel electrochemical method. *ChemPhysChem, 5*(1), 68–75.

McIntosh, C. M., Esposito, E. A., Boal, A. K., Simard, J. M., Martin, C. T., & Rotello, V. M. (2001). Inhibition of DNA transcription using cationic mixed monolayer-protected gold clusters. *Journal of the American Chemical Society, 123*(31), 7626–7629.

Mody, V. V., Singh, A., & Wesley, B. (2013). Basics of magnetic nanoparticles for their application in the field of magnetic fluid hyperthermia. *European Journal of Nanomedicine, 5*(1), 11–21.

Mody, V. V., Siwale, R., Singh, A., & Mody, H. R. (2010). Introduction to metallic nanoparticles. *Journal of Pharmacy and Allied Sciences, 2*(4), 282.

Mohamed, M. M., Fouad, S. A., Elshoky, H. A., Mohammed, G. M., & Salaheldin, T. A. (2017). Antibacterial effect of gold nanoparticles against Corynebacterium pseudotuberculosis. *International Journal of Veterinary Science and Medicine, 5*(1), 23–29.

Morens, D. M., & Fauci, A. S. (2013). Emerging infectious diseases: Threats to human health and global stability. *PLoS Pathogens, 9*(7), e1003467.

Morille, M., Passirani, C., Vonarbourg, A., Clavreul, A., & Benoit, J. P. (2008). Progress in developing cationic vectors for nonviral systemic gene therapy against cancer. *Biomaterials, 29*(24–25), 3477–3496.

Mubarak Ali, D., Thajuddin, N., Jeganathan, K., & Gunasekaran, M. (2011). Plant extract mediated synthesis of silver and gold nanoparticles and its antibacterial activity against clinically isolated pathogens. *Colloids and Surfaces b: Biointerfaces, 85*(2), 360–365.

Murphy, C. J., Gole, A. M., Stone, J. W., Sisco, P. N., Alkilany, A. M., Goldsmith, E. C., & Baxter, S. C. (2008). Gold nanoparticles in biology: Beyond toxicity to cellular imaging. *Accounts of Chemical Research, 41*(12), 1721–1730.

Nam, J., La, W. G., Hwang, S., Ha, Y. S., Park, N., Won, N., Jung, S., Bhang, S. H., Ma, Y. J., Cho, Y. M., & Jin, M. (2013). PH-responsive assembly of gold nanoparticles and "spatiotemporally concerted" drug release for synergistic cancer therapy. *ACS Nano, 7*(4), 3388–3402.

National Institutes of Health (US). (2007). Biological sciences curriculum study NIH curriculum supplement series. In Understanding Emerging and Re-emerging Infectious Diseases; National Institutes of Health (US). Available online: https://www.ncbi.nlm.nih.gov/books/NBK20370/. Retrieved April 20, 2017.

Nikoobakht, B., & El-Sayed, M. A. (2003). Preparation and growth mechanism of gold nanorods (NRs) using seed-mediated growth method. *Chemistry of Materials, 15*(10), 1957–1962.

Paciotti, G. F., Myer, L., Weinreich, D., Goia, D., Pavel, N., McLaughlin, R. E., & Tamarkin, L. (2004). Colloidal gold: A novel nanoparticle vector for tumor-directed drug delivery. *Drug Delivery, 11*(3), 169–183.

Pack, D. W., Hoffman, A. S., Pun, S., & Stayton, P. S. (2005). Design and development of polymers for gene delivery. *Nature Reviews Drug Discovery, 4*(7), 581–593.

Patra, C. R., Bhattacharya, R., Wang, E., Katarya, A., Lau, J. S., Dutta, S., Muders, M., Wang, S., Buhrow, S. A., Safgren, S. L., & Yaszemski, M. J. (2008). Targeted delivery of gemcitabine to pancreatic adenocarcinoma using cetuximab as a targeting agent. *Cancer Research, 68*(6), 1970–1978.

Pissuwan, D., Valenzuela, S. M., & Cortie, M. B. (2006). Therapeutic possibilities of plasmonically heated gold nanoparticles. *TRENDS in Biotechnology, 24*(2), 62–67.

Pokharkar, V., Bhumkar, D., Suresh, K., Shinde, Y., Gairola, S., & Jadhav, S. S. (2011). Gold nanoparticles as a potential carrier for transmucosal vaccine delivery. *Journal of Biomedical Nanotechnology, 7*(1), 57–59.

Polak, P., & Shefi, O. (2015). Nanometric agents in the service of neuroscience: manipulation of neuronal growth and activity using nanoparticles. *Nanomedicine: Nanotechnology, Biology and Medicine, 11*(6), 1467–1479.

Prasad, B. L., Stoeva, S. I., Sorensen, C. M., & Klabunde, K. J. (2002). Digestive ripening of thiolated gold nanoparticles: The effect of alkyl chain length. *Langmuir, 18*(20), 7515–7520.

Putnam, D. (2006). Polymers for gene delivery across length scales. *Nature Materials, 5*, 439–451.

Putzu, A., de Carvalho, C. M. P. D., de Almeida, J. P., Belletti, A., Cassina, T., Landoni, G., & Hajjar, L. A. (2018). Perioperative statin therapy in cardiac and non-cardiac surgery: A systematic review and meta-analysis of randomized controlled trials. *Annals of Intensive Care, 8*, 95.

Rai, A., Prabhune, A., & Perry, C. C. (2010). Antibiotic mediated synthesis of gold nanoparticles with potent antimicrobial activity and their application in antimicrobial coatings. *Journal of Materials Chemistry, 20*(32), 6789–6798.

Rietwyk, S., & Peer, D. (2017). Next-generation lipids in RNA interference therapeutics. *ACS Nano, 11*(8), 7572–7586.

Rosi, N. L., Giljohann, D. A., Thaxton, C. S., Lytton-Jean, A. K., Han, M. S., & Mirkin, C. A. (2006). Oligonucleotide-modified gold nanoparticles for intracellular gene regulation. *Science, 312*(5776), 1027–1030.

Saha, K., Agasti, S. S., Kim, C., Li, X., & Rotello, V. M. (2012). Gold nanoparticles in chemical and biological sensing. *Chemical Reviews, 112*(5), 2739–2779.

Satnami, M. L., Chandraker, K., Vaishanav, S. K., & Nagwanshi, R. (2015). Interaction of thiolated amino acids and peptide onto the gold nanoparticle surface: Radical scavenging activity. *Indian Journal of Chemistry Section, 54A*(10), 1206–1214.

Saylor, C., Dadachova, E., & Casadevall, A. (2009). Monoclonal antibody-based therapies for microbial diseases. *Vaccine, 27*, G38–G46.

Schroeder, A., Heller, D. A., Winslow, M. M., Dahlman, J. E., Pratt, G. W., Langer, R., Jacks, T., & Anderson, D. G. (2012). Treating metastatic cancer with nanotechnology. *Nature Reviews Cancer, 12*(1), 39–50.

Schäffler, M., Sousa, F., Wenk, A., Sitia, L., Hirn, S., Schleh, C., Haberl, N., Violatto, M., Canovi, M., Andreozzi, P., Salmona, M., Bigini, P., Kreyling, W. G., & Krol, S. (2014). Blood protein coating of nanoparticles as a potential tool for organ targeting. *Biomaterials, 35*, 3455–3466.

Scott, A. W., Garimella, V., Calabrese, C. M., & Mirkin, C. A. (2017). Universal Biotin–PEG-linked gold nanoparticle probes for the simultaneous detection of nucleic acids and proteins. *Bioconjugate Chemistry, 28*(1), 203–211.

Shah, M., Badwaik, V. D., & Dakshinamurthy, R. (2014b). Biological applications of gold nanoparticles. *Journal of Nanoscience and Nanotechnology, 14*(1), 344–362.

Shah, M., Badwaik, V., Kherde, Y., Waghwani, H. K., Modi, T., Aguilar, Z. P., Rodgers, H., Hamilton, W., Marutharaj, T., Webb, C., & Lawrenz, M. B. (2014a). Gold nanoparticles: Various methods of synthesis and antibacterial applications. *Frontiers in Bioscience-Landmark, 19*(8), 1320–1344.

Shao, J., Griffin, R. J., Galanzha, E. I., Kim, J. W., Koonce, N., Webber, J., Mustafa, T., Biris, A. S., Nedosekin, D. A., & Zharov, V. P. (2013). Photothermal nanodrugs: The potential of TNF-gold nanospheres for cancer theranostics. *Scientific Reports, 3*(1), 1293.

Siddiqi, K. S., & Husen, A. (2016). Engineered gold nanoparticles and plant adaptation potential. *Nanoscale Research Letters, 11*(400), 1–10. https://doi.org/10.1186/s11671-016-1607-2

Siddiqi, K. S., & Husen, A. (2017). Recent advances in plant-mediated engineered gold nanoparticles and their application in biological system. *Journal of Trace Elements in Medicine and Biology, 40*, 10–23. https://doi.org/10.1016/j.jtemb.2016.11.012

Sivaraman, S. K., Kumar, S., & Santhanam, V. (2011). Monodisperse sub-10 nm gold nanoparticles by reversing the order of addition in Turkevich method–the role of chloroauric acid. *Journal of Colloid and Interface Science, 361*(2), 543–547.

Sun, L., Liu, D., & Wang, Z. (2008). Functional gold nanoparticle-peptide complexes as cell-targeting agents. *Langmuir, 24*(18), 10293–10297.

Sun, Y., & Xia, Y. (2002). Shape-controlled synthesis of gold and silver nanoparticles. *Science, 298*(5601), 2176–2179.

Verma, A., Simard, J. M., Worrall, J. W., & Rotello, V. M. (2004). Tunable reactivation of nanoparticle-inhibited β-galactosidase by glutathione at intracellular concentrations. *Journal of American Chemical Society, 126*(43), 13987–13991.

Wang, F., Wang, Y. C., Dou, S., Xiong, M. H., Sun, T. M., & Wang, J. (2011). Doxorubicin-tethered responsive gold nanoparticles facilitate intracellular drug delivery for overcoming multidrug resistance in cancer cells. *ACS Nano, 5*(5), 3679–3692.

Waters, C. A., Mills, A. J., Johnson, K. A., & Schiffrin, D. J. (2003). Purification of dodecanethiol derivatized gold nanoparticles. *Chemical Communications, 4*, 540–541.

Xu, L., Liu, Y., Chen, Z., Li, W., Liu, Y., Wang, L., Liu, Y., Wu, X., Ji, Y., Zhao, Y., & Ma, L. (2012). Surface-engineered gold nanorods: Promising DNA vaccine adjuvant for HIV-1 treatment. *Nano Letters, 12*(4), 2003–2012.

Yafout, M., Ousaid, A., Khayati, Y., & El Otmani, I. S. (2021). Gold nanoparticles as a drug delivery system for standard chemotherapeutics: A new lead for targeted pharmacological cancer treatments. *Scientific African, 11*, e00685.

Yang, Y. C., Chen, C. N., & Chi, L. L. (2016). Inventors; GNT Biotech and Medicals Corp, assignee. Transcranial burst electrostimulation apparatus and its applications. United States patent application US 14/800,997. 2016 Jan 21.

Zamani, P., Fereydouni, N., Butler, A. E., Navashenaq, J. G., & Sahebkar, A. (2019). The therapeutic and diagnostic role of exosomes in cardiovascular diseases. *Trends in Cardiovascular Medicine, 29*(6), 313–323.

Zare, D., Khoshnevisan, K., Barkhi, M., & Tahami, H. V. (2014). Fabrication of capped gold nanoparticles by using various amino acids. *Journal of Experimental Nanoscience, 9*(9), 957–965.

Zhan, Q., Qian, J., Li, X., & He, S. (2009). A study of mesoporous silica-encapsulated gold nanorods as enhanced light scattering probes for cancer cell imaging. *Nanotechnology, 21*(5), 055704.

Zhang, X. D., Wu, D., Shen, X., Liu, P. X., & Fan, F. Y. (2015). The cellular uptake of gold nanoparticles in cancer cells is mediated by impaired autophagic flux rather than by the classic endocytic pathway. *Nanoscale, 7*(41), 17090–17103.

Zhang, G., Yang, Z., Lu, W., Zhang, R., Huang, Q., Tian, M., Li, L., Liang, D., & Li, C. (2009). Influence of anchoring ligands and particle size on the colloidal stability and in vivo biodistribution of polyethylene glycol-coated gold nanoparticles in tumor-xenografted mice. *Biomaterials, 30*(10), 1928–1936.

Zhao, X., Huang, Q., & Jin, Y. (2015). Gold nanorod delivery of LSD1 siRNA induces human mesenchymal stem cell differentiation. *Materials Science and Engineering, c: Materials for Biological Applications, 54*, 142–149.

Zharov, V. P., Mercer, K. E., Galitovskaya, E. N., & Smeltzer, M. S. (2006). Photothermal nanotherapeutics and nanodiagnostics for selective killing of bacteria targeted with gold nanoparticles. *Biophysical Journal, 90*, 619–662.

Use of Green Synthesized Platinum Nanoparticles for Biomedical Applications

Addisu Tamir Wassie, D. P. Pandey, Archana Bachheti, Azamal Husen, Deepti Singh Vashishth, and Rakesh Kumar Bachheti

Abstract Platinum nanoparticles (PtNPs) are extremely promising for biological applications due to their exceptional physicochemical characteristics. To address the drawbacks of the conventional physical and chemical procedures for the production of PtNPs, inexpensive and environmentally friendly alternatives have been devised. Several biogenic components, namely plant extracts, bacteria, fungus, and algae, were used in the green synthesis of PtNPs, and their biomedical effects were evaluated. The fabrication conditions, energy need, and reaction time were all decreased by using green methods to create PtNPs. The state-of-the-art techniques employed for the environmentally friendly production of PtNPs, the characterization of PtNPs, and their claimed biological uses are highlighted in this chapter.

A. T. Wassie · R. K. Bachheti (✉)
Department of Industrial Chemistry, College of Applied Sciences, Addis Ababa Science and Technology University, P. O. Box 16417 Addis Ababa, Ethiopia
e-mail: rkbachheti@gmail.com

A. T. Wassie
Department of Chemistry, College of Natural Sciences, Wollo University, P. O. Box 1145 Dessie, Ethiopia

D. P. Pandey
Department of Chemistry Government Degree College Dehradun Shahar, Suddhowala, Dehradun 248007, India

A. Bachheti · Deepti Singh Vashishth
Department of Environment Science, Graphic Era (Deemed to Be University), Dehradun, India

A. Husen
Department of Biotechnology, Smt. S.S. Patel Nootan Science & Commerce College, Sankalchand Patel University, Visnagar 384315, Gujarat, India

Department of Biotechnology, Graphic Era (Deemed to be University), Dehradun 248002, Uttarakhand, India

Laboratory of Bio Resource Management, Institute of Tropical Forestry and Forest Product, University Putra Malaysia, 43400 UPM Serdang, Selangor, Malaysia

R. K. Bachheti
Department of Allied Sciences, Graphic Era Hill University, Society Area, Clement Town, Dehradun 248002, India

R. K. Bachheti et al. (eds.), *Metal and Metal-Oxide Based Nanomaterials*, Smart Nanomaterials Technology, https://doi.org/10.1007/978-981-99-7673-7_13

Keywords Platinum nanoparticles · Green synthesis · Biomedical applications · Characterization

1 Introduction

Recently, nanoscale materials have drawn a lot of interest because of their potential to revolutionize a number of technologies and industries, including transportation, medicine, food safety, homeland security, energy, information technology, agriculture, environmental science, photonics, and catalysis (Bachheti et al., 2021, 2022; Bachheti et al., 2020a, 2020b; Husen & Iqbal, 2019; Husen & Siddiqi, 2023; Husen et al., 2019; 2023a, 2023b; Husen, 2022, 2023a, 2023b). Nanoparticles have distinct physicochemical features that allow them to interact with biological systems successfully due to their small size and high surface area-to-volume ratio. For several biomedical applications, such as medication transport, imaging, diagnostics, and therapy, nanoparticles have shown promise (Bachheti et al., 2019; Bloch et al., 2021; Painuli et al., 2020). Due to their wide range of uses in fuel cells, water gas shift reactions, organic catalysis, the petrochemical industry, electronics, selective CO oxidation, biosensors, optics, automobiles, and pharmaceuticals, platinum nanoparticles (PtNPs) are among the most relevant and promising metal nanoparticles (Jeyaraj et al., 2019). In addition to these because of their superior physicochemical qualities, biocompatibility, low toxicity, and catalytic activity, PtNPs have drawn more and more interest in biomedical research among the many types of nanoparticles (Siddiqi & Husen, 2016). Numerous biomedical uses of PtNPs have been investigated, including cancer treatment, imaging, and diagnosis. PtNPs have been demonstrated to specifically target cancer tissues and cells, increasing the therapeutic effectiveness of chemotherapy medications while lowering their toxicity to healthy cells. PtNPs have an inherent antibacterial, antioxidant, and catalytic activity that allows them to lower intracellular reactive oxygen species levels and hinder the inflammation-causing pathways downstream (Bloch et al., 2021).

There are different synthesis methods for PtNPs with their advantages and drawbacks. Even if different methods are used for the synthesis of PtNPs, they are broadly divided into two major categories: bottom-up and top-down methods. In top-down methods, metals are used in high amounts for the synthesis of PtNPs via mechanical breakdown with the major advantage of resulting in uniform shape and size (Dhand et al., 2015). Numerous physical methods are used for the synthesis of PtNPs such as lithography, sputtering thermal decomposition, milling process, and chemical etching grouped under top-down methods of synthesis. In the bottom-up method molecules and atoms are assembled to produce a kind of NPs. Chemical vapor deposition, sol–gel processing, laser pyrolysis, and plasma spraying are involved in the bottom-up synthesis method (Naseer et al., 2020).

The bottom-up and top-down synthesis methods of PtNPs are furtherly grouped into three which are physical, chemical, and biological methods as shown in Fig. 1. The advantages of the physical method synthesis of PtNPs include the use of nontoxic chemicals, high-speed process, purity of products, and uniform size, and shape. While its disadvantage includes high cost, high dilution, radiation exposure, requiring high temperature, pressure, and energy, producing high waste, less productivity, difficulty in size and shape tunability, lower stability, altered surface chemistry and physico-chemical properties of nanoparticles (Bloch et al., 2021). While the chemical production of PtNPs, however, requires the interaction of atoms and smaller molecules, where the precursor metal ions are transformed into the appropriate nanoparticles. Precursors that are soluble in water are utilized to speed up the reduction of metal monomers. The micro-emulsion, polyol synthesis, sol–gel process, pyrolysis, and hydrothermal are a few examples of different chemical synthesis methods PtNps. Chemical synthesis is efficient in terms of cost, versatility, yield, controllability, thermal stability, and dispersion, but it is also risky for both people and the environment due to its low purity and usage of dangerous chemicals. The synthesis of PtNPs has lately attracted a lot of interest in biological (green) synthesis due to its benefits of being straightforward, using non-toxic raw materials, easy to use, eco-friendly, and biocompatibility (Naseer et al., 2020).

The possible environmental and health risks associated with conventional procedures underline the significance of green synthesis techniques for PtNP production. In order to eliminate or reduce the use of hazardous chemicals and solvents, green synthesis techniques use natural materials as reducing and stabilizing agents, such as plant extracts, microorganisms, or enzymes (Naseer et al., 2020; Thakkar et al., 2010). Scalability, eco-friendly, cost-effectiveness, monodispersed, greater yields, rapid synthesis, lesser toxicity, and low waste production are just a few benefits that green synthesis techniques have over conventional ones (Jeyaraj et al., 2019). PtNPs in particular offer a lot of potential in a variety of biomedical applications. Due to this green synthesis techniques are used to synthesize PtNPs in a cost-efficient, environmentally friendly manner that enables their conversion to therapeutic usage. This chapter's goal is to provide a summary of how green synthetic platinum is used and characterized.

2 Green Synthesis of Platinum Nanoparticles

In order to decrease and stabilize platinum ions, green synthesis techniques for PtNPs use natural resources such as plant extracts, bacteria, fungi, and algae. These sources contain bioactive substances such as phenolic compounds, terpenoids, and flavonoids that function as stabilizing and reducing agents (Yang et al., 2017). Aqueous synthesis, microwave-assisted synthesis, and sonochemical synthesis are just a few of the different ways the green synthesis process which can be carried out (Singh et al., 2016). Compared to conventional approaches, green synthesis techniques have a number of benefits, such as improved yields, scalability, and lower toxicity. The employment of toxic chemicals and solvents in conventional synthesis techniques like chemical reduction and electrochemical reduction leads to the production of hazardous wastes and byproducts. Green synthesis techniques are

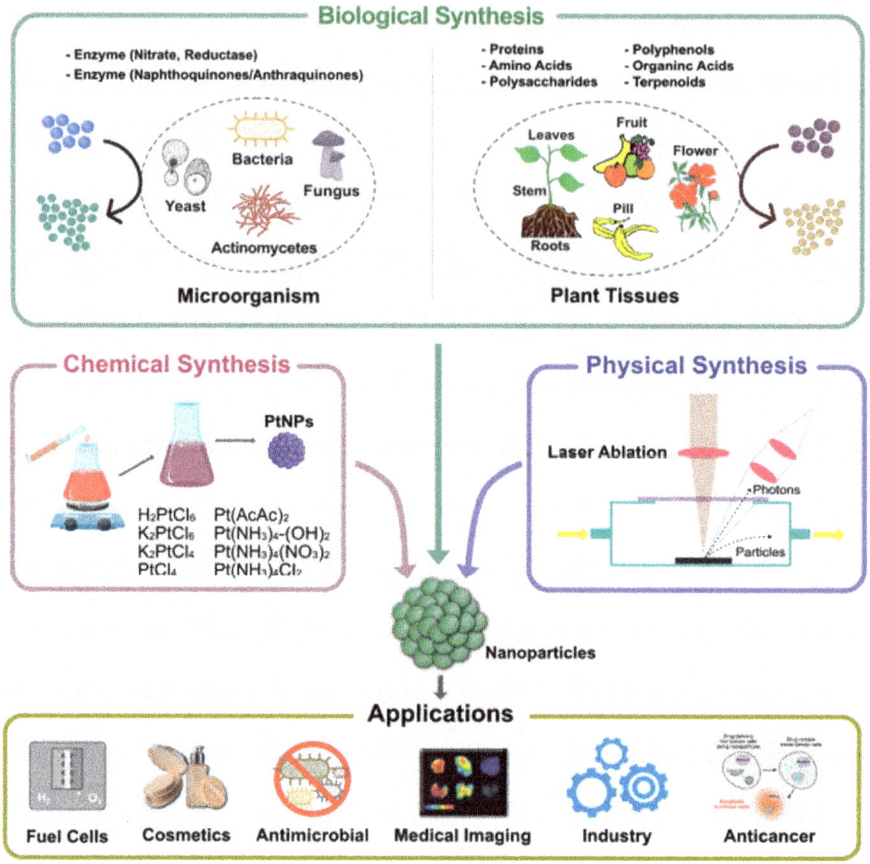

Fig. 1 Synthesis methods and applications of PtNPs (Jeyaraj et al., 2019)

more sustainable and environmentally friendly because they don't employ as many hazardous chemicals and solvents. A regulated mixture of a platinum precursor and a stabilizing and reducing agent, usually a plant extract, is required to create green platinum nanoparticles. Typically, the reaction takes place at ambient temperature and with normal atmospheric pressure. The size, shape, and stability of the nanoparticles can be controlled by adjusting reaction parameters such as pH, temperature, and reaction time (Siddiqi & Husen, 2016).

The green manufacture of ptNPs using gum ghatti (*Anogeissus latifolia*), a naturally occurring polymer derived from the bark of the Indian tree, is discussed in this research paper along with its use in dye-sensitized solar cells. Gum ghatti was utilized in the production of PtNPs as a reducing agent and stabilizer, producing stable nanoparticles with an average size of 5–10 nm. Several methods, including UV–visible spectroscopy, transmission electron microscopy, X-ray diffraction, and Fourier-transform infrared spectroscopy, were used to characterize the synthesized nanoparticles. Platinum nanoparticle-based dye-sensitized solar cells performed better than cells without PtNPs in the study's evaluation of photocurrent and photovoltage, which were also measured. According to the findings, gum ghatti may be a viable option for the environmentally friendly production of PtNPs and their use in renewable energy systems (Kora et al. 2018; Pal et al., 2019).

Selvi and coworkers (2020) used the aqueous extract of *T. arjuna* bark as a reducing agent in order to synthesize PtNPs. The synthesis was done at standard conditions and UV–Vis, TEM, and XRD analyses were used to characterize the synthesized PtNPs. The synthesized PtNPs showed outstanding antibacterial activity against gram-negative and gram-positive bacteria (Selvi et al., 2020). As different study shows, PtNPs were synthesized by using various plant materials such as *Punica granatum* peel, and *Azadirachta indica* plant extracts as reducing agents. The synthesis was done at room temperature, and UV–Vis, TEM, and XRD analyses were used to characterize the produced PtNPs. The synthesized PtNPs have shown outstanding antibacterial and antioxidant activity as well as cytotoxicity against cancer cells (Fahmy et al., 2020; Şahin et al., 2018). The formation of PtNPs for a range of biomedical applications can be done safely, sustainably, and inexpensively using green synthesis processes. Natural resources are employed to reduce and stabilize the use of dangerous chemicals and solvents, making them more ecologically friendly and sustainable.

3 Characterization of Platinum Nanoparticles

Maximizing the usage of NPs in various therapeutic applications requires an understanding of their physicochemical properties. The physical and chemical properties of PtNPs are crucial for their biological applications. PtNPs have distinct physicochemical properties due to their large surface area, high reactivity, and catalytic activity. It is possible to alter the surface chemistry of nanoparticles by coating them with different molecules, such as polymers, peptides, and antibodies. And also, the size and structure of PtNPs have a considerable impact on their biological applications. The smaller nanoparticles have higher surface area-to-volume ratios, which enhance their reactivity and cell absorption. The shape of nanoparticles can have an impact on both the biodistribution and cellular absorption of those particles (Xiang et al., 2018). Therefore, it is essential to characterize Pt NPs in order to know the most of their uses in various medicinal applications. The common once are the shape, size, and surface chemistry of nanoparticles which have a significant impact on their biological activity and interactions with biological systems.

The various techniques which are used to determine synthesized PtNPs are: Energy Dispersive X-Ray Spectroscopy (EDS), X-Ray Photoelectron Spectroscopy (XPS), UV–Visible Spectroscopy (UV), X-Ray Fluorescence (XRF), Hydrophobic Interaction Chromatography (HIC), Khan et al. (2021), Transmission Electron Microscopy (TEM), Dynamic Light Scattering (DLS), Fourier Transform Infrared Spectroscopy (FTIR) and Powder X-Ray Diffraction (XRD) are common examples

(Aygun et al., 2020; Şahin et al., 2018). For the successful synthesis of Pt NPs, plant extracts and biological derivatives were used. The particle was then characterized utilizing a variety of advanced techniques (Figs. 2, 3 and 4). Figure 2 displays the UV–vis spectra of PtNPs and Cordyceps flower extract (CE). Liu and coworkers study confirmed that the formation of PtNPs was because of the maximum wavelength of CE and PtNPs in the absorption spectra was 260 nm. The fact that CE was applied to the surface of PtNPs as a stabilizer during the reduction of platinum ions to platinum atoms may be the cause of the comparable absorption peaks (Liu et al., 2022).

Aygun and coworkers employed XRD, TEM, and XPS techniques to ascertain the crystal structure, particle morphology, and size of Pt NPs. The synthesized Pt NPs are displayed in Fig. 3 TEM image and it can be seen that the biogenic Pt NPs had spherical shapes with an average particle size of 3.47 ± 1.31 nm (Fig. 3b). The atomic lattice fringe of biogenic Pt nanoparticles is 0.22 nm, according to the HR-TEM data. The findings were generally in line with the body of literature. Figure 3c illustrates how the elemental energy loss spectroscopy (TEM–EELS) measurements revealed the presence of Pt in the produced biogenic nanoparticles. Additionally, the generated biogenic nanoparticles' XRD pattern had unique peaks. They were crystallized with a face-centered cubic structure and had four distinct diffraction peaks at approximately 39.9, 46.2, 67.3, and 82.1°, which can be assigned to Pt (111), (200), (220), and (311) based on the spectra shown in Fig. 4a. By employing the plant extract, it was evident that the platinum nanoparticles reduced the Pt(IV) ions (Aygun et al., 2020).

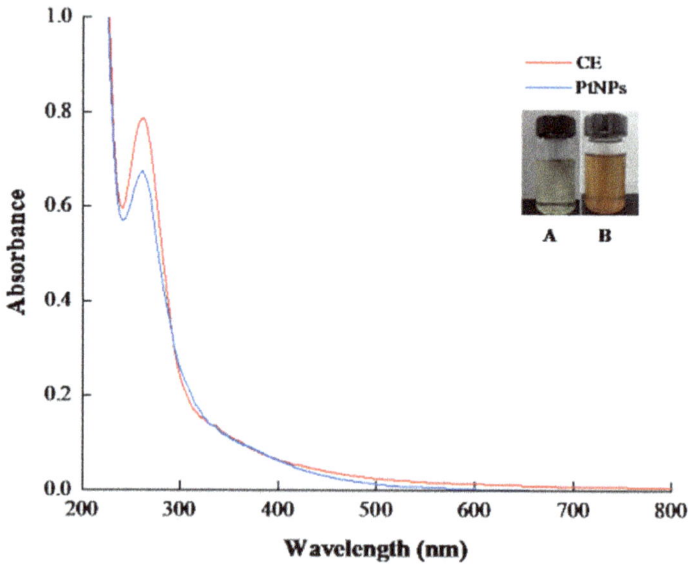

Fig. 2 UV–vis spectra of CE and PtNPs. **a** a mixture of Hexachloroplatinic acid and CE; **b** a mixture of the remaining CE and the synthetic PtNPs (adapted from: Liu et al., 2022)

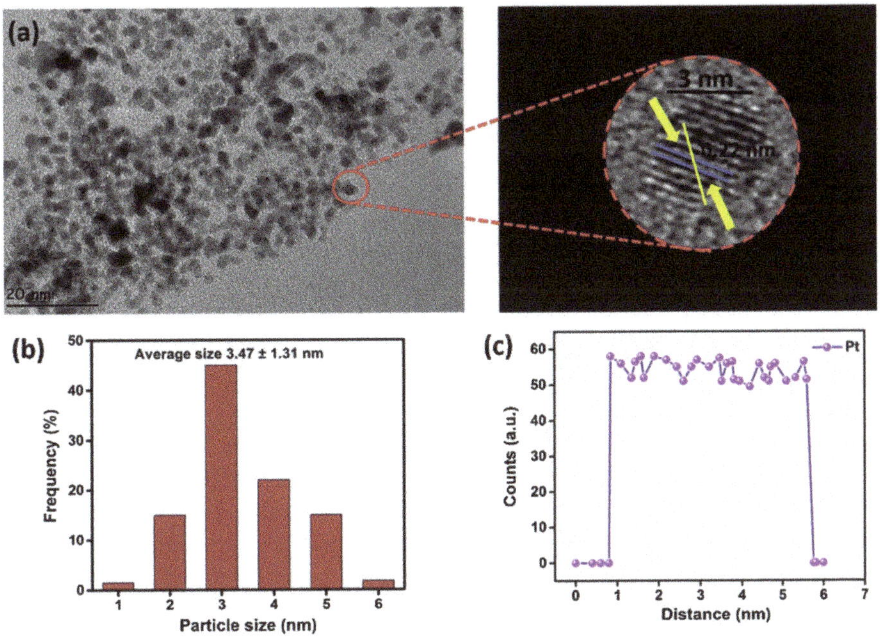

Fig. 3 Representative TEM image and HRTEM image (**a**), particle size histogram (**b**), EELS line profile (**c**) of biogenic PtNPs

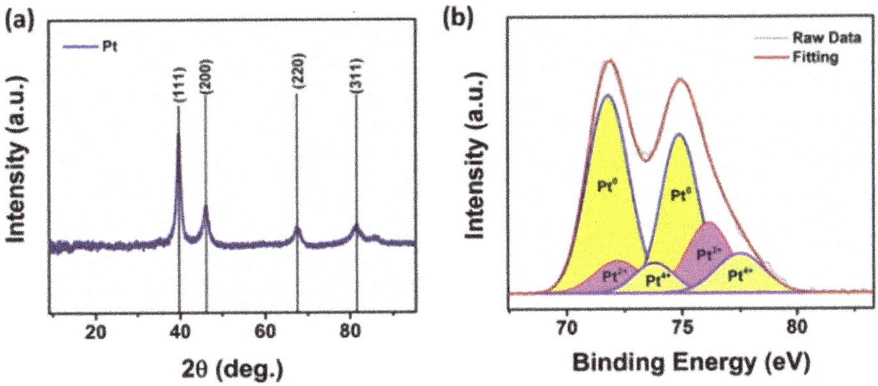

Fig. 4 XRD (**a**) and XPS (**b**) of PtNPs synthesized from black cumin seed extracts (Aygun et al., 2020)

4 Biomedical Applications of Platinum Nanoparticles

Due to their distinctive physicochemical characteristics, PtNPs have generated a great deal of attention in biomedical applications. Numerous biomedical applications (Fig. 5), such as cancer treatment, imaging and diagnosis, and drug delivery, have been investigated for PtNPs. One of the most thoroughly researched uses of PtNPs is cancer therapy. By causing apoptosis and preventing cancer cell proliferation, PtNPs have been demonstrated to exhibit powerful anticancer action. By enhancing drug transport and lowering toxicity to healthy tissues, PtNPs can also overcome the drawbacks of traditional platinum-based chemotherapy (Jeyaraj et al., 2019). The green synthesis of platinum nanoparticles from plants and their applications are shown in Table 1.

The use of PtNPs in imaging and medical diagnosis has also been investigated. PtNPs are well suited for application in a variety of imaging modalitics, including computed tomography (CT), magnetic resonance imaging (MRI), and optical imaging, as they have been found to have good optical and magnetic properties (Chen et al., 2017). Additionally, targeting moieties can be added to platinum nanoparticles to enable them to bind only to cancer cells, enhancing the sensitivity and specificity of imaging and diagnosis. Another interesting use for PtNPs is drug delivery. To improve their cellular absorption and target particular cells or tissues, PtNPs can be functionalized with a variety of compounds. By encapsulating medications inside of the nanoparticles and shielding them from degradation and clearance, PtNPs can also be employed to increase the bioavailability and effectiveness of medications (Bloch et al., 2021).

Şahin and coworkers studied the utilization of biosynthesized monodisperse PtNPs from *Punica granatum* crusts as anti-tumor agents on the human breast cancer cell line, MCF-7. After synthesis, they characterized obtained PtNPs by using the UV–vis spectrum, XRD, TEM, FESEM, and FTIR. The efficiency of synthesized PtNPs was evaluated by flow cytometry, cell viability, propidium iodide staining test, and comet tests on the MCF-7 cancer cell line. The cell existence percentage was

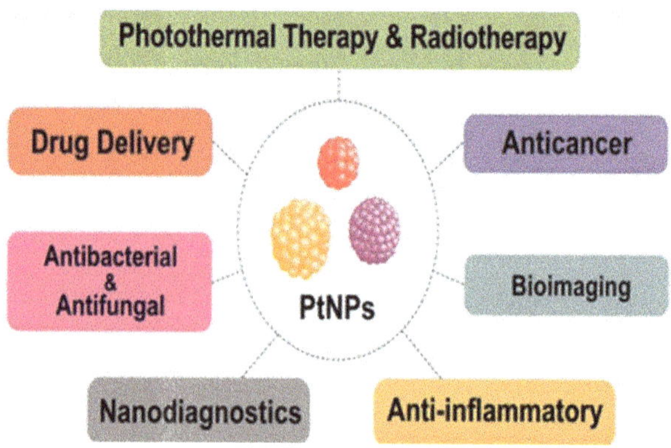

Fig. 5 Multifarious applications of platinum nanoparticles (Jeyaraj et al., 2019)

Table 1 Green synthesis of platinum nanoparticle fromplant material and biomedical aplications

Plant name	Family name	Plant part used	Major Phytochemicals	Type of extract	Biomedical application	References
Atriplex halimus	Amaranthaceae	Leaves	Alkaloids, phenolic acids, flavonoids, tannins, and saponins	Crude	For potential antimicrobial, and antioxidant activities	Eltaweil et al. (2022)
Barleria prionitis	Acanthaceae	Leaves	Phenolic acids, flavonoids, and phytosterols	Crude	Activity against human breast adenocarcinoma (mcf-7) cell lines	Rokade et al. (2017)
Centella asiatica	Apiaceae	Leaves	Flavonoids	Pure flavonoid	Antibacterial and anticancer activity	Kavitha (2021)
Combretum erythrophyllum	Combretaceae	Leaves	Flavonoids, carbohydrates, phenols alkaloids, and sterols	Crude	Antibacterial activities	Fanoro et al. (2021)
Cordyceps	Cordycipitaceae	Flower	carbohydrates, terpenoids and flavonoids	Crude	Antioxidant and antibacterial activities	Liu et al. (2022)
Gloriosa superba	Colchicaceae	Tuber	Tannins, flavonoids, phenols, saponins, alkaloids, steroids, glycosides, and terpenoids	Crude	Anticancer activity against mcf-7 cells	Rokade et al. (2018)
Nigella sativa	Ranunculaceae	Seed	Alkaloids, terpenoids, saponins, and tannins	Crude	Antimicrobial and anticancer agents	Aygun et al. (2020)
Nymphaea tetragona	Nymphaeaceae	Flower	Flavonoids	Crude	Palpitation of heart, blood purifier and aphrodisiac activities	(Zhang et al. 2023)
Ocimum sanctum	Lamiaceae	Leaves	Anthocyanins, tannins, phenolic acids, phytosterols, and policosanols	Crude	Cancer therapy	Prabhu and Gajendran (2017)
Polygonum salicifolium	Polygonaceae	Leaves	Flavonoids, glycosides, terpenoids, and alkaloids	Crude	Antimicrobial, antioxidant, and antidiabetic activities	Hosny et al. (2022)

(continued)

Table 1 (continued)

Plant name	Family name	Plant part used	Major Phytochemicals	Type of extract	Bicmedical application	References
Punica granatum	Punicaceae	Fruit peel	Saponins, phenols, flavonoids, steroids, quinones, cardinalities, tannins and terpenoids	Cude	Inhibition of proliferation of human breast cancer cell	Şahin et al. (2018)
Prosopis farcta	Fabaceae	Fruits	Steroids, alkaloids, phenolics, terpenes tannins, and flavonoids	Crude	Contrast agents for computed tomography imaging applications	Jameel et al. (2021)
Prunus x yedoensis	Rosaceae	Gum	Flavonoids, phenolic acids steroids, and terpenes,	Crude	Antifungal activity	Velmurugan et al. (2016)
Peganum harmala	Nitrariaceae	Seed	Alkaloids	Pure alkaloid	Antioxidant activity and antitumor activities	Fahmy et al. (2021)
Salix tetraspeama	Salicaceae	Leaves	Terpenoids, polysaccharides, proteins, tannins, and polyphenols	Crude	Antioxidant and Anticancer activities	Ramachandiran et al. (2021)
Saudi's dates (ajwa and barni)	Arecaceae	Fruit	Flavonoids, amino acids, sugar, organic acid, and phenol	Cude	Cancer cell treatment, antibacterial effect	Al-Radadi (2019)
Tragia involucrata	Euphorbiaceae	Leaves	Tannins, flavonoids, phenols, steroids, amino acids, carbohydrates, and protein	Crude	Antibacterial, antioxidant, and anticancer activities	Selvi et al. (2020)
Water hyacinth	Pontederiaceae	Fresh leaves	Phenolics, flavonoids, and glutathione	Crude	Not identified	Leo and Oluwafemi (2017)
Xanthium strumarium	Asteraceae	Leaves	Phenyl propenoids, triterpenoids, sesquiterpenoids, and lignanoids	Crude	Antibacterial, antifungal, and cytotoxic activities	Kumar et al. (2019)

resoluted by 3-(4,5-dimethylthiazol-2-yl)-2,5-diphenyltetrazolium bromide assay. After 48 h of incubation, the biosynthesized monodisperse PtNPs suppressed MCF-7 proliferation with an IC_{50} of 17.84 μg/ml. By fragmenting molecular DNA, the monodisperse PtNPs caused apoptosis, as shown by propidium iodide staining (Şahin et al., 2018).

Gutiérrez and coworkers' study shows that PtNPs have antioxidant activity because they reduce oxidative stress by removing intracellular reactive oxygen species. It has also been demonstrated that certain chemicals that have been conjugated to nanoparticle surfaces have a higher bioavailability, which increases their ability to enter cells. Similar to functionalization, they make attractive candidates for antibacterial drugs due to their inhibitory impact against Gram-positive and Gram-negative microbes (de la Rosa et al., 2022). In conclusion, PtNPs have numerous potential biomedical uses, such as cancer treatment, imaging and diagnosis, and medication administration. To fully comprehend the potential of PtNPs in these applications and to enhance their characteristics for clinical usage, more study is required.

5 Toxicity and Safety Considerations

Before being used in biomedical applications, platinum nanoparticles' toxicity, and safety must be thoroughly assessed, just like with any possible medicinal drug. The potential toxicity of platinum nanoparticles has been the subject of numerous research, with varying degrees of success. Platinum nanoparticles have been shown in several studies to cause cytotoxicity and genotoxicity in cells and tissues. According to other research, platinum nanoparticles can cause tissue damage and cell death by inducing oxidative stress, inflammation, and cell death (Wang et al., 2014). The toxicity of platinum nanoparticles is also extremely dependent on their size, shape, surface chemistry, and dose, which must be taken into consideration. Particles having sharp edges or irregular forms have been found to be more cytotoxic than smooth, spherical particles, while smaller platinum nanoparticles have been demonstrated to be more poisonous than bigger particles It is essential to carefully manage the synthesis and functionalization of platinum nanoparticles in order to assure their safety for biological applications. According to studies (Parveen et al., 2016), green synthesis techniques produce nanoparticles that are less hazardous than those made using conventional techniques. Platinum nanoparticles can be made more biocompatible and less poisonous by being functionalized with targeted molecules or biocompatible polymers (Wang et al., 2014). Platinum nanoparticle toxicity and safety can be assessed using a variety of techniques, such as in vivo animal studies and in vitro cell-based experiments. Imaging methods like CT or MRI can be used to track the biodistribution and clearance of platinum nanoparticles In vivo in order to assess their toxicity. Finally, before using platinum nanoparticles in biomedical applications, it is important to carefully assess their potential toxicity. However, platinum nanoparticles can be made more dependable and biocompatible for use in a range of therapeutic and diagnostic applications with proper synthesis and functionalization.

6 Future Directions and Conclusions

Even while the use of PtNPs in medical applications shows significant probability, there are quite a number of limitations and challenges that need to be resolved. A critical limitation that demands careful study and modification is the potential toxicity of the PtNPs. Another challenge is the targeted delivery of the PtNPs to certain tissues or cells, which is affected by factors like method of administration, particle size, and surface chemistry. There are a few intriguing areas for this field's upcoming study and improvement despite these barriers. One area of focus is the development of more effective targeting strategies, such as the use of certain ligands or antibodies to drive the PtNPs to cancer cells. Another area of study is the development of functionalization and synthesis procedures to produce PtNPs with improved biocompatibility and reduced toxicity. PtNPs may also help other therapeutic medicines, including immune checkpoint inhibitors or chemotherapeutic medications, work more effectively while reducing their negative effects. The environmentally friendly production of PtNPs enables the development of safe and efficient medicines for a number of biomedical applications. While there are still problems to be overcome, it is projected that the ongoing research and development will lead to significant breakthroughs in this field over the coming few years.

References

Al-Radadi, N. S. (2019). Green synthesis of platinum nanoparticles using Saudi's Dates extract and their usage on the cancer cell treatment. *Arabian Journal of Chemistry, 12*(3), 330–349.

Aruna, K., & Kavitha, K. (2021). Synthesis and characterization of platinum nanoparticles (Pt-NPs) from Centella Asiatica L leaf extract. *Annals of the Romanian Society for Cell Biology*, 16–22.

Aygun, A., Gülbagca, F., Ozer, L. Y., Ustaoglu, B., Altunoglu, Y. C., Baloglu, M. C., Atalar, M. N., Alma, M. H., & Sen, F. (2020). Biogenic platinum nanoparticles using black cumin seed and their potential usage as antimicrobial and anticancer agent. *Journal of Pharmaceutical and Biomedical Analysis, 179*, 112961.

Bachheti, R. K., Konwarh, R., Gupta, V., Husen, A., & Joshi, A. (2019). Green synthesis of iron oxide nanoparticles: Cutting edge technology and multifaceted applications. In *Nanomaterials and plant potential* (pp. 239–259). Springer. https://doi.org/10.1007/978-3-030-05569-1_9

Bachheti, R. K., Fikadu, A., Bachheti, A., & Husen, A. (2020). Biogenic fabrication of nanomaterials from flower-based chemical compounds, characterization and their various applications: A review. *Saudi Journal of Biological Sciences, 27*(10), 2551–2562.

Bachheti, R. K., Godebo, Y., Bachheti, A., Yassin, M. O., & Husen, A. (2020a). Root-based fabrication of metal/metal-oxide nanomaterials and their various applications. In A. Husen & M. Jawaid (Eds.), *Nanomaterials for agriculture and forestry applications* (pp.135–166). Elsevier. https://doi.org/10.1016/B978-0-12-817852-2.00006-8

Bachheti, A., Bachheti, R. K., Abate, L., & Husen, A. (2022). Current status of Aloe-based nanoparticle fabrication, characterization and their application in some cutting-edge areas. *South African Journal of Botany, 147*, 1058–1069.

Bachheti, R. K., Abate, L., Bachheti, A., Madhusudhan, A., & Husen, A. (2021). Algae-, fungi-, and yeast-mediated biological synthesis of nanoparticles and their various biomedical applications. In *Handbook of greener synthesis of nanomaterials and compounds* (pp. 701–34). Elsevier. https://doi.org/10.1016/B978-0-12-821938-6.00022-0

Bloch, K., Pardesi, K., Satriano, C., & Ghosh, S. (2021). Bacteriogenic platinum nanoparticles for application in nanomedicine. *Frontiers in Chemistry, 9*, 624344.

Chen, Q., Liang, C., Sun, X., Chen, J., Yang, Z., Zhao, H., Feng, L., & Liu, Z. (2017). H_2O_2-responsive liposomal nanoprobe for photoacoustic inflammation imaging and tumor theranostics via in vivo chromogenic assay. *Proceedings of the National Academy of Sciences, 114*(21), 5343–5348.

Dhand, C., Dwivedi, N., Loh, X. J., Ying, A. N. J., Verma, N. K., Beuerman, R. W., Lakshminarayanan, R., & Ramakrishna, S. (2015). Methods and strategies for the synthesis of diverse nanoparticles and their applications: a comprehensive overview. *RSC Advances, 5*(127), 105003–105037.

Eltaweil, A. S., Fawzy, M., Hosny, M., Abd El-Monaem, E. M., Tamer, T. M., & Omer, A. M. (2022). Green synthesis of platinum nanoparticles using Atriplex halimus leaves for potential antimicrobial, antioxidant, and catalytic applications. *Arabian Journal of Chemistry, 15*(1), 103517.

Fahmy, S. A., Fawzy, I. M., Saleh, B. M., Issa, M. Y., Bakowsky, U., & Azzazy, H. M. E. S. (2021). Green synthesis of platinum and palladium nanoparticles using Peganum harmala L. seed alkaloids: Biological and computational studies. *Nanomaterials, 11*(4), 965.

Fahmy, S. A., Preis, E., Bakowsky, U., & Azzazy, H. M. E. S. (2020). Platinum nanoparticles: green synthesis and biomedical applications. *Molecules, 25*(21), 4981.

Fanoro, O. T., Parani, S., Maluleke, R., Lebepe, T. C., Varghese, R. J., Mgedle, N., Mavumengwana, V., & Oluwafemi, O.S. (2021). Biosynthesis of smaller-sized platinum nanoparticles using the leaf extract of combretum erythrophyllum and its antibacterial activities. *Antibiotics, 10*(11), 1275.

Gutiérrez de la Rosa, S. Y., Muñiz Diaz, R., Villalobos Gutiérrez, P. T., Patakfalvi, R., & Gutiérrez Coronado, Ó. (2022). Functionalized platinum nanoparticles with biomedical applications. *International Journal of Molecular Sciences, 23*(16), 9404.

Hosny, M., Fawzy, M., El-Fakharany, E. M., Omer, A. M., Abd El-Monaem, E. M., Khalifa, R. E., & Eltaweil, A. S. (2022). Biogenic synthesis, characterization, antimicrobial, antioxidant, antidiabetic, and catalytic applications of platinum nanoparticles synthesized from Polygonum salicifolium leaves. *Journal of Environmental Chemical Engineering, 10*(1), 106806.

Husen, A., Rahman, Q. I., Iqbal, M., Yassin, M. O., & Bachheti, R. K. (2019). Plant-mediated fabrication of gold nanoparticles and their applications. In *Nanomaterials and plant potential* (pp. 71–110). Springer. https://doi.org/10.1007/978-3-030-05569-1_3

Husen, A. (2022). *Engineered nanomaterials for sustainable agricultural production, soil improvement and stress management*. Elsevier Inc. https://doi.org/10.1016/C2021-0-00054-7

Husen, A. (2023a). *Nanomaterials from agricultural and horticultural products*. Springer Nature Singapore Pte Ltd.

Husen, A. (2023b). *Nanomaterials and nanocomposites exposures to plants (Response, Interaction, Phytotoxicity and Defense Mechanisms)*. Springer Nature Singapore Pte Ltd. https://doi.org/10.1007/978-981-99-2419-6

Husen, A., Bachheti, R. K., & Bachheti, A. (2023a). *Nanomaterials for environmental and agricultural sectors*. Springer Nature Singapore Pte Ltd. https://doi.org/10.1007/978-981-99-2874-3

Husen, A., Bachheti, R. K., & Bachheti, A. (2023b). *Current trends in green nano-emulsions (Food, Agriculture and Biomedical Sectors)*. Springer Nature Singapore Pte Ltd.

Husen, A., & Iqbal, M. (2019). *Nanomaterials and plant potential*. Springer International Publishing AG. https://doi.org/10.1007/978-3-030-05569-1

Husen, A., & Siddiqi, K. S. (2023). *Advances in smart nanomaterials and their applications*. Elsevier Inc. https://doi.org/10.1016/C2021-0-02202-1

Jameel, M. S., Aziz, A. A., Dheyab, M. A., Mehrdel, B., Khaniabadi, P. M., & Khaniabadi, B. M. (2021). Green sonochemical synthesis platinum nanoparticles as a novel contrast agent for computed tomography. *Materials Today Communications, 27*, 102480.

Jeyaraj, M., Gurunathan, S., Qasim, M., Kang, M. H., & Kim, J. H. (2019). A comprehensive review on the synthesis, characterization, and biomedical application of platinum nanoparticles. *Nanomaterials, 9*(12), 1719.

Khan, M. A. R., Al Mamun, M. S., & Ara, M. H. (2021). Review on platinum nanoparticles: Synthesis, characterization, and applications. *Microchemical Journal, 171*, 106840.

Kora, A. J., & Rastogi, L. (2018). Green synthesis of palladium nanoparticles using gum ghatti (*Anogeissus latifolia*) and its application as an antioxidant and catalyst. *Arabian Journal of Chemistry, 11*(7), 1097–1106.

Kumar, P. V., Kala, S. M. J., & Prakash, K. S. (2019). Green synthesis derived Pt-nanoparticles using Xanthium strumarium leaf extract and their biological studies. *Journal of Environmental Chemical Engineering, 7*(3), 103146.

Leo, A. J., & Oluwafemi, O. S. (2017). Plant-mediated synthesis of platinum nanoparticles using water hyacinth as an efficient biomatrix source–An eco-friendly development. *Materials Letters, 196*, 141–144.

Liu, L., Jing, Y., Guo, A., Li, X., Li, Q., Liu, W., & Zhang, X. (2022). Biosynthesis of platinum nanoparticles with cordyceps flower extract: characterization, antioxidant activity and antibacterial activity. *Nanomaterials, 12*(11), 1904.

Naseer, A., Ali, A., Ali, S., Mahmood, A., Kusuma, H. S., Nazir, A., Yaseen, M., Khan, M. I., Ghaffar, A., Abbas, M., & Iqbal, M. (2020). Biogenic and eco-benign synthesis of platinum nanoparticles (Pt NPs) using plants aqueous extracts and biological derivatives: environmental, biological and catalytic applications. *Journal of Materials Research and Technology, 9*(4), 9093–9107.

Painuli, S., Semwal, P., Bacheti, A., Bachheti, R. K., & Husen, A. (2020). Nanomaterials from nonwood forest products and their applications. In A. Husen & M. Jawaid (Eds.), *Nanomaterials for agriculture and forestry applications* (pp. 15–40). Elsevier. https://doi.org/10.1016/B978-0-12-817852-2.00002-0

Pal, P., Singh, S. K., Mishra, S., Pandey, J. P., & Sen, G. (2019). Gum ghatti based hydrogel: Microwave synthesis, characterization, 5-Fluorouracil encapsulation and 'in vitro'drug release evaluation. *Carbohydrate Polymers, 222*, 114979.

Parveen, K., Banse, V., & Ledwani, L. (2016). Green synthesis of nanoparticles: Their advantages and disadvantages. In *Paper Presented at the AIP Conference Proceedings*.

Prabhu, N., & Gajendran, T. (2017). Green synthesis of noble metal of platinum nanoparticles from Ocimum sanctum (Tulsi) Plant-extracts. *IOSR Journal of Biotechnology and Biochemistry, 3*, 107–112.

Ramachandiran, D., Elangovan, M., & Rajesh, K. (2021). Structural, optical, biological and photocatalytic activities of platinum nanoparticles using salixtetrasperma leaf extract via hydrothermal and ultrasonic methods. *Optik, 244*, 167494.

Rokade, S. S., Joshi, K. A., Mahajan, K., Patil, S., Tomar, G., Dubal, D. S., Parihar, V. S., Kitture, R., Bellare, J.R., & Ghosh, S. (2018). Gloriosa superba mediated synthesis of platinum and palladium nanoparticles for induction of apoptosis in breast cancer. *Bioinorganic Chemistry and Applications*.

Rokade, S. S., Joshi, K. A., Mahajan, K., Tomar, G., Dubal, D. S., Singh, V., Kitture, R., Bellare, J., & Ghosh, S., (2017). Novel anticancer platinum and palladium *nanoparticles from Barleria prionitis. Global Journal of Nanomedicine, 2*(5), 555600.

Şahin, B., Aygün, A., Gündüz, H., Şahin, K., Demir, E., Akocak, S., & Şen, F. (2018). Cytotoxic effects of platinum nanoparticles obtained from pomegranate extract by the green synthesis method on the MCF-7 cell line. *Colloids and Surfaces B: Biointerfaces, 163*, 119–124.

Selvi, A. M., Palanisamy, S., Jeyanthi, S., Vinosha, M., Mohandoss, S., Tabarsa, M., You, S., Kannapiran, E., & Prabhu, N. M. (2020). Synthesis of Tragia involucrata mediated platinum nanoparticles for comprehensive therapeutic applications: Antioxidant, antibacterial and mitochondria-associated apoptosis in HeLa cells. *Process Biochemistry, 98*, 21–33.

Siddiqi, K. S., & Husen, A. (2016). Green synthesis, characterization and uses of palladium/platinum nanoparticles. *Nanoscale Research Letters, 11*, 1–13.

Singh, P., Kim, Y. J., Zhang, D., & Yang, D. C. (2016). Biological synthesis of nanoparticles from plants and microorganisms. *Trends in Biotechnology, 34*(7), 588–599.

Thakkar, K. N., Mhatre, S. S., & Parikh, R. Y. (2010). Biological synthesis of metallic nanoparticles. *Nanomedicine: Nanotechnology, Biology and Medicine, 6*(2), 257–262.

Velmurugan, P., Shim, J., Kim, K., & Oh, B. T. (2016). Prunus× yedoensis tree gum mediated synthesis of platinum nanoparticles with antifungal activity against phytopathogens. *Materials Letters, 174*, 61–65.

Wang, X., Ji, Z., Chang, C. H., Zhang, H., Wang, M., Liao, Y. P., Lin, S., Meng, H., Li, R., Sun, B., & Winkle, L.V. (2014). Use of coated silver nanoparticles to understand the relationship of particle dissolution and bioavailability to cell and lung toxicological potential. *Small, 10*(2), 385–398.

Xiang, Z. P., Deng, H. Q., Peljo, P., Fu, Z. Y., Wang, S. L., Mandler, D., Sun, G. Q., & Liang, Z. X. (2018). Electrochemical dynamics of a single platinum nanoparticle collision event for the hydrogen evolution reaction. *Angewandte Chemie International Edition, 57*(13), 3464–3468.

Yang, C., Wang, M., Zhou, J., & Chi, Q. (2017). Bio-synthesis of peppermint leaf extract polyphenols capped nano-platinum and their in-vitro cytotoxicity towards colon cancer cell lines (HCT 116). *Materials Science and Engineering: C, 77*, 1012–1016.

Zhang, Y., Cheng, S., Jia, H., Zhou, J., Xi, J., Wang, J., Chen, X., & Wu, L. (2023). Green synthesis of platinum nanoparticles by Nymphaea tetragona flower extract and their skin lightening, antiaging effects. *Arabian Journal of Chemistry, 16*(1), 104391.

Role of Zinc Oxide Nanomaterials for Photocatalytic Degradation of Environmental Pollutants

Addisu Tamir Wassie, Ibrahim Nasser Ahmed, Archana Bachheti, Azamal Husen, and Rakesh Kumar Bachheti

Abstract The environment is seriously exposed to numerous pesticides and chemical compounds produced by quick industrialization and pharmaceutical firms. The ability of new photocatalysts based on zinc oxide nanomaterials to absorb these organic and inorganic contaminants from the environment is very promising. The photocatalysts have many outstanding qualities, including excellent stability, nontoxicity, and photocatalytic degradation potential. However, several drawbacks, like poor affinity, a large band gap, particle agglomeration, and recapture difficulties, are also connected to using these photocatalysts. Therefore, optimization is necessary to raise their effectiveness while ensuring they are affordable and long-lasting. This chapter highlights the current advances in photocatalytic principles and mechanisms for

A. T. Wassie · I. N. Ahmed · R. K. Bachheti (✉)
Department of Industrial Chemistry, College of Applied Sciences, Addis Ababa Science and Technology University, P.O. Box 16417 Addis Ababa, Ethiopia
e-mail: rkbacheti@gmail.com

A. T. Wassie
Department of Chemistry, College of Natural Sciences, Wollo University, P.O. Box 1145 Dessie, Ethiopia

I. N. Ahmed · R. K. Bachheti
Nanotechnology Center of Excellence, Department of Industrial Chemistry, Addis Ababa Science and Technology University, Addis Ababa, Ethiopia

A. Bachheti
Department of Environment Science, Graphic Era (Deemed to Be University), Dehradun, India

A. Husen
Department of Biotechnology, Smt. S. S. Patel Nootan Science & Commerce College, Sankalchand Patel University, Visnagar 384315, Gujarat, India

Department of Biotechnology, Graphic Era (Deemed to be University), Dehradun 248002, Uttarakhand, India

Laboratory of Bio Resource Management, Institute of Tropical Forestry and Forest Product, University Putra Malaysia, 43400 UPM Serdang, Selangor, Malaysia

R. K. Bachheti
Department of Allied Sciences, Graphic Era Hill University (G.E.H.U), Society Area, Clement Town, Dehradun, India

R. K. Bachheti et al. (eds.), *Metal and Metal-Oxide Based Nanomaterials*, Smart Nanomaterials Technology, https://doi.org/10.1007/978-981-99-7673-7_14

environmental pollutants degradation, synthesis and characterization of ZnO nanomaterials, photocatalytic degradation of environmental pollutants by ZnO nanomaterials, and different enhancement strategies for ZnO photocatalysis that increase the elimination efficiency of ZnO-based photocatalysts.

Keywords Zinc oxide · Nanomaterial · Photocatalytic degradation · Pollutants · Environment

1 Introduction

The growth of nanotechnology has opened up advanced options and given people worldwide a broader perspective on what they may accomplish by manipulating materials at the nanoscale (Cruz et al., 2020; Zcbeaman et al., 2023). Nanoscience and technology have made enormous strides in recent years, making them appear as promising options for energy generation and environmental cleanup. Over the years, nanomaterials (NMs) have created a wide range of new opportunities for manufacturing and industrial applications (Husen & Siddiqi, 2023). Compared to bulk materials, nanomaterials have many advantages, such as a better capability for light absorption, more reactive sites, and more surface area (Bachheti et al., 2019; Husen & Siddiqi, 2023; Husen et al., 2019; Husen & Iqbal, 2019; Painuli et al., 2020). Some applications of nanomaterials include wastewater cures and the elimination of dangerous pollutants in the atmosphere (Raja & Husen, 2020; Raja et al., 2023). Toxins are released as industrialization advances, and dangerous chemicals are released into the atmosphere. In this context, various organic and hazardous industrial pollutants were extensively treated using techniques like immobilization, biological and chemical oxidation, and ignition (Khan & Khan, 2019). Nowadays, semiconductor-driven heterogeneous photocatalysis operating at ambient circumstances has earned the paramount signature in solar energy alteration, wastewater decontamination, and the inactivation of various dangerous microbes as one of the practical techniques in innovative oxidation processes. This adaptable green strategy, essential for the degradation of numerous harmful pollutants, denotes the whole change of organic carbon to CO_2 without explicitly exiting any additional hazardous by-products (Kumar & Rao, 2017).

Due to its versatility, nanostructured photocatalysis is an appealing and exceptional alternative for the environment. Some of these features include the requirement of less harsh conditions, such as low temperature, the lack of a secondary waste controlling problem, and the manufacture of hydrogen. At the same time, the air is being purified, the complete decomposition of target pollutants, antibacterial activities, and the improvement of oxidation in the presence of sunlight or UV light. Many conventional strategies are anticipated to treat environmental pollutants, including precipitation, membrane filtration, and other biological techniques, but these strategies have their drawbacks. Even if they are economical and environmentally benign, biological approaches like microalgae cannot remove or significantly lower chemical oxygen demand values caused by dyes and other less biodegradable materials. Despite being straightforward and effective, chemical treatments produce sludge, require higher chemical concentrations to handle waste, and consume large amounts of chemicals. Despite being secure and very effective, conventional photocatalysts like membrane technology have high maintenance costs and are either nonrecyclable or challenging to recover (Chen et al., 2020).

Photocatalysis intermediated through NMs is the most popular and economical method for eliminating environmental contaminants. This is because NMs have the exceptional ability to alter the characteristics of materials in their magnetic, electrical, and optical capabilities. Researchers from all across the world have recently concentrated their findings on environmentally friendly and maintainable characteristics. Compared to other environmental concerns, contamination of wastewater is one of the main dangers resulting from manufacturing activities. The innovative growth of technologies for purifying wastewater outflow from diverse sectors is a topic of many investigations. The ability of photocatalysis to degrade contaminants into non-toxic, biodegradable compounds makes it a popular method for treating water. Cheap photocatalysts are progressively entering the academic world (Eyasu et al., 2013; Sivaraman et al., 2022). Under the right light irradiation conditions, photocatalysis can safely oxidize modest amounts of organic contaminants, including dyes, insecticides, and volatile organic compounds, and transform them into harmless chemicals (Hekmatara & Haghani, 2019). A photocatalyst is a flexible method that may be used for various activities, including hydrogen generation, the degradation of different organic pollutants in wastewater, and the cleansing of hydrogen. Compared to other approaches, photocatalysis is swiftly growing and gaining more consideration from researchers due to its numerous advantages, including accessibility and appealing efficiency. Photocatalysis is an innovative strategy for tackling issues with energy and the environment (Saravanan et al., 2017).

ZnO photocatalysts have gained important applicability in treating environmental pollutants due to their versatile and tuneable structure-electronic properties. The present study in heterogeneous photocatalysis is rapidly moving towards emphasizing ZnO as a suitable semiconductor to various competing metal oxides such as Fe_2O_3, Bi_2O_3, WO_3, and TiO_2 (Sushma & Kumar, 2017). ZnO is the leading material for energy applications as a complement to the extensively studied TiO_2 because ZnO has a large band gap of 3.37 eV and displays high exciton binding energy (60 meV) amongst known semiconductors at 25 °C. Additionally, ZnO nanomaterial exhibits greater photoactivity than TiO_2 because it produces and separates photoinduced electrons and holes more effectively. ZnO nanostructures have a significant excitonic binding energy and a large band gap, which makes them a viable choice for light-emitting diodes and lasers in the UV spectrum. Other numerous applications of ZnO nanomaterials are photocatalysts (Arabi et al., 2018), gas sensors, solar cell nanogenerators, and photodetectors are some examples (Kumar et al., 2019a, 2019b).

Additionally, ZnO nanomaterials have a great absorbance capacity of a significant portion of the solar spectrum, which makes them effective at removing organic pollutants from surfaces when exposed to visible light. In addition, ZnO is a more sensible and affordable nanomaterial than the others (Hekmatara et al., 2019). So, ZnO is praised for its exceptional physicochemical and optoelectronic properties, favorable band edge positions, straightforward preparation techniques, high exciton binding energy and electron mobility, and ability to be structurally stabilized by supporting numerous conducting substrates. In contrast to TiO_2, ZnO absorbs more light fractions and may be made using a variety of physical and chemical techniques with varied morphologies. ZnO is produced at less cost, up to 75% less than TiO_2 and Al_2O_3 NPs. It has been recommended that ZnO be utilized in heterogeneous photocatalysis due to ZnO's benefits over TiO_2.

Furthermore, a few research findings also point to the dynamic performance of ZnO above TiO_2 for the breakdown of organic contaminants under light illumination conditions (Lee et al., 2016). So, this chapter aims to provide a brief overview of photocatalytic principles and mechanisms of ZnO nanomaterials, factors influencing photocatalytic efficiency, synthesis and characterization of ZnO nanomaterials, enhancement strategies for ZnO photocatalysis, and application of ZnO NP photocatalytic degradation on organic and inorganic pollutants. Finally, limitations, challenges, and future perspectives are stated at the end of this chapter.

2 Photocatalytic Principles and Mechanisms

The photocatalyst term is a composite of the two Greek words "phos" and "katalyo." "Phos" means light, but "katalyo" means catalyst, which is decomposition or disintegration. This is an ancient method originally used in such actions using TiO_2. This catalytic oxidation reaction absorbs light by adding catalyst substances (semiconductor nanomaterials) and maintains that state even after the reaction is complete. The process involves both photochemical and catalytic materials, which can be used as a preliminary stage to eliminate hazardous, non-biodegradable, and persistent compounds from the environment by improving biodegradability. Conduction and valence bands in semiconductors have a tiny energy gap, and visible or ultraviolet light can drive electrons to move from the valence band to the conduction band. For instance, ZnO has a bandgap of about 3.37 eV with 400 nm absorption wavelength, and the semiconductor surface temperature increases with continued exposure to light (Arun et al., 2023). Increasing this temperature causes the contaminants to break down using dissolved O_2 and OH radicals. Contaminants adsorbed on the active sites of the photocatalyst. After the electron excitation, the photocatalytic decomposition of pollutants begins. When electrons are excited from the valence band to the conduction band by light, it leaves a positive gap (hole) in the valence band. Electrons in the conduction band react with O_2 to form hydroperoxyl radicals ($HO_2\cdot$) or superoxide ions (O_2^-). At a similar time, H_2O is oxidized using valence band holes to produce H^+ and OH radicals (OH). These produced radicals are accountable for breaking down contaminants to form CO_2 and H_2O. Once the decomposition process is finished, desorption events release the damaged molecules into the H_2O. The general overall process is shown in equation a-d. Furthermore, the reactions that may occur during the photocatalytic process can be elucidated as follows (Puri & Gupta, 2023):

$$h\upsilon + \text{Photocatalyst} \rightarrow h^{+} + e^{-}(\text{Electron and hole generation}) \quad (1)$$

$$H_2O + h^{+} \rightarrow H^{+} + {}^{\cdot}OH \quad (2)$$

$$O_2 + e^{-} \rightarrow O_2^{-\cdot} \quad (\text{superoxide radical generation}) \quad (3)$$

$$OH + O_2^{-\cdot} \rightarrow HO_2 \quad (\text{hydroperoxyl radical generation}) \quad (4)$$

Generally, the ZnO photocatalysis mechanisms are divided into diverse stages as follows (Mutalib & Jaafar, 2022).

(a) Adsorption of organic contaminants onto the ZnO's surface
(b) Diffusion of the organic contaminants onto the surface of ZnO
(c) At the adsorbed phase, the redox reactions take place
(d) The products desorption
(e) At the area of the interface, product clearance occurred.

3 Key Factors Influencing Photocatalytic Efficiency of ZnO Nanomaterials

There are some limitations when using ZnO as a semiconductor for photocatalytic decomposition, for example, photocatalytic efficiency, slow decomposition rate, aggregation, and recovery of nanoparticles problems. These limitations can be overcome by altering concentration, particle size, and dopant type to increase affinity. Temperature, light intensity, pH, photocatalyst dosage, and treatment time are important parameters for the fruitful execution of the procedure. The light intensity has a significant effect on the degradation rate of photocatalysts. The intensity increases, and the faster the photocatalytic rate occurs until all active sites are filled. For instance, the decomposition of isopropanol increased to 56% as the UV light intensity increased from 0 to 3 mW/cm^2. Moreover, extra light is unlikely to subsidize more significant degradation (Cheng et al., 2018).

Another issue that has been found to significantly impact the decomposition of persistent organic contaminants (Fig. 1) is the optimal pH value. Further research has shown that an acidic pH is optimal for decomposing organic contaminants. The photocatalyst's positively charged (acidic pH) surface can attract negatively charged organic contaminants. In contrast, positively charged dyes are better decomposed at alkaline pH because alkalinity negatively charges the photocatalytic surface. For instance, Shubha and coworkers investigated the influence of various pH levels on eliminating methylene blue (MB). The outcomes demonstrated that a lower pH (pH =

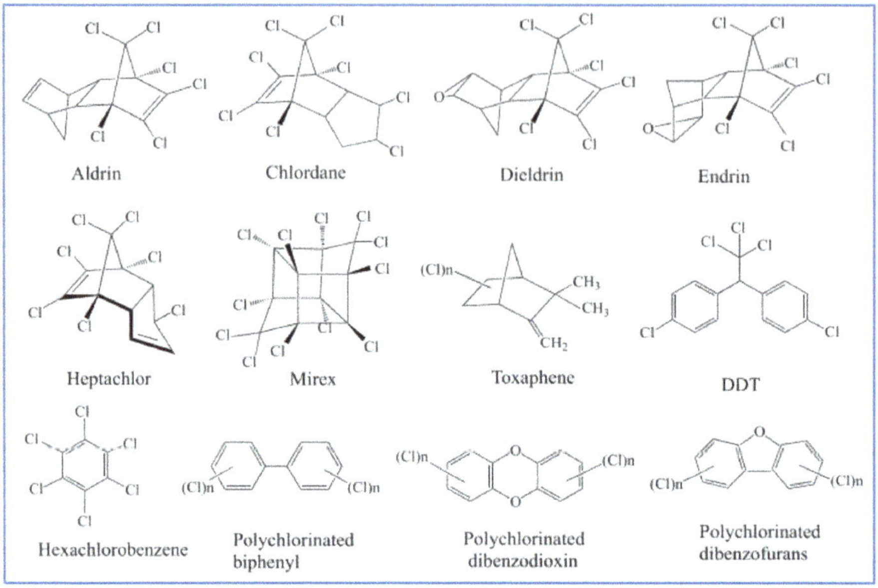

Fig. 1 The most common persistent organic pollutants

4) resulted in just 52% decomposition of the dye, whereas an advanced pH value (pH = 10) caused 98% deterioration. Utilizing the optimum catalyst quantity concentration for the photodegradation of pollutants is a crucial issue that must be considered. A more significant catalyst dose makes a more active surface area for pollutant absorption availability. However, excessive photocatalyst concentration causes it to aggregate, which inhibits solar, UV, and visible light from penetrating the effluent (Shubha et al., 2022). According to a study by Caliskan and coworkers, 1 g/L of ZnO photocatalyst was the ideal concentration to employ while attempting to degrade azodye. A concentration greater than 1 gL^{-1} causes photons to scatter, resulting in limited absorption (Çalışkan et al., 2017).

4 Synthesis and Characterization of ZnO Nanomaterials

Nanotechnology is a fast-evolving area concerned with synthesizing new materials at the nanoscale. In other words, nanotechnology aims to synthesize, characterize, and manipulate matter from 1 to 100 nm in size. The study community has been eager to create nanoparticles for the past few decades using green, economical, and environmentally acceptable synthesis methods. This is because green sources have several functions acting as steadying and reducing agents to help synthesize size and shape-controlled nanoparticles (Pillai et al., 2020). The biosynthesis of metal nanoparticles serves as an illustrative example. Biosynthetic nanotechnology has significant applications in biomedical science, food, cosmetics, drugs, and gene delivery, healthcare, chemical industries, electronics, energy science, mechanics, and space industries. These are just a few examples of the areas.

Moreover, by decreasing the size of materials, it is possible to attain a wide diversity of advanced physicochemical features and a wide range of potential applications, including material science and biomedical applications (Sivasankarapillai et al., 2019). ZnO NPs have gained intense scientific curiosity because of their wide range of potential uses, including photocatalysts and intelligent UV sensors (Sosna-Głębska et al., 2020), targeted medicine delivery (Fahimmunisha et al., 2020), biosensors, environmental cleanup (Singh et al., 2018), and antioxidant activity (Lingaraju et al., 2016). The capacity to choose between different nanoparticle morphologies based on the biological source used while maintaining increased stability is one of the unique features of ZnO NP biosynthesis. Traditional top-down and bottom-up methodologies can categorize the traditional chemical and physical techniques utilized in nano synthesis. The "top-down preparation procedure approach" of NMs uses high-energy lasers, lithographic processes, and mechanical and thermal energy to create nanoscale materials. Chemical vapor deposition and atomic layer deposition, in another way, "bottom-up synthesis approach" and are gas-phase techniques, whereas metal salt reduction, template synthesis, sol–gel procedures, and electrodeposition are liquid-phase techniques (Pillai et al., 2020).

5 Various Synthesis Methods for ZnO Nanomaterials

There are several common methods to prepare ZnO NPs, including the wet chemical route, vapor phase procedure, hydrothermal method, precipitation method, and sonochemical approach. However, along with the time-consuming process control, those procedures always call for using harmful chemicals and expensive equipment. Therefore, a straightforward, environmentally friendly approach to making ZnO NPs is highly desired. Due to its ease, affordability, and nontoxicity, the biosynthesis of NPs has recently received much interest. It was established that plant extracts were used to make metal NPs more stable than conventional ones. Plant extracts are suitable materials to scale up and produce well-dispersed metal NPs in an environmentally friendly (Zheng et al., 2015). To assist in making NPs, biological synthesis techniques used either plant extract or microorganisms. These green practices differ from chemical processes in many ways, which can be attributed to them (Jeevanandam et al., 2016). The possibility of using biosynthesized NPs directly in living systems or biomedical applications is the main advantage. This is because these NPs are less hazardous than those produced by physicochemical methods. The stabilizing effects of the biocomponents used in engineering and creating a corona by modifying the NPs' surfaces, which makes them more suited for use in living systems, are potential additional advantages (Singh et al., 2016).

Due to some characteristics, like feasibility, the use of easily accessible plants, and the wide range of ZnO NP morphologies synthesis (depending on plant extract utilized), plant-mediated synthesis is one of the more alluring biological approaches (Ghaseminezhad et al., 2012). By employing *Ixoracoccinea* leaf extract, Pillai and coworkers could create spherical ZnO NPs (80–130 nm) (Pillai et al., 2020). Lingaraju and coworkers made ZnO NPs by using stem extracts (aqueous) of *Ruta-graveolen* that have been shown to exhibit antibacterial and antioxidant properties (Lingaraju et al., 2016). Using *Passiflora caerulea* leaf extract as a starting material, Santhoshkumar and coworkers synthesized ZnO NPs, which showed promising results against bacteria that cause urinary tract infections (Santhoshkumar et al., 2017). Ismail and coworkers also prepared ZnO NPs by chemical sol–gel synthesis approach from dihydrated zinc acetate and NaOH as reacted ingredients, and they used deionized water for dissolving at 25 °C with energetic stirring for 3 h by adding 0.5% Polyethylene glycol, which was used as a surfactant. The resulting ZnO NPs showed superior antibacterial activities on gram-negative as well as gram-positive bacteria (Ismail et al., 2019).

Khalafi and coworkers also developed a new environmentally friendly process to produce stable, safer, high-purity ZnO NPs using *Chlorella* microalgae extract. They synthesized using zinc nitrate and *Chlorella* microalgae extracts under encompassing conditions. The Chlorella microalgae extract stabilizes and reduces the new ZnO NPs in this formulation. Figure 2 shows a potential procedure for the biogenic production of ZnO NPs using aqueous Chlorella extract algae-derived carbohydrates as a model (Khalafi et al., 2019).

Other studies also show that semiconductor ZnO nanomaterials have been produced using different approaches, such as vapor phase-based and solution-based

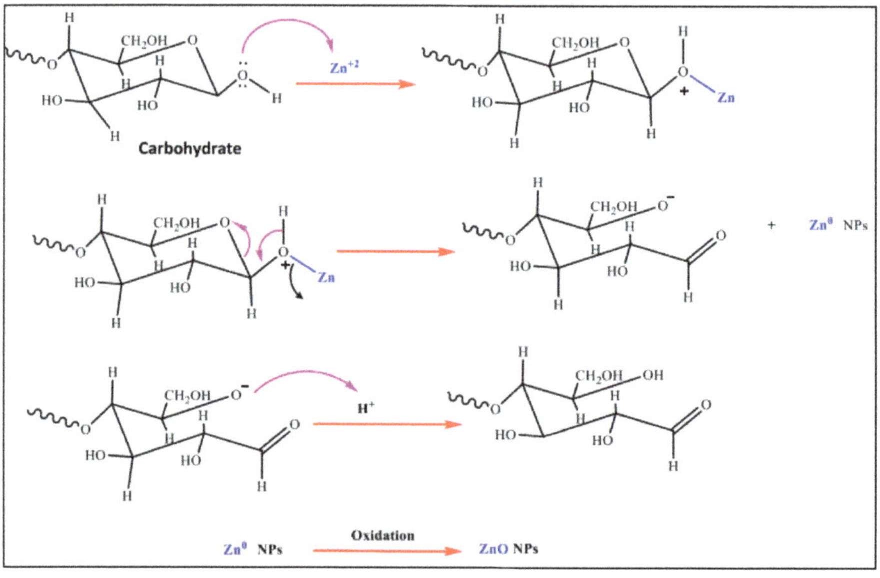

Fig. 2 Proposed schematic biogenic synthesis mechanism of ZnO NPs

synthesis methods. It is important to note that many factors, like crystallinity and particle size, which impact a material's band gap energy and transfer of charge, can be considerably influenced by the synthesis processes used. To create very effective catalytic materials, controlled synthesis techniques are crucial. In addition to the size and crystallinity, the morphology of nanocomposites is a key factor in influencing photocatalytic activity. According to reports, solution-based synthesis (mainly involving the sol–gel method, hydrothermal, flux method, solvothermal, microemulsion, microwave, and wet chemical route) techniques are the best way to alter the precursors, types of solvents, and conditions of reaction, like temperature and heating duration to regulate the size and shape of nanostructures (Kumar et al., 2020). ZnO nanostructure material can be classified into zero-dimensional, one-dimensional, two-dimensional, and three-dimensional, representing 0D,1D, 2D, and 3D, respectively. The 0D furtherly partitioned into quantum dots groups, 1D into elongated groups, 2D into planar groups, and 3D into neat structures. 1D ZnO groups include nanowires, nanorods, nanotubes, nanofibers, and nanoneedles but 2D arrays are nanosheets, and also 3D arrays are nanoflowers. 1D ZnO has pervasive potential applications in optoelectronics, gas sensing, and electronics. Desai and coworkers' study showed that ZnO nanowires were subjected to mechanical strength testing. They were found to have exceptional fracture strain, making them suitable for nanoscale sensors and actuators (Ong et al., 2018).

6 Characterization Techniques for ZnO Nanomaterials

Synthesized ZnO NP materials can be characterized by XRD, SEM, FT-IR, TEM, EDX, and SAED to confirm the formation of ZnO NPs. Suresh and coworkers synthesized biocompatible ZnO NPs using $Zn(NO_3)_2$ and the therapeutic plant *Costus pictus D. Don* leaf extracts. They subsequently characterized particles using various tools such as the FTIR spectrum in Fig. 3 shows the ZnO NPs absorb at 441.5–665.5 cm^{-1}, and the hydroxyl stretching performs at 3372.37 cm^{-1} with a precise broadband peak which shows the occurrence of the aliphatic carboxylic acid compound. The two peaks at 1112.64 cm^{-1} attributed to carbon–fluorine stretching represent mono- and poly-fluorinated compounds, and moderate absorbance levels between 1468.12 and 1384.65 cm^{-1} indicate the presence of aromatic rings. The absorption peak at 1047.70 cm^{-1} relates to the C–O stretching of saturated primary alcohols. A distinct level of doublet absorption detected between 2337.850 and 2427.510 cm^{-1} indicates the existence of C-H stretching frequencies in aromatic aldehydes. The band detected at 1590.61 cm^{-1} resembles the flavonoid carbonyl group. Finally, their FTIR analysis approves that phenolic compounds in flavonoids can bind better with metals, suggesting that phenolic groups can form metal NPs to inhibit aggregation and calm the medium. This proposes that biomolecules can play a dual role in forming and stabilizing ZnO NPs in aqueous media.

According to structural study using XRD, this study also shows pure hexagonal phase structures of ZnO NPs have formed, as shown in Fig. 3. The majority of

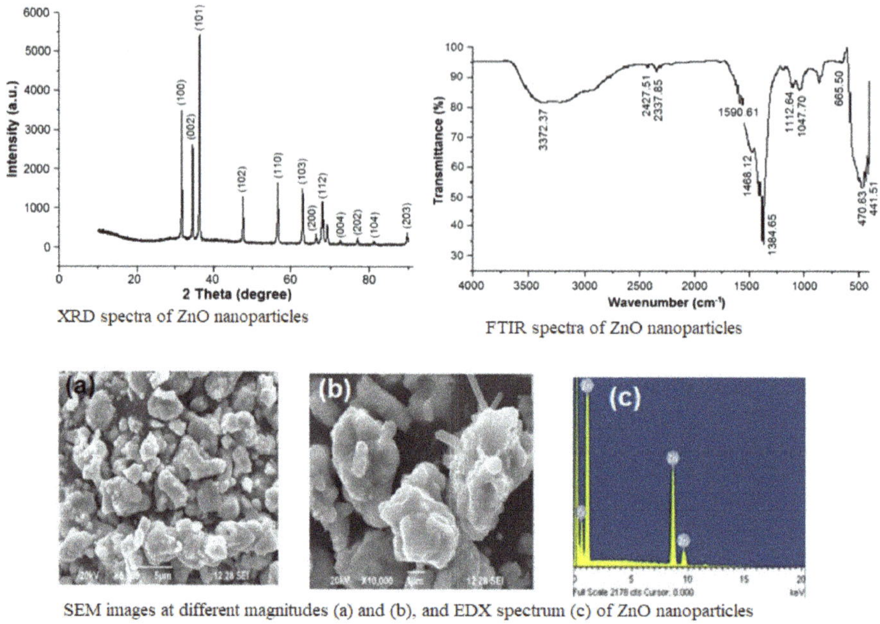

Fig. 3 FT-IR and XRD spectra and SEM, and EDX image of ZnO NPs. Adapted from Suresh et al. (2018)

ZnO crystallites appear to be hexagonal based on the surface morphologies of ZnO NPs, as seen under SEM. The presence of zinc and oxygen is confirmed by EDX analysis. Individual ZnO NP agglomerations have been seen in the SEM images. A closer inspection reveals that the agglomerated lump contains numerous aggregates of nanoparticles. Figure 3a shows some individual crystals that are visible and particles that appear to be agglomerated. Some particles in Fig. 3b appear rod-shaped, hexagonal, and elongated. Figure 3c displayed the EDX results, which further determine the presence of ZnO NPs (Suresh et al., 2018).

SAED and TEM patterns display the morphologies and sizes of NPs that have been analyzed. At various nanoscale sizes, the images in Fig. 4 demonstrate that the NPs are spherical, rod-shaped, or hexagonal. Halo rings indicate that NPs may contain amorphous components. The existence of an amorphous phase, which lacks the orientation of crystallography, is thought to be the cause of the occurrence of spherical particles. Biomolecules and hexagonal molecules cover the ZnO NPs. Similar hexagonal molecules have been seen in synthesizing ZnO NPs in the Euphorbia jatropha plant (Geetha et al., 2016).

The face-centered cubic ZnO lattice planes and Debye-Scherer rings at (100), (002), (101), (102), (110), (103), (200), (112), (201), (004), (202), (104), and (203) in the SAED pattern indicate that the biogenic NPs visible in the TEM images are nano-crystalline (Fig. 4a–c). The biomolecules impacted ZnO NP formation. The particle sizes estimated by the TEM study and the XRD analysis agree. Figure 5

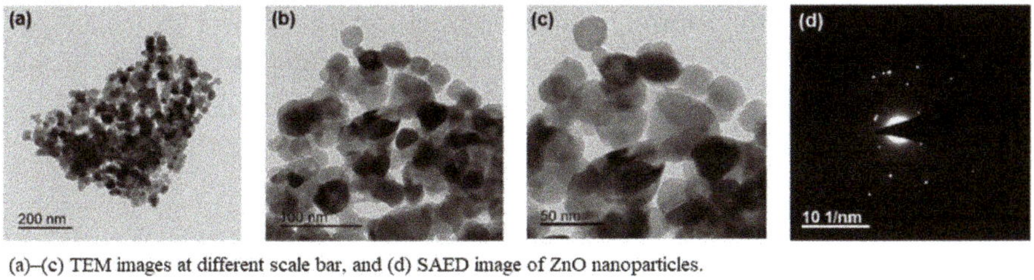

Fig. 4 Different scale bar TEM images (**a**)–(**c**), and ZnO NPs SAED image (**d**). Adapted from Suresh et al. (2018)

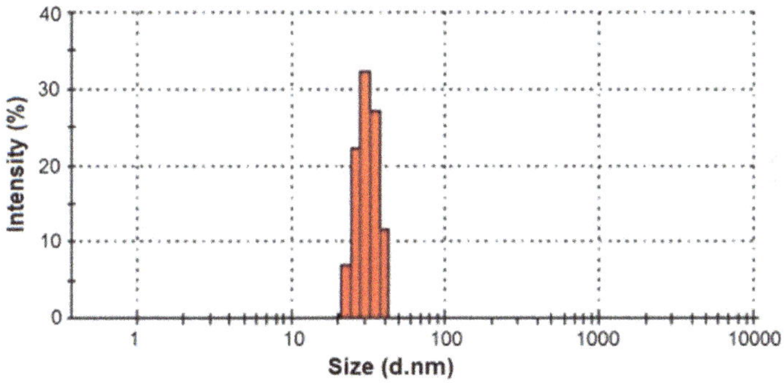

Fig. 5 ZnO NP size distribution found in SAED. Adapted from Suresh et al. (2018)

illustrates the outcomes of determining the average size of the particles and the size distribution of the synthesized ZnONPs using a particle size analyzer. The size distribution of the ZnONPs in the final particle size ranges from 20 to 80 nm. The results of the SEM study are consistent with 40 nm, an average particle size.

7 Photocatalytic Degradation of Environmental Pollutants by ZnO Nanomaterials

7.1 Degradation of Organic Pollutants

Chemical substances known as organic pollutants are resilient to environmental deterioration and may not be eliminated by treatment procedures. They may damage both people and animals if they are present. Metal oxide semiconductor-based photocatalysts have the potential for CO_2 abatement, fuel generation from water splitting, and environmental remediation. ZnO nanostructures are prospective photocatalyst possibilities for use in photodegradation because of inexpensive, non-toxic, and more actual in absorption throughout a significant portion of the solar spectrum (Sanakousar et al., 2022). Some common organic and inorganic pollutants' degradations using ZnO NMs are reviewed.

7.1.1 Degradation of Organic Dyes

Organic dyes, some commonly known as shown in Fig. 6, are synthetic aromatic substances that may have mutagenic and carcinogenic effects. Organic dyes, categorized as cationic or anionic dyes, are typical pollutants from the textile, plastics, and paper industries. The major technique for removing organic dyes from water is adsorption. In the adsorption of organic dyes from H_2O, nanomaterials with high surface-to-volume ratios have been used extensively (Lu & Astruc, 2020). Using ZnO NPs for dye degrading with lower toxicity and phytotoxicity of the textile pollutants is a promising green chemical method.

Nonetheless, the photostability of the ZnO NPs during dye photolysis can be limited by UV light irradiation photo corrosion. The photogenerated electron–hole pair recombination rate, which is very common, considerably lowers the photocatalytic activity of ZnO (Dihom et al., 2022). So, structural amendment or inclusion of

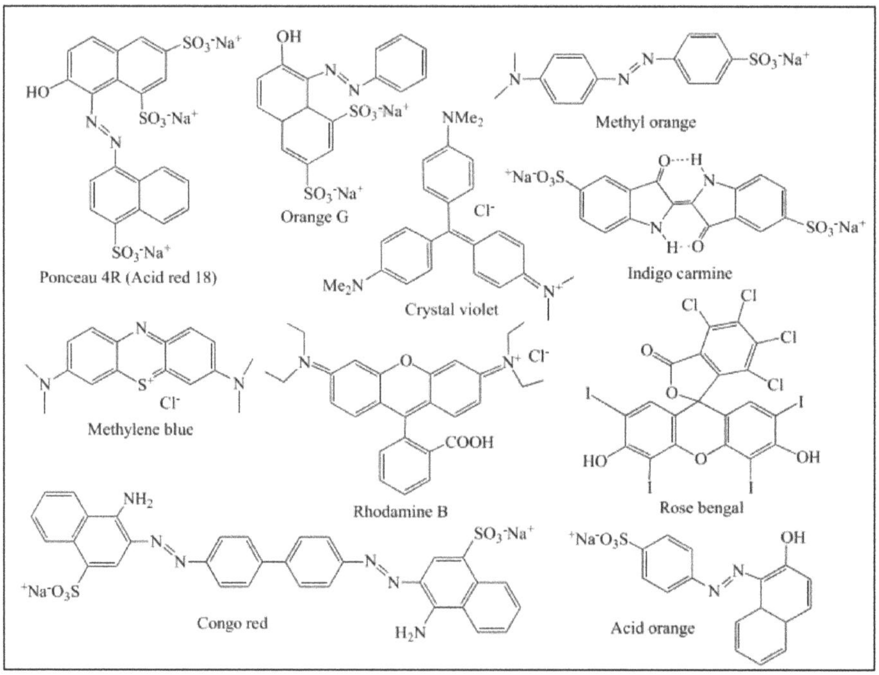

Fig. 6 Chemical structures of some common organic dyes

other elements into ZnO is a promising strategy for improving the breakdown efficiency of dye contaminants. Isa and coworkers engineered ZnO nanoparticles into micro flower design to maximize methyl orange breakdown efficiency. As a result, employing 150 mg of ZnO catalyst, 99% elimination of the 10 mg/L MO dye was obtained within 300 min of the reaction. In another study, Xiaoke and coworkers described the effective breakdown of rapid acid black dye employing a ZnO catalyst immobilized over polyurethane. As a result, the ZnO amendment technique improved the catalytic performance of ZnO under UV and solar irradiation by removing the acid black dye in 75 min for UV and solar 240 min, respectively (Mutalib & Jaafar, 2022).

Taha and coworkers tested the ZnO nanostructures' photocatalytic activity by watching MB organic contaminant dye photodegrade after being given 60 min to reach equilibrium (Fig. 7). Measurements of MB's absorbance were at λmax = 662 (Kappadan et al., 2016), and various illumination times under UV and visible radiation were used to track the degradation process. As seen in Fig. 8a, b for the S2 sample and Fig. 9a, b for the S1 sample, a steady decrease in absorbance is associated with the produced sample's increased catalytic performance in the UV and visible parts of the spectrum. As observed by them, the ZnO photocatalyst had better degrading efficiency than the $BaTiO_3$NPs, which produced results that were comparable but required a lower MB concentration (Taha et al., 2019).

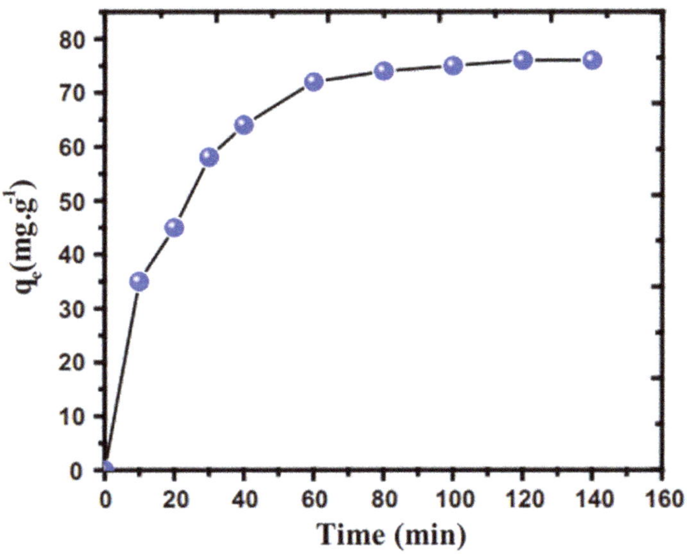

Fig. 7 Equilibrium of adsorption MB onto the ZnO NPs. Adapted from Taha et al. (2019)

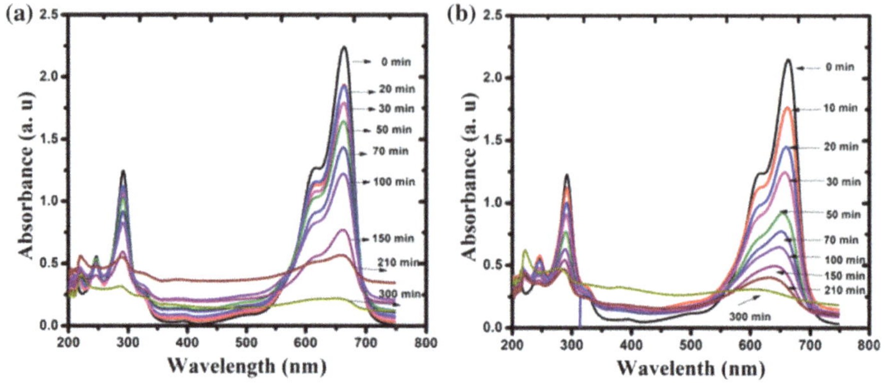

Fig. 8 1 MB absorbance for S2 at various times in visible (**a**) and UV (**b**) light. Adapted from Taha et al. (2019)

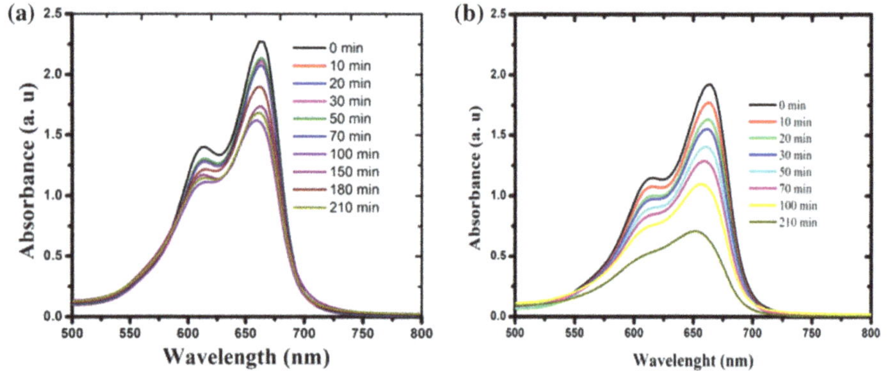

Fig. 9 The absorbance of MB at various time intervals (**a**) and (**b**) under visible and UV irradiation for S1, respectively. Adapted from Taha et al. (2019)

7.1.2 Degradation of Pesticides

ZnO has been demonstrated to degrade agricultural contaminants in various synthetic pesticides and herbicides efficiently. For instance, Niranjani and Anchana have proved the viability of chemically generated ZnO in the photolysis of phorate which is a very hazardous organophosphate insecticide frequently employed in the majority of impoverished countries (Niranjani & Anchana-Devi, 2017). The conducted practical tests showed that the ZnO photocatalyst could decompose phorate at a very high rate in a 120 min reaction under sunlight. One of the most widely used insecticides, mineralized imidacloprid, is also effective when combined with ZnO (Yari et al., 2019). 88% of Imidacloprid was efficiently destroyed under ideal reaction conditions. Most earlier investigations concurred that the degradation efficiency is significantly influenced by operating parameters like pesticide concentration, pH, catalyst dose, and the kind of synthetic pesticide itself (Khan & Pathak, 2020; Premalatha & Miranda, 2019). ZnO has been suggested as the most favorable catalyst, even more, viable than its much-studied counterparts, for removing agricultural pollutants due to its surface characteristics, such as surface defects and oxygen vacancies, which also significantly affect photocatalytic ability. To increase the degradation rate, the addition of peroxide as oxidizing agent can facilitate the dissociation of pesticides into harmless end products (Khan & Pathak, 2020).

7.1.3 Degradation of Analgesics and Antibiotics

Like other organic contaminants, the degradation rate of various analgesics and antibiotics is related to temperature, time, light sources, and pH (Majumder et al., 2020). Some of the antibiotics and analgesics effectively degraded with photocatalyst of ZnO are diclofenac, ibuprofen, ampicillin, ciprofloxacin, ofloxacin, and paracetamol (Eskandari et al., 2018; Malakootian et al., 2019; Kumar et al., 2019a, 2019b). From the variations of degradation performance reported by Eskandari and coworkers, it can be concluded that the molecular content differences between the used drugs determine their degradability by ZnO, in addition to structure and size factors. In most cases, without any modification, ZnO shows a removal efficiency of 10–60% when irradiated with light in the UV range (Eskandari et al., 2018; Mirzaei et al., 2016). However, with adjustments such as doping metals, heterostructure creation, or integration of metal–organic framework, the photocatalysts of ZnO can work with good speed under solar irradiation (Akkari et al., 2018; Kumar et al., 2019a, 2019b; Mitra et al., 2018; Ritika et al., 2018). The band gap may be tuned by adding metal to ZnO, which also enhances the defect states of the acceptor and donor, making the catalyst more visible and light-active. Furthermore, the formation of heterostructures by incorporating metal oxides enhances the optoelectric characteristics of ZnO, which is also endorsed by the increased photoreactivity in less intense energy circumstances and sunlight (Majumder et al., 2020).

In addition to the organic contaminants listed above, ZnO can also cleanse wastewater with trace amounts of stimulants, including morphine, coffee, ketamine, and amphetamine. Ethalil and coworkers have created Mg–ZnO–Al_2O_3 for the UV-induced breakdown of caffeine. The amazing degradation rate (98.9%) attained in 70 min using a 0.3 g/L catalyst dose suggests that ZnO catalyst may help break down stimulant compounds (Elhalil et al., 2018). However, another research has that including a ZnO catalyst allows for the precise removal of ketamine, morphine, and methamphetamine from wastewater. The elimination effectiveness of medications can reach up to 90–100% in 30 min (Mutalib & Jaafar, 2022).

7.2 *Degradation of Inorganic Pollutants*

ZnO nanoparticles also remove inorganic pollutants, the main class of pollutants released from chemical and related industries like fertilizers, pharmaceuticals, and refineries. Heavy metals and other inorganic contaminants are trace elements, metals, inorganic salts, mineral acids, metal compounds, and metals with organic compounds such as complexes, cyanides, and sulfates. If all of these concentrations exceed the permissible limit values, they can pollute the environment. These inorganic contaminants may not be decomposable and remain in the environment. Many inorganic contaminants have harmful effects on community health as well as on aquatic flora and animals (Wasewar et al., 2020).

A viable substitute and upcoming method for eliminating heavy metals from water is photocatalysis. Bao and coworkers report that the ZnO photocatalytic process simultaneously lowers heavy metals to their corresponding elemental form and transforms them into non-toxic ionic states (Bao et al., 2019; Shukor et al., 2019). Polyaniline/ZnO nanosheets were fabricated as surface-hybridized to remove Cr(VI) ions from water with a UV light source. The high surface areas of ZnO, with OH groups on the surface, help as the active sites for the adsorption and uptake of the positively charged ions. In aqueous media, chromium often occurs in two ionic forms (Cr(III) and Cr(VI)). These oxidation states dictate the degree of toxicity and reactivity of chromium. Comparatively speaking, Cr(VI) is much high harmful than Cr(III). Polyaniline and ZnO, which resemble flowers, were hybridized to improve the photoreduction of dangerous Cr(VI) ion to benign Cr(III) ion with great repeatability. The high dispersion of the active terminal OH groups with amine groups of Polyaniline enhances the photocatalytic abilities of ZnO nanoparticles. Cr (VI) elimination was pH reliant with an ideal 4–7 pH range. Salehi and coworkers used PANI/ZnO and GO/ZnO composites to remove chromate ions under visible light. Due to its superior dye adsorption and charge separation capabilities, GO/ZnO composite demonstrated better photoactivity than PANI/ZnO (Salehi-Babarsad et al., 2020).

On the other hand, Le and coworkers conducted comparative research to eliminate the heavy metal ions by using two different light sources. They removed the heavy metal ions by depositing them on the surface of ZnO by forming metal/metal oxide-linked ZnO hybrid particles (Le et al., 2019). After being subjected for one hour to UV light irradiation, ZnO particles had a remarkable capacity to remove ions of Pb(II), Cu(II), as well also Ag(I) with above 85% efficiency, but the poor elimination ions of Cr(VI), Ni(II), Cd(II) and Mn(II) ions that are less than 15%. Under visible light, up to ~40, ~60, and 100% of Cr(VI), Pb(II), and Cu(II), respectively, were removed, but the other materials had removal efficiencies of less than 10%. Because of inadequate energy of visible light to create electron holes for photo redox, removal performances were lowered. It was suggested that a physical absorption mechanism was responsible for the elimination of these metal ions. Therefore, the light source and the types of metal ions are the primary factors in the removal of heavy metals (Abdullah et al., 2022).

8 Enhancement Strategies for ZnO Photocatalysis

ZnO has a poor quantum yield and a high rate of photogenerated electron–hole recombination, which restrict its photocatalytic activity. The quantum yield is decreased during this recombination process, and the energy is wasted. Therefore, the electron–hole recombination route should be minimized to assist efficient photocatalysis. Doping with metals and non-metals may reduce recombination by widening the charge gap between electrons and holes. The possibility of electron–hole recombination, which would prevent the photocatalytic system from functioning, can be reduced by dopants' ability to trap electrons. Due to variations in band gap energy and surface area, doped ZnO degrades pollutants quicker than undoped ZnO by enhancing the photocatalytic degradation of dyes. Particle size reduces, which energetically contributes to or correlates with the photocatalytic reaction, increases the band gap, and a bigger surface area predicts a strong redox reaction. Additionally, the entire process degrades at a faster pace (Gnanasekaran et al., 2017). Therefore, doping is a simple and effective method (Yousefi et al., 2015). Alkaline earth metals (Mg, Sr) (Chithira & John, 2020), transition metals (Fe, Cu) (Isai & Shrivastava, 2019), noble metals (Au, Ag, Pd) (Ahmed et al., 2019; Güy et al., 2016), non-metals (S, C, N) (Byzynski et al., 2017), rare earth metals (Eu, Nd, Er) (Raza et al., 2016), Co-doping (Zn_2SnO_4, $ZnFeO_4$) (Jain et al., 2020; Mishra et al., 2019) and semiconductor coupling ($Zn_2SnO_4V_2O_5$) (Raja et al., 2018) were used as dopants to enhance the photocatalytic activity.

8.1 *Surface Modification of ZnO Nanomaterials for Improved Photocatalytic Performance*

In the process of dye photodegradation, which is responsible for the charge transfer procedure and dye adsorption, the surface characteristics of TiO_2 and TiO_2-coated ZnO play a significant role. Rhodamine B (RhB) dye solution degradation under UV-A irradiation was investigated as a function of time in order to shed light on the photocatalytic activity of TiO_2/ZnO nanocomposite (TZN) films with various numbers of ZnO deposition cycles. In addition, for comparison purposes, a blank photodegradation test without a catalyst was performed in the presence of UV-A irradiation, which showed that the RhB dye has good photostability under UV-A irradiation. However, under the same experimental conditions, the photodegradation of the RhB dye solution increased over the TiO_2 & TZN film catalysts, resulting in the formation of % hydroxyl radicals due to the UV irradiation can react with the surface O_2 of the catalyst surface to give % hydroxyl radicals (Çırak et al., 2019) since the electron–hole production model might explain how the dye solution degraded across the metal oxide composite catalysts. The catalyst surface absorbs photons with energies equal to or greater than its band gap energy, producing holes (h+) and electron pairs (e−) that move toward the catalyst surface (Wang et al., 2016);

subsequently, during photoirradiation, h+ and e− interact with surface hydroxyl and water species to produce superoxide radicals (O_{2-}.%) and % hydroxyl radicals. As a result, when dye molecules are exposed to UV light, these reactive oxygen species degrade the dye molecules (Chen et al., 2017; Ranjith et al., 2017). According to the photodegradation results, the photocatalytic activity of TZN films improved with time parallel to the number of ZnO deposition cycles, with TNZ8 films showing the greatest performance. Pure TiO_2 was found to have a 65% RhB photodegradation efficiency, but the TNZ8 film showed a 95% RhB degradation (Çırak et al., 2019).

8.2 *Hybridization of ZnO with Other Materials for Synergistic Effects*

Even though ZnO has received much consideration as a photocatalyst in latest years, it is still challenging to create inexpensive and Vi-light-sensitive ZnO-based photocatalysts that can be applied widely. To solve such difficulty, Xu and coworkers synthesize CuO/ZnO NCs by using a hydrothermal technique, and they examined the effect of the CuO concentration on the photocatalytic characteristics of the nanocomposites. They tested the photocatalytic performance of the CuO/ZnO NCs under the illumination of a Xe lamp by using MB dyes as a simulated pollutant. The CuO/ZnO NCs photocatalytic activity is significantly increased compared to pure ZnO. The sample was made from the precursor solution with the molar ratio Zn^{2+}: Cu^{2+} was 2:1. They concluded that band coupling, which enhances the separation efficiency of photogenerated electrons and holes, and the CuO/ZnO coupling system's higher solar energy utilization efficiency are the two leading causes of the improvement in photocatalytic activity (Xu et al., 2017).

Cerrato and colleagues synthesized ZnO-based photocatalysts doped with various lanthanide elements (La, Pr, Yb, Er, and Ce) to enhance ZnO photocatalytic efficiency. The dopants, introduced in minimal quantities (1% molar), exhibited different behaviors based on their oxidation number and *f* orbital electron count. While most elements integrated into the ZnO matrix, only Ce produced a new oxide phase. ZnO–La was chosen for evaluating OH radical establishment in water suspension due to its strong light contact in the solid state. Using 5,5-Dimethyl-1-Pyrroline-N-Oxide as a spin trap molecule to measure OH radical concentration, they observed that the doped sample was three times more active under UV light compared to naked ZnO. This suggests that lanthanide-doped ZnO holds promise for photocatalytic degradation of organic contaminants (Cerrato et al., 2018).

9 Limitations of ZnO Nanomaterials for Photocatalysis Applications

A good photocatalyst material should have a suitable band gap structure that allows it to absorb various solar energy wavelengths, as well as a maximum redox potential, strong photostability, cheap cost, and adequate amount. Low absorption and photo corrosion are two problems that most traditional photocatalyst materials experience. Despite ZnO's many advantages as a photocatalyst, it has disadvantages, such as a high band gap energy (3.37 eV) that is related to the UV region of the solar energy spectrum (~5%), so it can only use a small portion of it, and quick photoinduced charge recombination that lowers the photocatalytic activity because charge carriers cannot reach the catalyst surface (Puri & Gupta, 2023). As the equation $ZnO + 2h^+ \rightarrow Zn^{2+} + 1/2O_2$ illustrates, photo corrosion of ZnO caused by photogenerated holes during a photocatalytic reaction results in less stability. Additionally, the ZnO dissolves in both strongly acidic and basic environments to produce Zn^{2+} and $Zn(OH)_4$, respectively. Therefore, the photocatalytic process using ZnO as the catalyst is only favorable at moderate pH conditions. Also, ZnO's propensity to aggregate during photocatalytic reactions might hinder activity. Therefore, it can be concluded that ZnO NP has several limitations and that intensive research has been concentrated on overcoming these issues. So, it is evident that despite significant advancements in the field of photocatalysis over the past 10 years, many obstacles still need to be addressed before this method can be extensively used, especially for industrial applications (Kumar et al., 2020).

10 Future Challenges and Prospects

Current research suggests that using visible-light-driven photocatalysis can degrade organic contaminants more effectively than using traditional techniques. Future advancements in this technology's efficiency will necessitate major efforts to solve many problems. The first step is learning more about the mechanisms of degradation and the interactions between photocatalysts and other organic contaminants. Second, the photocatalysts employed in the processes significantly impact the photocatalytic reactions' effectiveness. While it is possible to make high-quality photocatalysts using currently available techniques (with small particle sizes, vast surface areas, and prolonged carrier lifetimes), more investigation is required. The next phase for the researchers is to investigate the effectiveness and caliber of photocatalysts thoroughly. Additionally, there are operational characteristics like the photocatalyst recovery and loss during post-treatment and the photoactivity of recycled photocatalysts. More effort is also required to construct and test mathematical models to ascertain the quantum yield, kinetics, and ideal state of the process for photocatalytic processes for wastewater treatment.

11 Conclusion

ZnO nanostructures have emerged as potential photocatalyst candidates for solar-powered photodegradation processes of insistent organic and inorganic contaminants, from textile dyes to stimulant drugs. This is due to the low cost of production (75% lower compared to TiO_2 and Al_2O_3), nontoxicity, and the ability to absorb more of the solar spectrum compared to TiO_2. Various techniques have been tried to improve its catalytic reactivity and narrow the band gap for solar light reactivity. These techniques have been tried to improve the photoresponse of ZnO nanomaterials, such as hybridizing or structurally modifying. With the adjustments, the use of ZnO catalysts is unrestricted to the UV region, which is not ideal given the negative energy implications. ZnO properties such as particle size, 0D, 1D, 2D, and 3D dimensions, and morphology are influenced by synthesis methods. Various types of research indicated that the band gap energy played an important role in defining the photoactivity of ZnO in the application. ZnO can be improved as a photocatalyst in photodegradation applications by non-metal or metal doping, coupling with other semiconductors, and coupling with nanocarbon. These methods enhance their performance by altering the band gap energy, suppressing the rate of electron–hole pair recombination, boosting the production of OH radicals, raising the efficiency of charge separation, producing smaller particles with higher specific surfaces, and facilitating better dispersion in the medium. The decomposition of Rhodamine B by ZnO nanoparticles with the most significant specific surface area was demonstrated through experimentation.

References

Abdullah, F. H., Bakar, N. A., & Bakar, M. A. (2022). Current advancements in the fabrication, modification, and industrial application of zinc oxide as a photocatalyst in the removal of organic and inorganic contaminants in aquatic systems. *Journal of Hazardous Materials, 424*, 127416.

Ahmed, M. A., Abou-Gamra, Z. M., ALshakhanbeh, M. A., & Medien, H. (2019). Control synthesis of metallic gold nanoparticles homogeneously distributed on hexagonal ZnO nanoparticles for photocatalytic degradation of methylene blue dye. *Environmental Nanotechnology, Monitoring and Management, 12*, 100217.

Akkari, M., Aranda, P., Belver, C., Bedia, J., Amara, A. B. H., & Ruiz-Hitzky, E. (2018). Reprint of ZnO/sepiolite heterostructured materials for solar photocatalytic degradation of pharmaceuticals in wastewater. *Applied Clay Science, 160*, 3–8.

Arabi, M., Baizaee, S. M., Bahador, A., & Otaqsara, S. M. T. (2018). Rapid, controllable, one-pot and room-temperature aqueous synthesis of ZnO: Cu nanoparticles by a pulsed UV laser and its application for photocatalytic degradation of methyl orange. *Luminescence, 33*(3), 475–485.

Arun, J., Nachiappan, S., Rangarajan, G., Alagappan, R. P., Gopinath, K. P., & Lichtfouse, E. (2023). Synthesis and application of titanium dioxide photocatalysis for energy, decontamination, and viral disinfection: A review. *Environmental Chemistry Letters, 21*(1), 339–362.

Bao, C., Chen, M., Jin, X., Hu, D., & Huang, Q. (2019). Efficient and stable photocatalytic reduction of aqueous hexavalent chromium ions by polyaniline surface-hybridized ZnO nanosheets. *Journal of Molecular Liquids, 279*, 133–145.

Bachheti, R. K., Konwarh, R., Gupta, V., Husen, A., & Joshi, A. (2019). Green synthesis of iron oxide nanoparticles: Cutting edge technology and multifaceted applications. In *Nanomaterials and plant potential* (pp. 239–259). Springer. https://doi.org/10.1007/978-3-030-05569-1_9

Byzynski, G., Melo, C., Volanti, D. P., Ferrer, M. M., Gouveia, A. F., Ribeiro, C., Andrés, J., & Longo, E. (2017). The interplay between morphology and photocatalytic activity in ZnO and N-doped ZnO crystals. *Materials and Design, 120*, 363–375.

Çalışkan, Y., Yatmaz, H. C., & Bektaş, N. (2017). Photocatalytic oxidation of highly concentrated dye solutions enhanced by hydrodynamic cavitation in a pilot reactor. *Process Safety and Environmental Protection, 111*, 428–438.

Cerrato, E., Gionco, C., Berruti, I., Sordello, F., Calza, P., & Paganini, M. C. (2018). Rare earth ions doped ZnO: Synthesis, characterization, and preliminary photoactivity assessment. *Journal of Solid State Chemistry, 264*, 42–47.

Chen, M., Bao, C., Cun, T., & Huang, Q. (2017). One-pot synthesis of ZnO/oligoaniline nanocomposites with improved removal of organic dyes in water: Effect of adsorption on photocatalytic degradation. *Materials Research Bulletin, 95*, 459–467.

Chen, D., Cheng, Y., Zhou, N., Chen, P., Wang, Y., Li, K., Huo, S., Cheng, P., Peng, P., Zhang, R., & Wang, L. (2020). Photocatalytic degradation of organic pollutants using TiO_2-based photocatalysts: A review. *Journal of Cleaner Production, 268*, 121725.

Cheng, H. Y., Chang, K. C., Lin, K. L., & Ma, C. M. (2018). Study on isopropanol degradation by UV/TiO_2 nanotube. In *Paper Presented at the AIP Conference Proceedings.*

Chithira, P. R., & John, T. T. (2020). Correlation among oxygen vacancy and doping concentration in controlling the properties of cobalt-doped ZnO nanoparticles. *Journal of Magnetism and Magnetic Materials, 496*, 165928.

Çırak, B. B., Caglar, B., Kılınç, T., Karadeniz, S. M., Erdoğan, Y., Kılıç, S., Kahveci, E., Ekinci, A. E., & Çırak, Ç. (2019). Synthesis and characterization of ZnO nanorice decorated TiO_2 nanotubes for enhanced photocatalytic activity. *Materials Research Bulletin, 109*, 160–167.

Cruz, D. M., Mostafavi, E., Vernet-Crua, A., Barabadi, H., Shah, V., Cholula-Díaz, J. L., Guisbiers, G., & Webster, T. J. (2020). Green nanotechnology-based zinc oxide (ZnO) nanomaterials for biomedical applications: a review. *Journal of Physics: Materials, 3*(3), 034005.

Dihom, H. R., Al-Shaibani, M. M., Mohamed, R. M. S. R., Al-Gheethi, A. A., Sharma, A., & Khamidun, M. H. B. (2022). Photocatalytic degradation of disperse azo dyes in textile wastewater using green zinc oxide nanoparticles synthesized in plant extract: A critical review. *Journal of Water Process Engineering, 47*, 102705.

Elhalil, A., Elmoubarki, R., Farnane, M., Machrouhi, A., Sadiq, M., Mahjoubi, F. Z., Qourzal, S., & Barka, N. (2018). Photocatalytic degradation of caffeine as a model pharmaceutical pollutant on Mg-doped ZnO–Al_2O_3 heterostructure. *Environmental Nanotechnology, Monitoring and Management, 10*, 63–72.

Eskandari, M., Goudarzi, N., & Moussavi, S. G. (2018). Application of low-voltage UVC light and synthetic ZnO nanoparticles to photocatalytic degradation of ciprofloxacin in aqueous sample solutions. *Water and Environment Journal, 32*(1), 58–66.

Eyasu, A., Yadav, O. P., & Bachheti, R. K. (2013). Photocatalytic degradation of methyl orange dye using Cr-doped ZnS nanoparticles under visible radiation. *International Journal of ChemTech Research, 5*(4), 1452–1461.

Fahimmunisha, B. A., Ishwarya, R., AlSalhi, M. S., Devanesan, S., Govindarajan, M., & Vaseeharan, B. (2020). Green fabrication, characterization and antibacterial potential of zinc oxide nanoparticles using Aloe socotrina leaf extract: A novel drug delivery approach. *Journal of Drug Delivery Science and Technology, 55*, 101465.

Geetha, M. S., Nagabhushana, H., & Shivananjaiah, H. N. (2016). Green-mediated synthesis and characterization of ZnO nanoparticles using Euphorbia Jatropa latex as a reducing agent. *Journal of Science: Advanced Materials and Devices, 1*(3), 301–310.

Ghaseminezhad, S. M., Hamedi, S., & Shojaosadati, S. A. (2012). Green synthesis of silver nanoparticles by a novel method: Comparative study of their properties. *Carbohydrate Polymers, 89*(2), 467–472.

Gnanasekaran, L., Hemamalini, R., Saravanan, R., Ravichandran, K., Gracia, F., Agarwal, S., & Gupta, V. K. (2017). Synthesis and characterization of metal oxides (CeO_2, CuO, NiO, Mn_3O_4, SnO_2, and ZnO) nanoparticles as photocatalysts for degradation of textile dyes. *Journal of Photochemistry and Photobiology B: Biology, 173*, 43–49.
Güy, N., Çakar, S., & Özacar, M. (2016). Comparison of palladium/zinc oxide photocatalysts prepared by different palladium doping methods for congo red degradation. *Journal of Colloid and Interface Science, 466*, 128–137.
Hekmatara, S. H., Mohammadi, M., & Haghani, M. (2019). Novel water-soluble, copolymer-capped zinc oxide nanorods with high photocatalytic activity for degradation of organic pollutants from water. *Chemical Physics Letters, 730*, 345–353.
Husen, A., Rahman, Q. I., Iqbal, M., Yassin, M. O., & Bachheti, R. K. (2019). Plant-mediated fabrication of gold nanoparticles and their applications. In *Nanomaterials and plant potential* (pp. 71–110). Springer. https://doi.org/10.1007/978-3-030-05569-1_3
Husen, A., & Iqbal, M. (2019) *Nanomaterials and plant potential*. Springer International Publishing AG. https://doi.org/10.1007/978-3-030-05569-1
Husen, A., & Siddiqi, K. S. (2023). *Advances in smart nanomaterials and their applications*. Elsevier Inc. https://doi.org/10.1016/C2021 0 02202 1
Isai, K. A., & Shrivastava, V. S. (2019). Photocatalytic degradation of methylene blue using ZnO and 2% Fe–ZnO semiconductor nanomaterials synthesized by sol-gel method: A comparative study. *SN Applied Sciences, 1*, 1–11.
Ismail, A. M., Menazea, A. A., Kabary, H. A., El-Sherbiny, A. E., & Samy, A. (2019). The influence of calcination temperature on structural and antimicrobial characteristics of zinc oxide nanoparticles synthesized by Sol-Gel method. *Journal of Molecular Structure, 1196*, 332–337.
Jain, S., Shah, A. P., & Shimpi, N. G. (2020). Efficient photocatalytic degradation of organic dyes under visible light using zinc stannate (Zn_2SnO_4) nanorods prepared by microwave irradiation. *Nanostructures and Nano-Objects, 21*, 100410.
Jeevanandam, J., Chan, Y. S., & Danquah, M. K. (2016). Biosynthesis of metal and metal oxide nanoparticles. *ChemBioEng Reviews, 3*(2), 55–67.
Kappadan, S., Gebreab, T. W., Thomas, S., & Kalarikkal, N. (2016). Tetragonal $BaTiO_3$ nanoparticles: an efficient photocatalyst for the degradation of organic pollutants. *Materials Science in Semiconductor Processing, 51*, 42–47.
Khalafi, T., Buazar, F., & Ghanemi, K. (2019). Photosynthesis and enhanced photocatalytic activity of zinc oxide nanoparticles toward organosulfur pollutants. *Scientific Reports, 9*(1), 1–10.
Khan, S. H., & Pathak, B. (2020). Zinc oxide based photocatalytic degradation of persistent pesticides: A comprehensive review. *Environmental Nanotechnology, Monitoring and Management, 13*, 100290.
Khan, I., Saeed, K., & Khan, I. (2019). Nanoparticles: Properties, applications, and toxicities. *Arabian Journal of Chemistry, 12*(7), 908–931.
Kumar, S., Kumar, A., Kumar, A., & Krishnan, V. (2020). Nanoscale zinc oxide-based heterojunctions as visible light active photocatalysts for hydrogen energy and environmental remediation. *Catalysis Reviews, 62*(3), 346–405.
Kumar, S. G., & Rao, K. K. (2017). Comparison of modification strategies towards enhanced charge carrier separation and photocatalytic degradation activity of metal oxide semiconductors (TiO_2, WO_3 and ZnO). *Applied Surface Science, 391*, 124–148.
Le, A. T., Pung, S. Y., Sreekantan, S., & Matsuda, A. (2019). Mechanisms of removal of heavy metal ions by ZnO particles. *Heliyon, 5*(4), e01440.
Lee, K. M., Lai, C. W., Ngai, K. S., & Juan, J. C. (2016). Recent developments of zinc oxide based photocatalyst in water treatment technology: A review. *Water Research, 88*, 428–448.
Lingaraju, K., Raja Naika, H., Manjunath, K., Basavaraj, R. B., Nagabhushana, H., Nagaraju, G., & Suresh, D. (2016). Biogenic synthesis of zinc oxide nanoparticles using Ruta graveolens (L.) and their antibacterial and antioxidant activities. *Applied Nanoscience, 6*, 703–710.
Lu, F., & Astruc, D. (2020). Nanocatalysts and other nanomaterials for water remediation from organic pollutants. *Coordination Chemistry Reviews, 408*, 213180.

Majumder, S., Chatterjee, S., Basnet, P., & Mukherjee, J. (2020). ZnO based nanomaterials for photocatalytic degradation of aqueous pharmaceutical waste solutions–A contemporary review. *Environmental Nanotechnology, Monitoring and Management, 14*, 100386.

Malakootian, M., Yaseri, M., & Faraji, M. (2019). Removal of antibiotics from aqueous solutions by nanoparticles: A systematic review and meta-analysis. *Environmental Science and Pollution Research, 26*, 8444–8458.

Mirzaei, A., Chen, Z., Haghighat, F., & Yerushalmi, L. (2016). Removal of pharmaceuticals and endocrine disrupting compounds from water by zinc oxide-based photocatalytic degradation: a review. *Sustainable Cities and Society, 27*, 407–418.

Mishra, S., Sahu, T. K., Verma, P., Kumar, P., & Samanta, S. K. (2019). Microwave-assisted catalytic degradation of brilliant green by spinel zinc ferrite sheets. *ACS Omega, 4*(6), 10411–10418.

Mitra, P., Dutta, D., Das, S., Basu, T., Pramanik, A., & Patra, A. (2018). Antibacterial and photocatalytic properties of ZnO-9-aminoacridine hydrochloride hydrate drug nanoconjugates. *ACS Omega, 3*(7), 7962–7970.

Mutalib, A. A., & Jaafar, N. F. (2022). ZnO photocatalysts applications in abating the organic pollutant contamination: A mini-review. *Total Environment Research Themes*, 100013.

Niranjani, R., & Anchana-Devi, C. (2017). Photocatalytic degradation of pesticide phorate using zinc oxide nanoparticles. *International Journal of Academic Research and Development, 2*, 35–40.

Ong, C. B., Ng, L. Y., & Mohammad, A. W. (2018). A review of ZnO nanoparticles as solar photocatalysts: Synthesis, mechanisms, and applications. *Renewable and Sustainable Energy Reviews, 81*, 536–551.

Pillai, A. M., Sivasankarapillai, V. S., Rahdar, A., Joseph, J., Sadeghfar, F., Rajesh, K., & Kyzas, G. Z. (2020). Green synthesis and characterization of zinc oxide nanoparticles with antibacterial and antifungal activity. *Journal of Molecular Structure, 1211*, 128107.

Premalatha, N., & Miranda, L. R. (2019). Surfactant modified ZnO-Bi_2O_3 nanocomposite for degradation of lambda-cyhalothrin pesticide in visible light: A study of reaction kinetics and intermediates. *Journal of Environmental Management, 246*, 259–266.

Puri, N., & Gupta, A. (2023). Water remediation using titanium and zinc oxide nanomaterials through disinfection and photocatalysis process: A review. *Environmental Research*, 115786.

Painuli, S., Semwal, P., Bacheti, A., Bachheti, R. K., & Husen, A. (2020). Nanomaterials from nonwood forest products and their applications. In A. Husen & M. Jawaid (Eds.), Nanomaterials for agriculture and forestry applications (pp. 15–40). Elsevier. https://doi.org/10.1016/B978-0-12-817852-2.00002-0

Raja, V. R., Rosaline, D. R., Suganthi, A., & Rajarajan, M. (2018). Facile sonochemical synthesis of Zn_2SnO_4–V_2O_5 nanocomposite as an effective photocatalyst for degradation of Eosin Yellow. *Ultrasonics Sonochemistry, 44*, 310–318.

Raja, M. A., Ahmad, M. A., Daniyal, M., & Husen, A. (2023). Management of wastewater and other environmental issues using smart nanomaterials. In: A. Husen & K. S. Siddiqi (Eds.), *Advances in smart nanomaterials and their applications*. Elsevier Inc. https://doi.org/10.1016/B978-0-323-99546-7.00003-3

Raja, M. A., & Husen, A. (2020). Role of nanomaterials in soil and water quality management. In A. Husen & M. Jawaid (Eds.), Nanomaterials for agriculture and forestry applications (pp. 491–503). Elsevier Inc. https://doi.org/10.1016/B978-0-12-817852-2.00020-2

Ranjith Kumar, D., Ranjith, K. S., Haldorai, Y., Kandasami, A., & Rajendra Kumar, R. T. (2019). Nitrogen-implanted ZnO nanorod arrays for visible light photocatalytic degradation of pharmaceutical drug acetaminophen. *ACS Omega, 4*(7), 11973–11979.

Ranjith, K. S., Manivel, P., Rajendrakumar, R. T., & Uyar, T. (2017). Multifunctional ZnO nanorod-reduced graphene oxide hybrids nanocomposites for effective water remediation: Effective sunlight driven degradation of organic dyes and rapid heavy metal adsorption. *Chemical Engineering Journal, 325*, 588–600.

Raza, W., Faisal, S. M., Owais, M., Bahnemann, D., & Muneer, M. (2016). Facile fabrication of highly efficient modified ZnO photocatalyst with enhanced photocatalytic, antibacterial and anticancer activity. *RSC Advances, 6*(82), 78335–78350.

Ritika, Kaur, M., Umar, A., Mehta, S. K., Singh, S., Kansal, S. K., Fouad, H., & Alothman, O. Y. (2018). Rapid solar-light driven superior photocatalytic degradation of methylene blue using MoS_2-ZnO heterostructure nanorods photocatalyst. *Materials, 11*(11), 2254.

Salehi-Babarsad, F., Derikvand, E., Razaz, M., Yousefi, R., & Shirmardi, A. (2020). Heavy metal removal by using ZnO/organic and ZnO/inorganic nanocomposite heterostructures. *International Journal of Environmental Analytical Chemistry, 100*(6), 702–719.

Sanakousar, F. M., Vidyasagar, C. C., Jiménez-Pérez, V. M., & Prakash, K. (2022). Recent progress on visible-light-driven metal and non-metal doped ZnO nanostructures for photocatalytic degradation of organic pollutants. *Materials Science in Semiconductor Processing, 140*, 106390.

Santhoshkumar, J., Kumar, S. V., & Rajeshkumar, S. (2017). Synthesis of zinc oxide nanoparticles using plant leaf extract against urinary tract infection pathogen. *Resource-Efficient Technologies, 3*(4), 459–465.

Saravanan, R., Gracia, F., & Stephen, A. (2017). Basic principles, mechanism, and challenges of photocatalysis. In *Nanocomposites for visible light-induced photocatalysis* (pp. 19–40).

Satish Kumar, N., Reddy, M. S., Bheeram, V. R., Mukkamala, S. B., Raju Chowhan, L., & Chandrasekhara Rao, L. (2019). Zinc oxide nanoparticles as an efficient catalyst for the synthesis of novel di-spiroindolizidine bis oxindoles in an aqueous medium. *Environmental Chemistry Letters, 17*, 455–464.

Shubha, J. P., Sushma, N. V., Adil, S. F., Khan, M., Assal, M. E., Hatshan, M. R., & Shaik, B. (2022). ZnO/La_2O_3/NiO based ternary heterostructure nano-photocatalyst: Preparation, characterization and its application for the degradation of methylene blue. *Journal of King Saud University-Science, 34*(1), 101738.

Shukor, S. A. A., Hamzah, R., Bakar, M. A., Noriman, N. Z., Al-Rashdi, A. A., Razlan, Z. M., Shahriman, A. B., Zunaidi, I., & Khairunizam, W. (2019). Metal oxide and activated carbon as photocatalysts for wastewater treatment. In *Paper presented at the IOP conference series: Materials science and engineering*.

Singh, J., Dutta, T., Kim, K. H., Rawat, M., Samddar, P., & Kumar, P. (2018). 'Green'synthesis of metals and their oxide nanoparticles: applications for environmental remediation. *Journal of Nanobiotechnology, 16*(1), 1–24.

Singh, P., Kim, Y. J., Zhang, D., & Yang, D. C. (2016). Biological synthesis of nanoparticles from plants and microorganisms. *Trends in Biotechnology, 34*(7), 588–599.

Sivaraman, C., Vijayalakshmi, S., Leonard, E., Sagadevan, S., & Jambulingam, R. (2022). Current developments in the effective removal of environmental pollutants through photocatalytic degradation using nanomaterials. *Catalysts, 12*(5), 544.

Sivasankarapillai, V. S., Jose, J., Shanavas, M. S., Marathakam, A., Uddin, M. S., & Mathew, B. (2019). Silicon quantum dots: promising theranostic probes for the future. *Current Drug Targets, 20*(12), 1255–1263.

Sosna-Głębska, A., Sibiński, M., Szczecińska, N., & Apostoluk, A. (2020). UV–Visible silicon detectors with zinc oxide nanoparticles acting as wavelength shifters. *Materials Today: Proceedings, 20*, 25–29.

Suresh, J., Pradheesh, G., Alexramani, V., Sundrarajan, M., & Hong, S. I. (2018). Green synthesis and characterization of zinc oxide nanoparticle using insulin plant (Costus pictus D. Don) and investigation of its antimicrobial as well as anticancer activities. *Advances in Natural Sciences: Nanoscience and Nanotechnology, 9*(1), 015008.

Sushma, C., & Girish Kumar, S. (2017). Advancements in the zinc oxide nanomaterials for efficient photocatalysis. *Chemical Papers, 71*(10), 2023–2042.

Taha, K. K., Al Zoman, M., Al Outeibi, M., Alhussain, S., Modwi, A., & Bagabas, A. A. (2019). Green and sonogram synthesis of zinc oxide nanoparticles for the photocatalytic degradation of methylene blue in the water. *Nanotechnology for Environmental Engineering, 4*, 1–11.

Wang, Z., Liu, B., Xie, Z., Li, Y., & Shen, Z. Y. (2016). Preparation and photocatalytic properties of RuO_2/TiO_2 composite nanotube arrays. *Ceramics International, 42*(12), 13664–13669.
Wasewar, K. L., Singh, S., & Kansal, S. K. (2020). *Process intensification of treatment of inorganic water pollutants inorganic pollutants in water* (pp. 245–271). Elsevier.
Xu, L., Zhou, Y., Wu, Z., Zheng, G., He, J., & Zhou, Y. (2017). Improved photocatalytic activity of nanocrystalline ZnO by coupling with CuO. *Journal of Physics and Chemistry of Solids, 106*, 29–36.
Yari, K., Seidmohammadi, A., Khazaei, M., Bhatnagar, A., & Leili, M. (2019). A comparative study for the removal of imidacloprid insecticide from water by chemical-less UVC, UVC/TiO2 and UVC/ZnO processes. *Journal of Environmental Health Science and Engineering, 17*, 337–351.
Yousefi, R., Jamali-Sheini, F., Cheraghizade, M., Khosravi-Gandomani, S., Saaedi, A., Huang, N. M., Basirun, W. J., & Azarang, M. (2015). Enhanced visible-light photocatalytic activity of strontium-doped zinc oxide nanoparticles. *Materials Science in semiconductor processing, 32*, 152–159.
Zebeaman, M., Bachheti, R. K., Bachheti, A., Pandey, D. P., & Husen, A. (2023). Green and cost-effective nanoparticles synthesis from some frequently used medicinal plants and their various applications. In R. K. Bachheti & A. Bachheti (Eds.), *Secondary metabolites from medicinal plants* (pp. 287–304), CRC Press.
Zheng, Y., Fu, L., Han, F., Wang, A., Cai, W., Yu, J., Yang, J., & Peng, F. (2015). Green biosynthesis and characterization of zinc oxide nanoparticles using Corymbia citriodora leaf extract and their photocatalytic activity. *Green Chemistry Letters and Reviews, 8*(2), 59–63.

Application of Silver-Doped Nanomaterials for Wastewater Treatment

Versha Dixit, Himani Rawat, K. Aggarwal, Kundan Kumar Chaubey, Anish Kumar Pal, B. T. Manjunath, Shweta Sharma, and Rakesh Kumar Bachheti

Abstract Every living thing is strongly dependent on water to survive. A lack of water can lead to a wide range of problems. Without water, humankind may have a tough time surviving. The water supplies used by industries should be protected and treated in a variety of right ways in order to save, conserve, and protect the water supply. Treating wastewater is one of the most cost-effective and convenient ways to overcome water scarcity. Depending on the treatment system, water can be treated differently physically, chemically, biologically, or in combination. Nanotechnology in these systems allows wastewater treatment to be carried out at a significantly higher quality level than conventional systems, and most of their limitations can be solved. Nanomembranes used in membrane bioreactor technology are among

V. Dixit · B. T. Manjunath
Department of Life Sciences, CHRIST (Deemed to Be University), Bangalore, Karnataka 560029, India

H. Rawat
School of Basic & Applied Sciences, Shri Guru Ram Rai University, West Patel Nagar, Patel Nagar, Dehradun, Uttarakhand, India

K. Aggarwal
School of Biotechnology, Gautam Buddha University, Greater Noida, Uttar Pradesh, India

K. K. Chaubey (✉) · A. K. Pal
Division of Research and Innovation, School of Applied and Life Sciences, Uttaranchal University, Arcadia Grant, P.O. Chandanwari, Premnagar, Dehradun, Uttarakhand 248007, India
e-mail: kundan2006chaubey@gmail.com

K. K. Chaubey
Department of Biotechnology, School of Basic and Applied Sciences, Sanskriti University, Mathura 281401, Uttar Pradesh, India

S. Sharma
College of Biotechnology, DUVASU, Mathura, Uttara Pradesh, India

R. K. Bachheti
Department of Allied Sciences, Graphic Era Hill University, Society Area, Clement Town, Dehradun, India

Department of Industrial Chemistry, Addis Ababa Science and Technology University, P.O. Box 16417, Addis Ababa, Ethiopia

R. K. Bachheti et al. (eds.), *Metal and Metal-Oxide Based Nanomaterials*, Smart Nanomaterials Technology, https://doi.org/10.1007/978-981-99-7673-7_15

the best and most effective technologies currently available for water treatment because they are both appropriate and efficient. They can be applied to put off water shortages for several more years. Silver-doped nanoparticles have many properties, including antimicrobial, antibacterial, disinfectant, and many more. This chapter aims to present an in-depth analysis of the potential and efficacy of silver-doped nanoparticles for wastewater treatment.

Keywords Silver nanoparticles · Nanomembrane · Water pollution · Wastewater · Water treatment · Nanomaterials

1 Introduction

Water pollution is a severe and significant area of care and concern nowadays due to increasing pollution, industrial waste, and other compounds. The regular discharge of waste material into the water bodies makes it contaminated day by day. This contamination in water is damaging and harmful to aquatic life and human beings. Water is a vital substance for all the living organisms living on the motherland Earth and is an essential and precious resource for animal, plant, and human civilization. Access and availability of clean and safe water to everyone is one of the humanitarian goals and still a very challenging target to achieve. Worldwide around 780 million people still lack fresh and clean water for drinking (WHO, 2012). Therefore treating wastewater is a significant concern and needs immediate attention and solution. In both developing and developing countries, human activities play a significant role in exacerbating water. The increasing population and their demand increase the consumption rate of water and its need. Moreover, the global climate accentuates the uneven distribution of fresh and clean water supply.

In recent times if we compare the rate of supply of fresh water to the rate of treatment of wastewater, then the rate of treatment of wastewater is very low compared to the need. This untreated water has so many chemical compounds which are nondegradable and hence are present in water even after the treatment and consumption of such water, making humans sick and unhealthy. Another phenomenon called bioaccumulation through the food chain and food web causes very harmful and adverse negative effects on human and animal health along with environmental microflora (Kumar et al., 2015a, 2015b; Larramendy & Soloneski, 2015; Sidhu et al., 2019). The direct disposal and degradation of untreated wastewater in water bodies, including oceans, rivers, and lakes, causing the considering global problems and challenges and have a very adverse and harmful effect on the quantity and quality of wastewater (Naidoo & Olaniran, 2013). Overall, almost 38,354 million liters per day of wastewater are generated from the metropolitan cities of India. Around 13,468 MLD waste is generated, and India only can treat around 8080.8 MLD (Kaur et al., 2012). Water not only contains harmful compounds but also contain very harmful bacteria, protozoans, viruses, and many helminths as well, which are causative organism for many diseases and disorders such as campylobacteriosis, hepatitis, and lepidopterist (Kaur et al., 2018; Kumar et al., 2014a, 2014b; Lin & Ganesh, 2013; Singh et al., 2016, 2017a, 2017b). The government and people often make schemes and efforts to clean the water, which is becoming dirty daily by various anthropogenic enterprises (Kumar et al., 2013; Sahu, 2014). Various techniques, including filtration, coagulation, flocculation, carbon adsorption, many oxidation processes, membrane bioreactor, and many other advanced growth treatments, are used for treating wastewater (Altmann et al., 2014; Bollmann et al., 2016; Bonvin et al., 2016; Jegatheesan et al., 2016; Liang et al., 2014). All these mentioned processes and methods are capable of cleaning and removing the solid and particulate impurities from the water, which is considered waste, but all the following techniques are not capable of removing the endocrine disruptor chemicals from the wastewater (Arbabi et al., 2015; Gosavi & Sharma, 2014). So far, these methods and techniques still need to be more efficient in thoroughly treating wastewater (Luo et al., 2014). Thus the importance and need for validation and optimization in clearing and treating wastewater is paramount. Moreover, along with the technology development, a collaborative approach is needed. In addition to being eco-friendly and cost-effective, the technology was developed to remove all impurities from the water and ensure that it is free of contamination.

There is a class of materials known as nanomaterials in which one of the material's dimensions is between 1 and 100 nm in size (Bachheti & Bachheti, 2023; Stark et al., 2015). Several characteristics differentiate nanomaterials from other materials regarding their optical, electrical, and magnetic properties. There is strong evidence that nanomaterials possess high adsorption, reactivity, and catalytic activity levels compared to conventional materials (Khan et al., 2017). Nanoparticles have attracted researchers and scientists in the last few years because of their properties. Nanoparticles are used in many fields, such as biology, medicine, catalytic chemistry, sensing, agriculture, and forestry (Bachheti et al., 2019; Dauthal & Mukhopadhyay, 2016; Kamaly et al., 2016; Husen et al., 2019, 2023; Husen & Iqbal, 2019; Husen & Jawaid, 2020; Husen & Siddiqi, 2023; Husen, 2022, 2023a, 2023b; Taghiyari et al., 2022). In addition, nanoparticles also have other advantages in wastewater treatment due to their surface area (Raja & Husen, 2020; Raja et al., 2023).

The smaller size of the nanoparticles can be associated with a larger volume. Another unique property is their excellent adsorption reactivity and capacity (Bachhet et al., 2020; Mauter et al., 2018). It has been reported that there are many types of pollutants, such as inorganic anions, natural pollutants, bacteria, and organic solvents, which can cause the number of nanomaterials or nanoparticles to decompose (Mendez et al., 2017; Varjani et al., 2017). Several nanoparticles, such as zerovalent nanoparticles, carbon nanotubes, metal oxide nanoparticles, and nanocomposites, are being developed now (Prasad & Thirugnanasanbandham, 2019; Usmania et al., 2017).

Since silver is an excellent antimicrobial agent, silver-doped nanoparticles are widely used in wastewater. Silver nanoparticles are very effective against viruses, bacteria, fungi, and viruses. Silver-doped nanoparticles attack the cell membrane and increase its permeability of the cell membrane, producing free radicals (Le et al., 2012). Silver causes free radicals in the cell membrane and eventually induces cell apoptosis. Silver-doped nanoparticles also damage and ultimately destroy phosphorus and sulfur elements and compounds in DNA. In numerous cases, the disintegration of silver nanoparticles discharged the Ag positive particle, which makes contact with the thiol bunch shown within the gathering of essential proteins; this interaction debilitates the chemicals and hinders their ordinary life working (Aziz et al., 2014, 2015, 2016, 2019; Rudakiya & Pawar, 2017). Nanoparticle accumulation in the liquid media for a very long time decreases its efficiency and functioning. That is why in many cases, the silver-doped nanoparticles are immobilized on a suitable membrane for prolonged use. In situ remediation experimenters use the immobilized silver-doped nanoparticles on the cellulose fiber. The sheet shows the significant antibacterial properties and activities against the suspension culture of microbes, including *Enterococcus fecalis*, *Escherichia coli*, and many inactivated bacteria (Chitra & Annadurai, 2014; Peiris et al., 2017).

Many industries manufacturing silver-doped nanoparticles incorporate it in polyethersulfone (PES) or microfiltration membranes with different pore diameters (Ferreira et al., 2015). The activity of microorganisms was observed to be very much suppressed when used with membrane silver-doped nanoparticles. The PES silver-doped nanoparticles membrane shows strong antimicrobial properties and is held to be very useful in the treatment of wastewater.

The Ag NPs make a difference in murdering pathogens by actuating physical annoyance. Moreover, with a few oxidative pushes, through the devastation and disturbance of a particular microbial prepare through either oxidation or unsettling influence of an imperative and critical imperative cellular component and, in a few cases, disruption of cell film structure (Prasad, 2014; Prasad et al., 2018; Seo et al., 2014). Within the final two decades, the immobilization of Ag NPs on film and numerous ceramic materials has drawn consideration toward itself since of its potential properties and exercises. Moreover, broadly utilized in biofouling decrease and sensitization of squandered water treatment (Amin et al., 2014).

To extend its productivity and viability, numerous works are in advance, one of which is to utilize tall porosity channels made up of sawdust and clay, which are being used to evacuate *E.coli* and other destructive microscopic organisms and pathogens from the wastewater (Moosa & Muhsen, 2017). Nanotechnology is advertising jumping openings to create a sufficient water supply for another era. Nanotechnology is an emerging technology in wastewater treatment, allowing wastewater treatment plants that Rely on substantial foundations to achieve outstanding performance by using nanotechnology for wastewater treatment (Qu et al., 2013).

From a financial point of view, nanotechnology and nanoparticles permit the total utilization of exceptionally troublesome and challenging water assets and vitality preservation. In this manner, to utilize this innovation appropriately, the fetch ought to be taken care of carefully because of its competition with the conventional strategy of squandering water treatment advances (Crane et al., 2012).

Hence, there is a requirement for new and imaginative advances in wastewater treatment. One of the developing innovations for wastewater treatment is the utilization of nanomaterials. Nanomaterials have exciting properties due to their little measure and tall surface range, which make them profoundly productive in wastewater treatment. Nanomaterials have interesting properties due to their small measure and tall surface range, which make them profoundly productive in wastewater treatment. Silver-doped nanomaterials have shown promising results in expelling various poisons from wastewater, including heavy metals, natural compounds, and microorganisms. The silver particles within the nanomaterials have solid antimicrobial properties, which can successfully slaughter microscopic organisms and infections in wastewater. In addition, the tall surface zone of the nanomaterials gives an expansive contact area for toxin adsorption. The application of silver-doped nanomaterials for wastewater treatment is still within the inquiry and advancement stage. In any case, a few ponders have illustrated their potential for wastewater treatment. One of the noteworthy focal points of silver-doped nanomaterials is that they can be effortlessly synthesized utilizing diverse strategies, counting chemical, physical, and natural strategies.

This makes them exceedingly cost-effective and adaptable for mechanical wastewater treatment. Moreover, using silver-doped nanomaterials for wastewater treatment has a few natural benefits (Tsarev et al., 2016). The nanomaterials can expel toxins from wastewater without creating any nasty by-products. Besides, the nanomaterials can be effectively evacuated from the treated wastewater utilizing straightforward filtration strategies. In conclusion, applying silver-doped nanomaterials for wastewater treatment may be a promising innovation that can give an economic arrangement for the developing request for clean water. The unique properties of silver-doped nanomaterials make them exceedingly effective in poison expulsion from wastewater. However, a considerable amount of work still needs to be done to improve the blend and application of these nanomaterials for large-scale wastewater treatment applications.

2 Potential Application in Wastewater Treatment

Currently, accessible wastewater treatment advances of slurry can be improved by using nanoparticles in it. Connected within the powder form, the nano-adsorbents within the slurry have a noteworthy advantage because they increase the surfaces of the adsorbents.

Wastewater treatment is one of the world's foremost basic natural challenges today (Ju-Nam & Lead, 2016; Prasad et al., 2014, 2017). The developing populace and industrialization have increased water contamination, antagonistically impacting the environment and human well-being. In this manner, there is a requirement for inventive and viable strategies to treat wastewater. In later years, silver-doped nanomaterials developed as an effective instrument for wastewater treatment (Sharma et al., 2015).

Silver-doped nanomaterials are nanoparticles that contain silver particles, which have potent antimicrobial properties. These nanoparticles can be synthesized utilizing different strategies, such as chemical lessening, sol–gel, and aqueous strategies

(Fig. 2). The expansion of silver particles to nanomaterials improves their antimicrobial movement, making them successful in slaughtering microscopic organisms, infections, and other microorganisms.

The application of silver-doped nanomaterials for wastewater treatment has a few focal points. Firstly, these nanoparticles can evacuate many contaminants, counting natural and inorganic toxins, overwhelming metals, and pathogens. Furthermore, these nanoparticles are cost-effective and can be effortlessly created in expansive amounts. Thirdly, silver-doped nanomaterials can be effortlessly consolidated into existing wastewater treatment forms, making them a practical choice for mechanical applications.

The instrument of activity of silver-doped nanomaterials includes the discharge of silver particles, which are associated with the cell film of microorganisms, driving their inactivation (Nowack & Bucheli, 2007). The silver particles disturb the cell layer, driving the spillage of intracellular components coming about in cell passing. Moreover, silver-doped nanomaterials can produce responsive oxygen species, which assist in upgrading their antimicrobial action. In conclusion, silver-doped nanomaterials have monstrous potential for wastewater treatment. The utilization of these nanoparticles can lead to the advancement of inventive and maintainable strategies for treating wastewater, which can have a critical effect on the environment and human health (Liu et al., 2011a, 2011b). Be that as it may, using silver-doped nanomaterials raises concerns about their potential poisonous quality and natural impact (Ghosh et al., 2017; Mahgoub & Samaras, 2014). Subsequently, it is basic to conduct an assisted inquiry about the long-term impacts of these nanoparticles on the environment and human health (Hou et al., 2013).

3 Properties of Silver-Doped Nanoparticles

Nanosilver is widely used in many fields due to its unique properties. Adding additives further improves silver nanoparticles' properties, making them more effective. Silver-doped nanoparticles have many properties that make them suitable for many applications. One of the most important properties of silver-doped nanoparticles is their antibacterial properties (Fig. 3) (Shalin et al., 2012). Studies have shown that silver nanoparticles have strong antibacterial properties against many pathogens, including bacteria, viruses, and fungi (Le et al., 2012). The ability of the silver-doped nanoparticles to kill bacteria is further enhanced by adding chemicals like copper, gold, or platinum (Zhang et al., 2012). Both outstanding electrical and antibacterial capabilities can be found in silver-doped nanoparticles (Lee et al., 2008). Because of their features, they are perfect for electronic equipment, including sensors, circuits, and batteries. Silver nanoparticles' characteristics can be improved by adding additives, which increases their suitability for electrical applications (Aziz et al., 2014, 2015, 2016, 2019; Rudakiya & Pawar, 2017). Excellent catalytic capabilities can also be seen in silver-doped nanoparticles.

They are helpful as catalysts for various processes, such as oxidation, reduction, and hydrogenation. Silver nanoparticles can work more efficiently as catalysts by adding additives like palladium, platinum, or rhodium. Optical characteristics are another crucial aspect of silver-doped nanoparticles (Chitra & Annadurai, 2014; Peiris et al., 2017). They exhibit robust plasmon resonances that render them highly light-sensitive. They are advantageous in imaging applications in science, art, and medicine. Because of this, silver-doped nanoparticles possess various qualities that make them adaptable and effective in a wide range of applications. They are appropriate for use in medicine, electronics, and catalysis due to their antibacterial, electrical, catalytic, and optical capabilities (Prasad, 2014; Prasad et al., 2018; Seo et al., 2014). Researchers and engineers prefer silver-doped nanoparticles because adding dopants further improves these qualities.

4 Characterization of Silver-Doped Nanoparticles

Nanotechnology is very interested in silver-doped nanoparticles because they are attractive materials (Gonfa et al., 2023; Qu et al., 2013). Advanced methods must be used to study these items to comprehend their composition, structure, and characteristics. The development of novel applications and technology requires the characterization of silver-doped nanoparticles. Numerous techniques, such as spectroscopy, microscopy, and X-ray diffraction, can characterize silver-doped nanoparticles (Fig. 4). Spectroscopy is used to examine the optical characteristics of objects and provide details on their size, shape, and chemical composition (Gonfa et al., 2022; Liga et al., 2011).

Object shape and size distribution are investigated using microscopy methods, including transmission electron microscopy (TEM) and scanning electron microscopy (SEM). In order to ascertain the material's crystal structure, X-ray diffraction (XRD) is performed. Determining the chemical makeup of silver-doped nanoparticles is another characteristic (Glover et al., 2006). Energy-dispersive X-ray spectroscopy (EDS) and X-ray photoelectron spectroscopy (XPS) are used to do this. These concepts aid in comprehending the substance's electrical structure and chemical bonding. The size and form of silver-doped nanoparticles affect their characteristics. According to Wang et al. (2017), as particle size grows, the surface area to volume ratio also rises, boosting particle reactivity. Understanding these goods' structure, composition, and properties will be more accessible through this procedure, which is crucial for creating new applications and technologies. Many sectors may transform due to the exciting research on these materials (Mostafaii et al., 2017).

5 How Silver-Doped Nanoparticles Are Formed

Nanoparticles that have been doped with silver are helpful for research and development. These nanoparticles are employed in various fields, including nanoelectronics, medication delivery, and biological research. However, creating these nanoparticles can be difficult and requires specialized tools and knowledge. Chemical reduction is the most popular technique for creating silver-doped nanoparticles. This procedure adds a stabilizer while a silver salt and a reducing agent combine (Denali et al., 2017; Leyu et al., 2023). The silver ions are converted to silver atoms using a reducing agent and assembled to produce nanoparticles—Stabilizers aid in keeping items uniform in size and preventing clumping. Sodium borohydride is one of the most popular reducing agents to create silver-doped nanoparticles. Due to the compound's high reactivity, silver ions can be quickly reduced to create nanoparticles. Citrate, hydrazine, and ascorbic acid are additional reducing agents that may be employed (Sahara et al., 2014).

Surfactants or polymers are frequently utilized as stabilizers to create silver-doped nanoparticles. These chemicals aid in maintaining homogeneous nanoparticle size and shape by controlling the growth and aggregation of the particles (Yang & Yin, 2017). PVP (polyvinyl pyrrolidone), CTAB (cetyltrimethylammonium bromide), and SDS (sodium dodecyl sulfate) are a few of the stabilizers that are most frequently utilized. Techniques such as transmission electron microscopy (TEM), X-ray diffraction (XRD), and UV–Vis spectroscopy are frequently utilized when creating silver-doped nanoparticles. With the aid of these methods, researchers may identify the size, content, and optical and electrical characteristics of nanoparticles (Fig. 5).

In conclusion, a chemical reduction process involving specialized knowledge and tools is necessary to produce silver-doped nanoparticles. The most popular technique involves reducing and stabilizing agents, which produce nanoparticles of consistent size and quality by decreasing silver ions. Nanoparticles are characterized after synthesis using various techniques to ascertain their properties and suitability for various applications.

6 Advantage of Silver-Doped Nanoparticles in Wastewater Treatment

Due to their unique characteristics and capacity to remove pollutants from water, nanoparticles are helpful in various industries, including wastewater treatment. Silver-doped nanoparticles have emerged as one of the best and most efficient options for wastewater treatment. Due to their high surface area to volume ratio, these nanoparticles can cause more water pollution (Stark et al., 2015). The capacity of silver-doped nanoparticles to remove diverse contaminants is one of its key advantages in wastewater treatment (Fig. 6).

According to Mauter et al. (2018), these nanoparticles may remove germs, contaminants, and heavy metals from wastewater. They are unique in treating water with hazardous germs because they also have antibacterial characteristics. The durability of silver-doped nanoparticles provides an additional benefit. The antioxidant properties of these nanoparticles last for a very long time. Because of this, it may be recycled without losing its capacity to purge pollutants from wastewater (Prasad & Thirugnanasanbandham, 2019; Usmania et al., 2017).

Additionally, silver-doped nanoparticles are pretty inexpensive. It employs minimal material to cleanse vast amounts of wastewater and is relatively cheap to create. Due to this, they are an excellent option for wastewater treatment facilities and businesses that produce large amounts of wastewater (Méndez et al., 2017; Varjani et al., 2017). Silver-doped nanoparticles are efficient, cost-effective, and favorable to the environment. During processing, they do not create any waste or toxic by-products (Ghosh et al., 2017; Mahgoub & Samaras, 2014). It works well in delicate situations without endangering the ecosystem. In general, silver-doped nanoparticles are pretty beneficial for treating wastewater. Companies and municipalities wishing to upgrade their wastewater treatment procedures are an excellent option because of their capacity to remove contaminants, long-term safety, effectiveness, and environmental friendliness. As a result, they can be utilized in delicate and valuable locations without endangering the environment or having a detrimental effect on it. Overall, silver-doped nanoparticles have many benefits for treating wastewater. They are an excellent option for businesses and governments wishing to upgrade their wastewater treatment procedures because of their versatility in removing contaminants, long-term stability, cost-effectiveness, and environmental friendliness.

7 Different Types of Nanoparticles Used in Wastewater Treatment

Wastewater treatment is an essential step in maintaining a healthy and sustainable ecosystem. Numerous techniques, including the use of nanoparticles, have been developed to clean wastewater as technology has improved. According to Glover et al. (2006), nanoparticles are minuscule particles with distinct physical and chemical properties that range in size from 1 to 100 nm. Due to their substantial surface area, high reactivity, and capacity to adsorb contaminants, they have been demonstrated to be successful at removing pollutants from water (Cho et al., 2011). Many nanoparticle varieties are used to treat wastewater, each with unique qualities appropriate for particular purposes. Silver nanoparticles, which have great antibacterial qualities and aid in eradicating bacteria and viruses in wastewater, are among the most often utilized nanoparticles. To stop the growth of dangerous bacteria, these nanoparticles can be coated on pipelines and tanks or used as water filters (Kunduru et al., 2017). Metal nanoparticles are another form of nanoparticle utilized in wastewater treatment. Iron nanoparticles efficiently remove organic substances, heavy metals, and colors from water. They can be used in conjunction with other techniques like adsorption, coagulation, and visualization for best results.

Additionally, biodegradable metal nanoparticles are a green solution for wastewater treatment. Another nanoparticle form utilized in wastewater treatment is titanium dioxide (Nassar, 2012). Due to their high reactivity, they can be employed in photocatalytic processes to break down organic molecules in wastewater. Titanium dioxide nanoparticles can also eliminate Heavy metals and dyes (Fig. 7) (Sa & Premalatha, 2016).

As a result, how we handle water has altered due to employing nanoparticles in wastewater treatment. The many different kinds of nanoparticles utilized in wastewater treatment include titanium dioxide, iron, and silver. According to Dave and Chopuda (2014), each type of nanoparticle has distinctive characteristics that make it suited for a certain application. We anticipate discovering new applications for nanoparticles in wastewater treatment with continuing research and development.

8 Influence of Nanomaterials on Microbial Systems

In recent years, nanomaterials have grown in popularity across numerous industries. From energy to medicine, these seemingly little items are numerous and diverse (Amin et al., 2014; Prachi et al., 2013; Qu et al., 2013). As with any new technology, it is crucial to consider any potential effects on the environment and public health. The impact of nanoparticles on microbial systems is one area of interest. Microbial systems are crucial to many industries and play a significant environmental role. They are responsible for waste management, soil fertility, and cycling (Ahmed et al., 2012; Brady-Estévez et al., 2008). These systems are vulnerable to severe effects from any breach. According to research, Nanomaterials have beneficial and harmful effects on microbial life. However, promoting the development and activity of bacteria can enhance waste treatment and bioremediation (Bottino et al., 2001; Bae & Tak, 2005; Ebert et al., 2004; Maximus et al., 2010; Pendergast et al., 2010).

Additionally, they can be employed as antibiotics, which is quite helpful in medicine. On the other hand, microbiological species may also be poisoned by nanoparticles. They limit development and activity by rupturing cell membranes and interfering with cellular functions (Corsi et al., 2018). This has a significant environmental impact, especially if nanomaterials are widely used. More crucially, it is still vital to comprehend how nanoparticles affect microbial systems. More research is necessary to ascertain the long-term effects and potential hazards (Sa & Premalatha, 2016). When employing nanomaterials, monitoring them properly and considering how they will affect the environment and human health is crucial. When employing nanomaterials, proceeding cautiously and considering the potential effects on the environment and human health is imperative.

9 Future Prospective

The use of nanoparticles is the recent approach for wastewater treatment, a very effective and conventional technique with encouraging outcomes (Fig. 1). We can easily remove pollutants, including organic compounds and heavy metals, using nanoparticles. Nanoparticles have various properties, such as the large surface area to volume ratio, which help them quickly interact with toxins in wastewater and ultimately make them perfect for water filtration. Nanoparticles are small in size, and because of their small size, they can easily access areas of water sources that conventional methods cannot reach and remove toxins from them. They use nanoparticles for wastewater treatment to help improve the environment and human health because they remove pollutants and toxins from the water source. With the development of advanced research and techniques, scientists are focusing on developing novel materials and their application for wastewater treatment using nanoparticles. The use of nanoparticles in water treatment will become more important as demand for clean water increases. In conclusion, using nanoparticles to treat wastewater is a promising development with the potential to improve both the environment and human health. More research and advanced techniques are needed to address some of the most severe environmental problems. One of these challenges is wastewater treatment and purification, which can be accomplished by utilizing nanoparticles.

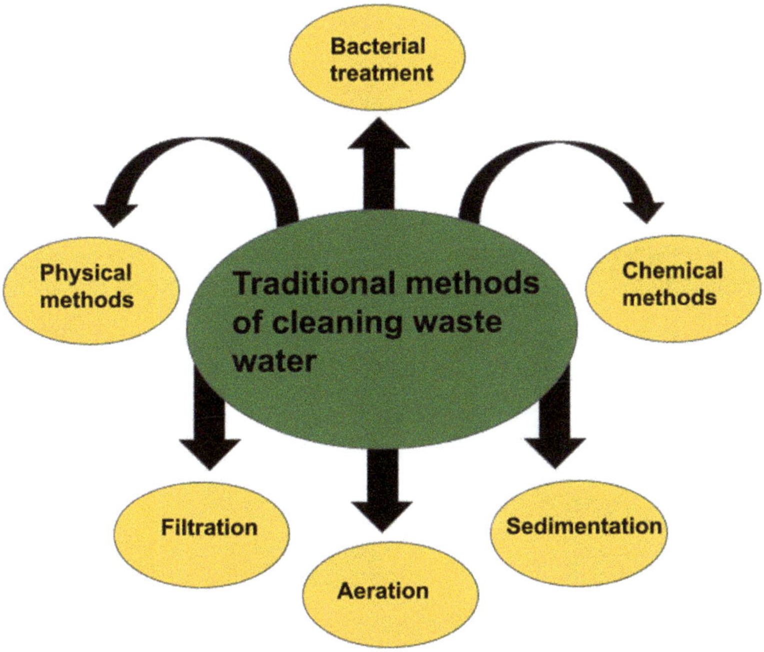

Fig. 1 Traditional methods of cleaning wastewater

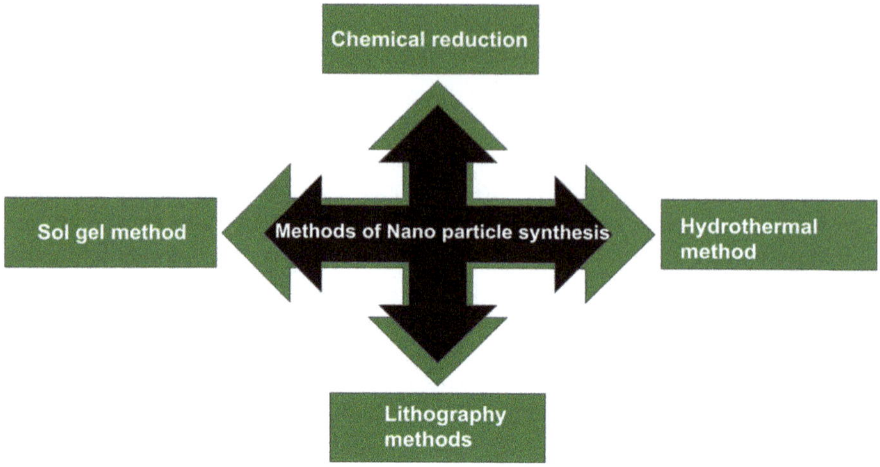

Fig. 2 Methods of nanoparticle synthesis

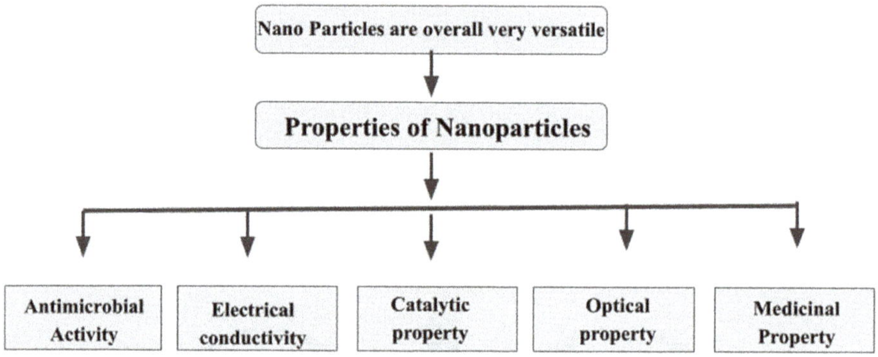

Fig. 3 Properties of nanoparticles

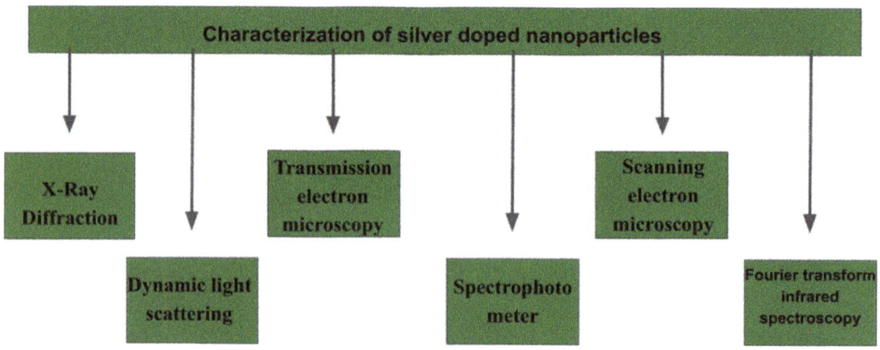

Fig. 4 Characterization of silver-doped nanoparticles

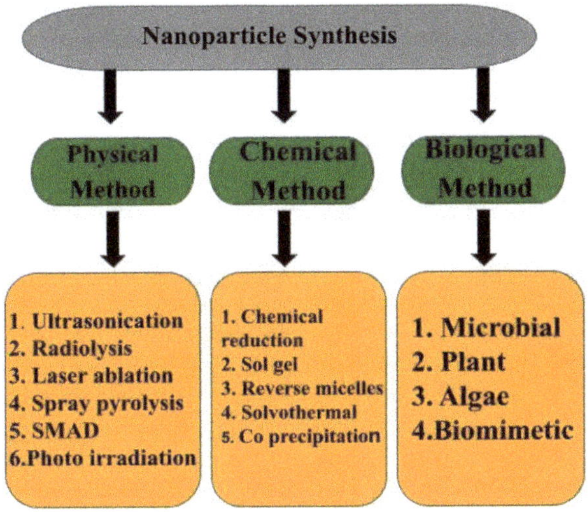

Fig. 5 Different methods used for synthesis of nanoparticles

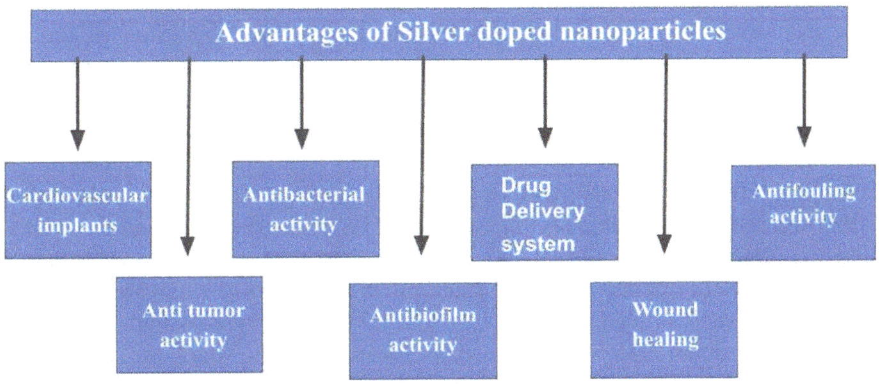

Fig. 6 Advantages of silver-doped nanoparticles

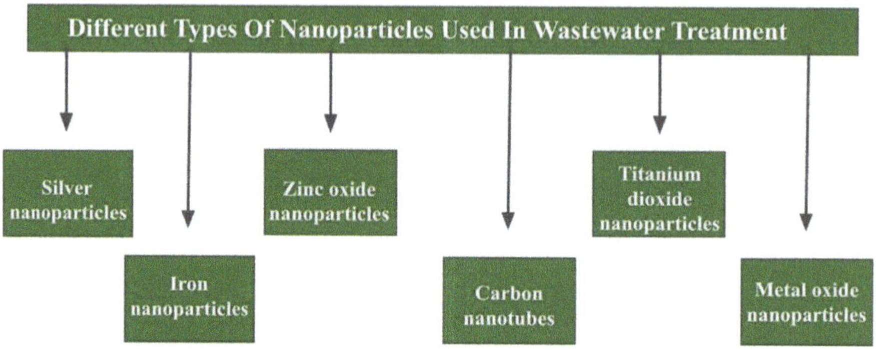

Fig. 7 Different types of nanoparticles used in wastewater treatment

10 Conclusion

The use of silver-doped nanoparticles for wastewater treatment is extensively studied. The main focus of this study is to evaluate the efficacy of silver-doped nanoparticles in eliminating bacteria and other contaminants from wastewater. Recent studies show that silver-doped nanoparticles are efficient in removing various pollutants, germs, viruses, and heavy metals such as lead, mercury, and arsenic. It has been demonstrated that it can also remove toxins, pesticides, and other impurities from wastewater. Conventional water purification techniques can become more efficient and conventional with the use of nanoparticles which can be made to target particular pollutants. Silver-doped nanoparticles are excellent for wastewater treatment as they utilize less energy and resources when compared to traditional methods. In order to fully comprehend the hazards and advantages of employing silver-doped nanoparticles in wastewater treatment, more research is required. Studies have revealed that heavy metals such as lead, cadmium, and mercury strongly associate with silver-doped nanoparticles. If these metals are not taken out of the water, they may result in severe health issues. Organic pollutants like oil and grease that could harm the environment if not addressed are also efficiently removed by nanoparticles.

Furthermore, silver-doped nanoparticles have antibacterial qualities that efficiently purge microorganisms from water. The ability of silver-doped nanoparticles to remove contaminants from wastewater at low concentrations is a significant benefit. This implies that even nanoparticles are capable of cleaning up water pollution. Additionally, using silver-doped nanoparticles is a successful technique for treating wastewater. Nanoparticles are cost-effective and can be easily produced on a large scale. Although they are cost-effective and have many advantages for wastewater treatment, their long-term environmental and human health consequences must be better understood. Silver products also affect aquatic life, and there are very few studies about their effect on aquatic animals and human health. There needs to be more studies about the effect of these chemicals on human health and the environment, mainly when this treated water is used for crop irrigation, drinking, or another purpose. As a result of the above discussion, using silver-doped nanoparticles is a cost-effective, efficient way to remove contaminants from wastewater. These nanoparticles have unique qualities to bind with various contaminants such as organic compounds, heavy metals, and microorganisms. However, their long-term and sustainable impact on the environment, plants, and human health is not clearly understood, so more research is required for the application and its long-term impact on the environment and human health.

References

Ahmed, N., Fessi, H., & Elaissari, A. (2012). Theranostic applications of nanoparticles in cancer. *Drug Discovery Today, 17*(17–18), 928–934.

Altmann, J., Ruhl, A. S., Zietzschmann, F., & Jekel, M. (2014). Direct comparison of ozonation and adsorption onto powdered activated carbon for micropollutant removal in advanced wastewater treatment. *Water Research, 55*, 185–193.

Amin, M. T., Alazba, A. A., & Manzoor, U. (2014). A review of removal of pollutants from water/wastewater using different types of nanomaterials. *Advances in Materials Science and Engineering, 2014*, 1–24.

Arbabi, M., Hemati, S., & Amiri, M. (2015). Removal of lead ions from industrial wastewater: A review of removal methods. *The International Journal of Epidemiology, 2*, 105–109.

Aziz, N., Pandey, R., Barman, I., & Prasad, R. (2016). Leveraging the attributes of Mucor hiemalis-derived silver nanoparticles for a synergistic broad-spectrum antimicrobial platform. *Frontiers in Microbiology, 7*, 1984. https://doi.org/10.3389/fmicb.2016.01984

Aziz, N., Fatma, T., Varma, A., & Prasad, R. (2014). Biogenic synthesis of silver nanoparticles using Scenedesmus abundans and evaluation of their antibacterial activity. *Journal of Nanoparticle Research*, Article ID 689419, https://doi.org/10.1155/2014/689419

Aziz, N., Faraz, M., Pandey, R., Sakir, M., Fatma, T., Varma, A., Barman, I., & Prasad, R. (2015). Facile algae-derived route to biogenic silver nanoparticles: Synthesis, antibacterial and photocatalytic properties. *Langmuir, 31*, 11605–11612. https://doi.org/10.1021/acs.langmuir.5b03081

Aziz, N., Faraz, M., Sherwani, M. A., Fatma, T., & Prasad, R. (2019). Illuminating the anticancerous efficacy of a new fungal chassis for silver nanoparticle synthesis. *Frontiers in Chemistry, 7*, 65. https://doi.org/10.3389/fchem.2019.00065

Bachheti, R. K., Fikadu, A., Bachheti, A., & Husen, A. (2020). Biogenic fabrication of nanomaterials from flower-based chemical compounds, characterization, and their various applications: A review. *Saudi Journal of Biological Sciences, 27*(10), 2551–2562.

Bachheti, R. K., Konwarh, R., Gupta, V., Husen, A., & Joshi, A. (2019). Green Synthesis of iron oxide nanoparticles: cutting edge technology and multifaceted applications. In *Nanomaterials and plant potential* (pp. 239–259). Springer. https://doi.org/10.1007/978-3-030-05569-1_9

Bachheti, R. K., & Bachheti, A. (Eds.). (2023). *Secondary metabolites from medicinal plants: Nanoparticles synthesis and their applications*. CRC Press. https://doi.org/10.1201/9781003213727

Bollmann, A. F., Seitz, W., Prasse, C., Lucke, T., Schulz, W., & Ternes, T. (2016). Occurrence and fate of amisulpride, sulpiride, and lamotrigine in municipal wastewater treatment plants with biological treatment and ozonation. *Journal of Hazardous Materials, 320*, 204–215.

Bonvin, F., Jost, L., Randin, L., Bonvin, E., & Kohn, T. (2016). Super-fine powdered activated carbon (SPAC) for efficient removal of micropollutants from wastewater treatment plant effluent. *Water Research, 90*, 90–99.

Brady-Estévez, A. S., Kang, S., & Elimelech, M. (2008). A single-walled-carbon-nanotube filter for removal of viral and bacterial pathogens. *Small, 4*(4), 481–484.

Central Pollution Control Board (2010) Status of water quality in India 2009. Central Pollution Control Board, Ministry of Environment and Forests, Government of India.

Chitra, K., & Annadurai, G. (2014). Antibacterial activity of pH-dependent biosynthesized silver nanoparticles against clinical pathogen. *BioMed Research International*, 725165.

Cho, M., Cates, E. L., & Kim, J. H. (2011). Inactivation and surface interactions of MS-2 bacteriophage in a TiO_2 photoelectrocatalytic reactor. *Water Research, 45*, 2104–2110.

Corsi, I., Winther-Nielse, M., Sethi, R., Punta, C., Della Torre, C., Libralato, G., & Cinuzzi, F. (2018). Ecofriendly nanotechnologies and nanomaterials for environmental applications: Key issue and consensus recommendations for sustainable and eco-safe nano remediation. *Ecotoxicology and Environmental Safety, 154*, 237–244.

Crane, R. A., & Scott, T. B. (2012). Nanoscale zero-valent iron: Future prospects for an emerging water treatment technology. *Journal of Hazardous Materials, 211*, 112–125.

Dauthal, P., & Mukhopadhyay, M. (2016). Noble metal nanoparticles: Plant-mediated synthesis, mechanistic aspects of synthesis, and applications. *Industrial and Engineering Chemistry Research, 55*, 9557–9577.

Dave, P. N., & Chopda, L. V. (2014). Application of iron oxide nanomaterials for the removal of heavy metals. *Journal of Nanotechnology, 2014*, 1–14.

Dinali, R., Ebrahiminezhad, A., Manley-Harris, M., Ghasemi, Y., & Berenjian, A. (2017). Iron oxide nanoparticles in modern microbiology and biotechnology. *Critical Reviews in Microbiology, 43*, 493–507.

Ferreira, A. M., Roque, É. B., Fonseca, F. V. D., & Borges, C. P. (2015). High flux microfiltration membranes with silver nanoparticles for water disinfection. *Desalination and Water Treatment, 56*, 3590–3598.

Ghosh, A., Nayak, A. K., & Pal, A. (2017). Nano-particle-mediated wastewater treatment: A review. *Current Pollution Reports, 3*, 17–30.

Glover, S., Gómez, L. A., Reyes, K., & Leal, M. T. (2006). A practical demonstration of water disinfection using TiO_2 films and sunlight. *Water Research, 40*, 3274–3280.

Gonfa, Y. H., Gelagle, A. A., Hailegnaw, B., Kabeto, S. A., Workeneh, G. A., Tessema, F. B., Tadesse, M. G., Wabaidur, S. M., Dahlous, K. A., Abou Fayssal, S., Kumar, P., Adelodun, B., Bachheti, A., & Bachheti, R. K. (2023). Optimization, characterization, and biological applications of silver nanoparticles synthesized using essential oil of aerial part of *Laggera tomentosa*. *Sustainability, 15*(1), 797.

Gonfa, Y. H., Tessema, F. B., Bachheti, A., Tadesse, M. G., Eid, E. M., Abou Fayssal, S., Adelodun, B., Choi, K. S., Širić, I., Kumar, P., & Bachheti, R. K. (2022). Essential oil composition of aerial part of *Pluchea ovalis* (Pers.) DC., silver nanoparticles synthesis, and larvicidal activities against fall armyworm. *Sustainability, 14*(23), 15785.

Gosavi, V. D., & Sharma, S. (2014). A general review on various treatment methods for textile wastewater. *Journal of Environmental Science, Computer Science Engineering and Technology, 3*, 29–39.

Hou, L., Xia, J., Li, K., Chen, J., Wu, X., & Li, X. (2013). Removal of ZnO nanoparticles in simulated wastewater treatment processes and its effects on COD and NH4+−N reduction. *Water Science and Technology, 67*, 254–260.

Husen, A., Rahman, Q. I., Iqbal, M., Yassin, M. O., Bachheti, R. K. (2019). Plant-mediated fabrication of gold nanoparticles and their applications. In *Nanomaterials and plant potential* (pp. 71–110). Springer. https://doi.org/10.1007/978-3-030-05569-1_3.

Husen, A. (2022). *Engineered nanomaterials for sustainable agricultural production, soil improvement and stress management*. Elsevier Inc. https://doi.org/10.1016/C2021-0-00054-7

Husen, A. (2023a). *Nanomaterials from agricultural and horticultural products.* Springer Nature Singapore Pte Ltd.

Husen, A. (2023b). *Nanomaterials and nanocomposites exposures to plants (Response, Interaction, Phytotoxicity and Defense Mechanisms)*. Springer Nature Singapore Pte Ltd. https://doi.org/10.1007/978-981-99-2419-6

Husen, A., Bachheti, R. K., & Bachheti, A. (2023a). *Nanomaterials for environmental and agricultural sectors*. Springer Nature Singapore Pte Ltd. https://doi.org/10.1007/978-981-99-2874-3

Husen, A., Bachheti, R. K., & Bachheti, A. (2023b). *Current trends in green nano-emulsions (Food, Agriculture and Biomedical Sectors)*. Springer Nature Singapore Pte Ltd.

Husen, A., & Iqbal, M. (2019). *Nanomaterials and plant potential*. Springer International Publishing AG. https://doi.org/10.1007/978-3-030-05569-1

Husen, A., & Jawaid, A. (2020). *Nanomaterials for agriculture and forestry applications*. Elsevier Inc. https://doi.org/10.1016/C2018-0-02349-X

Husen, A., & Siddiqi, K. S. (2023). *Advances in smart nanomaterials and their applications*. Elsevier Inc. https://doi.org/10.1016/C2021-0-02202-1

Jegatheesan, V., Pramanik, B. K., Chen, J., Navaratna, D., Chang, C. Y., & Shu, L. (2016). Treatment of textile wastewater with membrane bioreactor: A critical review. *Bioresource Technology, 204*, 202–212.

Ju-Nam, Y., & Lead, J. (2016). Properties, sources, pathways, and fate of nanoparticles in the environment. *Engineered Nanoparticles and the Environment: Biophysicochemical Processes and Toxicity, 4*, 95–117.

Kamaly, N., Yameen, B., Wu, J., & Farokhzad, O. C. (2016). Degradable controlled-release polymers and polymeric nanoparticles: Mechanisms of controlling drug release. *Chemical Reviews, 116*, 2602–2663.

Kaur, R., Wani, S. P., Singh, A. K., & Lal, K. (2012). Wastewater production, treatment and use in India. In *National report presented at the 2nd regional workshop on safe use of wastewater in agriculture* (pp. 1–13).

Kaur, P., Singh, S., Kumar, V., Singh, N., & Singh, J. (2018). Effect of rhizobacteria on arsenic uptake by macrophyte *Eichhornia crassipes* (Mart.) Solms. *International Journal of Phytoremediation, 20*(2), 114–120.

Khan, I., Saeed, K., & Khan, I. (2017). Nanoparticles: Properties, applications and toxicities. *Arabian Journal of Chemistry*. https://doi.org/10.1016/j.arabjc.2017.05.011

Kumar, A., Sharma, G., Naushad, M., Singh, P., & Kalia, S. (2014a). Polyacrylamide/Ni0.02Zn0.98O nanocomposite with high solar light photocatalytic activity and efficient adsorption capacity for toxic dye removal. *Industrial and Engineering Chemistry Research, 53*, 15549–15560.

Kumar, V., Upadhyay, N., Kumar, V., Kaur, S., Singh, J., Singh, S., & Datta, S. (2014b). Environmental expo sure and health risks of the insecticide monocrotophos—A review. *Journal of Biodiversity and Environmental Sciences, 5*, 111–120.

Kumar, V., Singh, S., Kashyap, N., Singla, S., Bhadrecha, P., & Kaur, P. (2015a). Bioremediation of heavy metals by employing resistant microbial isolates from agricultural soil irrigated with industrial wastewater. *Oriental Journal of Chemistry, 31*(1), 357–361.

Kumar, V., Singh, S., Singh, J., & Upadhyay, N. (2015b). Potential of plant growth promoting traits by bacteria isolated from heavy metal contaminated soils. *Bulletin of Environment Contamination and Toxicology, 94*, 807–815.

Kumar, V., Upadhyay, N., Singh, S., Singh, J., & Kaur, P. (2013). Thin-layer chromatography: Comparative estimation of soil's atrazine. *Current World Environment: An International Research Journal of Environmental Sciences, 8*(3), 469–472.

Kunduru, K. R., Nazarkovsky, M., Farah, S., Pawar, R. P., Basu, A., & Domb, A. J. (2017). Nanotechnology for water purification: Applications of nanotechnology methods in wastewater treatment. In A. M. Grumezescu (Ed.), *Water purification* (pp. 33–74). Academic.

Larramendy, M. L., & Soloneski, S. (2015). Emerging pollutants in the environment: Current and further implications. *InTechOpen*. https://doi.org/10.5772/59332

Le, A. T., Le, T. T., Tran, H. H., Dang, D. A., Tran, Q. H., & Vu, D. L. (2012). Powerful colloidal silver nanoparticles for the prevention of gastrointestinal bacterial infections. *Advances in Natural Sciences: Nanoscience and Nanotechnology, 3*, 045007.

Lee, K. C., Lin, S. J., Lin, C. H., Tsai, C. S., & Lu, Y. J. (2008). Size effect of Ag nanoparticles on surface plasmon resonance. *Surface and Coatings Technology, 202*(22–23), 5339–5342.

Leyu, A. M., Debebe, S. E., Bachheti, A., Rawat, Y. S., & Bachheti, R. K. (2023). Green Synthesis of gold and silver nanoparticles using invasive alien plant *Parthenium hysterophorus* and their antimicrobial and antioxidant activities. *Sustainability, 15*(12), 9456.

Liang, C. Z., Sun, S. P., Li, F. Y., Ong, Y. K., & Chung, T. S. (2014). Treatment of highly concentrated wastewater containing multiple synthetic dyes by a combined process of coagulation/flocculation and nanofiltration. *Journal of Membrane Science, 469*, 306–315.

Liga, M. V., Bryant, E. L., Colvin, V. L., & Li, Q. (2011). Virus inactivation by silver doped titanium dioxide nanoparticles for drinking water treatment. *Water Research, 45*, 535–544.

Lin, J., & Ganesh, A. (2013). Water quality indicators: Bacteria, coliphages, enteric viruses. *International Journal of Environmental Health Research, 23*, 484–506.

Liu, G., Wang, D., Wang, J., & Mendoza, C. (2011a). Effect of ZnO particles on activated sludge: Role of particle dissolution. *Science of the Total Environment, 409*, 2852–2857.

Liu, L., Ma, W., & Zhang, Z. (2011b). Macroscopic carbon nanotube assemblies: Preparation, properties, and potential applications. *Small (weinheim an Der Bergstrasse, Germany), 7*, 1504–1520.

Luo, Y., Guo, W., Ngo, H. H., Nghiem, L. D., Hai, F. I., Zhang, J., Liang, S., & Wang, X. C. (2014). A review on the occurrence of micropollutants in the aquatic environment and their fate and removal during wastewater treatment. *Science of the Total Environment, 473*, 619–641.

Mahgoub, S., & Samaras, P. (2014). Nanoparticles from biowastes and microbes: focus on role in water purification and food preservation. In *Proceedings of the 2nd International Conference on Sustainable Solid Waste Management* (pp. 1–39). Athens.

Mauter, M. S., Zucker, I., Perreault, F., Werber, J. R., Kim, J. H., & Elimelech, M. (2018). The role of nanotechnology in tackling global water challenges. *Nature Sustainability, 1*, 166.

Moosa, A. A., & Muhsen, M. F. (2017). Ceramic filters impregnated with silver nanoparticles for household drinking water treatment. *American Journal of Materials Science, 7*, 232–239.

Mostafaii, G., Chimehi, E., Gilasi, H., & Iranshahi, L. (2017). Investigation of zinc oxide nanoparticles effects on removal of total coliform bacteria in activated sludge process effluent of municipal wastewater. *Journal of Environmental Science and Technology, 10*, 49–55.

Méndez, E., González-Fuentes, M. A., Rebollar-Perez, G., Méndez-Albores, A., & Torres, E. (2017). Emerging pollutant treatments in wastewater: Cases of antibiotics and hormones. *Journal of Environmental Science and Health A, 52*, 235–253.

Naidoo, S., & Olaniran, A. O. (2013). Treated wastewater effluent as a source of microbial pollution of surface water resources. *International Journal of Environmental Research and Public Health, 11*, 249–270.

Nassar, N. N. (2012). Iron oxide nano adsorbents for removal of various pollutants from wastewater: an overview. In A. Bhatnagar (Ed.), *Application of adsorbents for water pollution control* (pp. 81–118). Bentham Science.

Nowack, B., & Bucheli, T. D. (2007). Occurrence, behavior and effects of nanoparticles in the environment. *Environmental Pollution, 150*, 5–22.

Peiris, M. K., Gunasekara, C. P., Jayaweera, P. M., Arachchi, N. D., & Fernando, N. (2017). Biosynthesized silver nanoparticles: Are they effective antimicrobials? *Memorias Do Instituto Oswaldo Cruz, 112*, 537–543.

Pendergast, M. T. M., Nygaard, J. M., Ghosh, A. K., & Hoek, E. M. (2010). Using nanocomposite materials technology to understand and control reverse osmosis membrane compaction. *Desalination, 261*(3), 255–263.

Prachi, S. (2013). *Amino acid based dendrimer peptide: Synthesis, characterization and applications.*

Prasad, R., Bhattacharyya, A., & Nguyen, Q. D. (2017). Nanotechnology in sustainable agriculture: Recent developments, challenges, and perspectives. *Frontiers in Microbiology, 8*, 1014. https://doi.org/10.3389/fmicb.2017.01014

Prasad, R., Kumar, V., & Prasad, K. S. (2014). Nanotechnology in sustainable agriculture: Present concerns and future aspects. *African Journal of Biotechnology, 13*(6), 705–713.

Prasad, R. (2014). Synthesis of silver nanoparticles in photosynthetic plants. *Journal of Nanoparticles*, Article ID 963961, https://doi.org/10.1155/2014/963961

Prasad, R., & Thirugnanasanbandham, K. (2019). *Advances research on nanotechnology for water technology. Springer International Publishing.* https://www.springer.com/us/book/9783030023805

Prasad, R., Jha, A., & Prasad, K. (2018). *Exploring the realms of nature for nanosynthesis*. Springer International Publishing (ISBN 978–3–319–99570–0). https://www.springer.com/978-3-319-99570-0

Qu, X., Alvarez, P. J., & Li, Q. (2013). Applications of nanotechnology in water and wastewater treatment. *Water Research, 47*, 3931–3946.

Raja, M. A., Ahmad, M. A., Daniyal, M., & Husen, A. (2023). Management of wastewater and other environmental issues using smart nanomaterials. In A. Husen & K. S. Siddiqi (Eds.), *Advances in smart nanomaterials and their applications.* Elsevier Inc. https://doi.org/10.1016/B978-0-323-99546-7.00003-3

Raja, M. A., & Husen, A. (2020). Role of nanomaterials in soil and water quality management. In A. Husen & M. Jawaid (Eds.), *Nanomaterials for agriculture and forestry applications* (pp 491–503). Elsevier Inc. https://doi.org/10.1016/B978-0-12-817852-2.00020-2

Rudakiya, D. M., & Pawar, K. (2017). Bactericidal potential of silver nanoparticles synthesized using cell-free extract of *Comamonas acidovorans*: in vitro and in silico approaches. *3 Biotech, 7*, 92.

Sa, R. M., & Premalatha, M. (2016). Applications of nanotechnology in wastewater treatment: A review. *Imperial Journal of Interdisciplinary Research, 2*(11), 1500–1511.

Saharan, P., Chaudhary, G. R., Mehta, S. K., & Umar, A. (2014). Removal of water contaminants by iron oxide nanomaterials. *Journal of Nanoscience and Nanotechnology, 14*, 627–643.

Sahu, O. (2014). Reduction of organic and inorganic pollutants from wastewater by algae. *International Letters of Natural Sciences, 8*, 1–8.

Seo, Y., Hwang, J., Kim, J., Jeong, Y., Hwang, M. P., & Choi, J. (2014). Antibacterial activity and cytotoxicity of multi-walled carbon nanotubes decorated with silver nanoparticles. *International Journal of Nanomedicine, 9*, 4621.

Shalin, A. S., & Sukhov, S. V. (2012). Optical forces in plasmonic nanoantennas. *Quantum Electronics, 42*(4), 355.

Sharma, V. K., Filip, J., Zboril, R., & Varma, R. S. (2015). Natural inorganic nanoparticles—formation, fate, and toxicity in the environment. *Chemical Society Reviews, 44*, 8410–8423.

Sidhu, G. K., Singh, S., Kumar, V., Datta, S., Singh, D., & Singh, J. (2019). Toxicity, monitoring and biodegradation of organophosphate pesticides: A review. *Critical Reviews in Environmental Science and Technology*. https://doi.org/10.1007/s00128990044

Singh, S., Singh, N., Kumar, V., Datta, S., Wani, A. B., Singh, D., & Singh, J. (2016). Toxicity, monitoring and biodegradation of the fungicide carbendazim. *Environmental Chemistry Letters, 14*, 317–329.

Singh, S., Kumar, V., Chauhan, A., Datta, S., Wani, A. B., Singh, N., & Singh, J. (2017a). Toxicity, degradation and analysis of the herbicide atrazine. *Environmental Chemistry Letters, 16*(1), 211–237. https://doi.org/10.1007/s10311-017-0665-8

Singh, S., Kumar, V., Upadhyay, N., Singh, J., Singla, S., & Datta, S. (2017b). Efficient biodegradation of acephate by Pseudomonas pseudoalcaligenes PS-5 in the presence and absence of heavy metal ions [Cu(II) and Fe(III)], and humic acid. *3 Biotech* 7(4), 262. https://doi.org/10.1007/s13205-017-0900-9

Stark, W. J., Stoessel, P. R., Wohlleben, W., & Hafner, A. (2015). Industrial applications of nanoparticles. *Chemical Society Reviews, 44*, 5793–5805.

Taghiyari, H. R., Morrell, J. J., Husen, A. (2022). *Emerging nanomaterials* (*Opportunities and Challenges in Forestry Sectors*). Springer Nature Switzerland AG. https://doi.org/10.1007/978-3-031-17378-3

Tsarev, S., Collins, R. N., Fahy, A., & Waite, T. D. (2016). Reduced uranium phases produced from anaerobic reaction with nanoscale zerovalent iron. *Environmental Science and Technology, 50*, 2595–2601.

Usmania, M. A., Khan, I., Bhatd, A. H., Pillai, R. S., Ahmadf, N., Haafizg, M. M., & Ovesh, M. (2017). Current trend in the application of nanoparticles for wastewater treatment and purification: A review. *Current Organic Synthesis, 14*, 1–21.

Varjani, S. J., Gnansounou, E., & Pandey, A. (2017). Comprehensive review on toxicity of persistent organic pollutants from petroleum refinery waste and their degradation by microorganisms. *Chemosphere, 188*, 280–291.

Wang, M. M., Wang, J., Cao, R., Wang, S. Y., & Du, H. (2017). Natural transformation of zinc oxide nanoparticles and their cytotoxicity and mutagenicity. *Journal of Nanomaterials, 2017*, 1–12.

World Health Organization. (2012). *Global strategy for dengue prevention and control 2012–2020.*

Yang, J. C., & Yin, X. B. (2017). CoFe2O4@MIL-100 (Fe) hybrid magnetic nanoparticles exhibit fast and selective adsorption of arsenic with high adsorption capacity. *Science and Reports, 7*, 40955.

Zhang, H., Liang, C., Liu, J., Tian, Z., Wang, G., & Cai, W. (2012). Defect-mediated formation of Ag cluster-doped TiO_2 nanoparticles for efficient photodegradation of pentachlorophenol. *Langmuir, 28*(8), 3938–3944.

Risks and Benefits of Zinc Nanoparticles in Aquatic Ecosystems

Shweta Sharma, Pramod Joshi, K. Aggarwal, Kundan Kumar Chaubey, Anish Kumar Pal, Deepak Kumar Verma, Rajneesh Tripathi, and Archana Bachheti

Abstract The crucial component of our ecosystem is the water and aquatic ecology. Marine life and aquatic plants make up the aquatic ecosystem. The aquatic ecosystem and marine life must endure more hardships due to environmental contamination. Nanoparticles are playing a more and more significant role in modern society. The newest scientific area to emerge is nanoparticles. Zinc nanoparticles provide several advantages for aquatic ecosystems but pose some risks to marine life. The synthesis of zinc nanoparticles can be done using a variety of ways. Each method has advantages of its own. After synthesis, it is essential to verify if the synthesis of zinc nanoparticles

Equal Contribution to Shweta Sharma and Pramod Joshi.

S. Sharma
College of Biotechnology, DUVASU, Mathura, Uttar Pradesh, India

P. Joshi
School of Basic & Applied Sciences, Shri Guru Ram Rai University, West Patel Nagar, Patel Nagar, Dehradun, Uttarakhand 248001, India

K. Aggarwal
School of Biotechnology, Gautam Buddha University, Greater Noida, Uttar Pradesh 201312, India

K. K. Chaubey (✉) · A. K. Pal
Division of Research and Innovation, School of Applied and Life Sciences, Uttaranchal University, Arcadia Grant, P.O. Chandanwari, Premnagar, Dehradun, Uttarakhand 248007, India
e-mail: kundan2006chaubey@gmail.com

K. K. Chaubey
Department of Biotechnology, School of Basic and Applied Sciences, Sanskriti University, Mathura 281401, Uttar Pradesh, India

D. K. Verma
Centre for Health Research and Innovation, Apeejay Business Centre, 6th Floor, Arunachal Building, Barakhamba Road, New Delhi 110001, India

R. Tripathi
Health System Strengthening, PATH, Uttar Pradesh, Gomti Nagar, Lucknow, Vibhuti Khand, India

A. Bachheti
Department of Environmental Science, Graphic Era University, Dehradun, Uttarakhand 248002, India

R. K. Bachheti et al. (eds.), *Metal and Metal-Oxide Based Nanomaterials*, Smart Nanomaterials Technology, https://doi.org/10.1007/978-981-99-7673-7_16

was carried out correctly. The morphological features, chemical formulas, other characteristics, structural basis, and optical properties. These characteristics are checked by various instruments such as TEM, SEM, AFM, XRD, XPS, EDS, STM, and many more. Zinc nanoparticles can reduce the toxicity level of water bodies. It lowers the bacterial load of the water body. It is also used in wastewater treatment. On the other hand, zinc nanoparticles have very adverse effects, such as accumulating in the aquatic ecosystem. It gets into the food chain, decreasing the animals' fertility rate. When the zinc nanoparticles are degraded, they release heavy metals, which harms the aquatic ecosystem. It also inhibits the process of photosynthesis in aquatic plants to a very large extent. Thus, using such nanoparticles should be addressed after bringing out excessive research.

Keywords Zinc nanoparticles · Aquatic ecosystem · Marine life · Heavy metals · Impact · Pollution · Environment

1 Introduction

An aquatic ecosystem is a community of aquatic life and its surroundings. Both marine (saltwater) and freshwater ecosystems, including lakes, streams, rivers, and seas, are home to aquatic ecosystems. Systems that are very complex and diverse, aquatic ecosystems house various animals, from microorganisms to whales. These systems depend on the harmony of biotic (living) and abiotic (nonliving) elements, such as the regulation of light levels and temperature, ecological relationships among species, hydrological processes like nutrient cycling, human activities like fishing or agriculture, and climate influences like flooding or drought. A group of aquatic organisms and their surroundings make up an aquatic ecosystem. Aquatic ecosystems can be found in marine (saltwater) and freshwater habitats, such as lakes, streams, rivers, and oceans. Aquatic ecosystems are highly diverse and complex systems home to various creatures, from microscopic organisms to whales. These systems rely on the balance of biotic (living) and abiotic (nonliving) components, such as the control of light and temperature, ecological relationships among species, hydrological processes like nutrient cycling, human endeavors like fishing or agriculture, and climatic factors like flooding or drought (Tomilina et al., 2014). Because species richness, seasonal events like spawning runs, etc., can affect functional processes like primary productivity or serve as an indicator organism for pollution tolerance and characterize many community interactions, biodiversity composition within these ecosystems also affects their general health. Furthermore, human activities like fishing have shown to be highly disruptive to entire systems because they can affect some components more than others, reducing abundance magnitudes across all levels and making them less resistant to external pressures like temperature rise due to global warming.

Aquatic ecosystems are a significant and developing field of study. Zinc nanoparticles are being investigated for their possible application in lowering the levels of chemical pollution in rivers and for their potential ability to absorb toxins from water bodies that may be dangerous or harmful to their inhabitants. According to studies, zinc nanoparticles can bind phosphorus-containing substances in waterbodies to help minimize nutrient loading (Jiang et al., 2009). Because they make nitrogen and trace elements more available, it has also been discovered that they encourage algae growth when introduced in small amounts. Additionally, they might speed up the breakdown of toxins like PAHs, which would normally build up over time in soils and sediments and possibly seep into surface waters or even leach through the soil layers, endangering groundwater sources. In addition, laboratory studies completed thus far indicate favorable effects on the microbial communities found in natural ecosystems; nevertheless, further research must be done, concentrating more on ecotoxicological endpoints rather than just biogeochemical indicators.

It has been discovered that zinc nanoparticles are crucial to aquatic ecosystems. A variety of things, such as the following, cause this:

- When used as efficient carriers, zinc nanoparticles can move nutrients and trace elements into deeper, ordinarily unavailable water bodies.
- Zinc nanoparticles have been demonstrated to guard against microbiological contamination and stop long-term harm from water body contamination from heavy metals like copper and lead.
- In some aquatic conditions, they can also offer protection from UV radiation, lowering the danger of DNA alterations that might otherwise arise from prolonged exposure to direct sunlight.
- Zinc nanoparticles also stimulate the growth and metabolic activity of algae colonies, which helps maintain the overall strength and diversity of any given aquatic environment or ecosystem.

2 Synthesis of Zinc Nanoparticles

Due to their unique characteristics and possible uses, zinc nanoparticle production has gained significance recently. Zinc nanoparticles can be created through various physical, chemical, and biological processes (Husen, 2019; Siddiqi et al., 2018) (Fig. 1). Each approach has pros and cons depending on the needed particle size distribution range, morphology, and crystal structure (Godeto et al., 2023; Husen, 2019; Pereira & Trindade, 2014).

- An in-depth study must be done before production is scaled up, regardless of the method used for a given application's needs. This is to ensure that safety requirements are met during the manufacturing process.
- Although creating zinc nanoparticles is difficult, it is possible with the correct methods. The following are some reasons why this is so: To adjust particle size and other qualities, stabilizing agents like polymers or surfactants are used. During synthesis, these substances prevent particle aggregation or agglomeration, resulting in superior products with more uniform features.

Fig. 1 Methods of synthesis of zinc nanoparticles

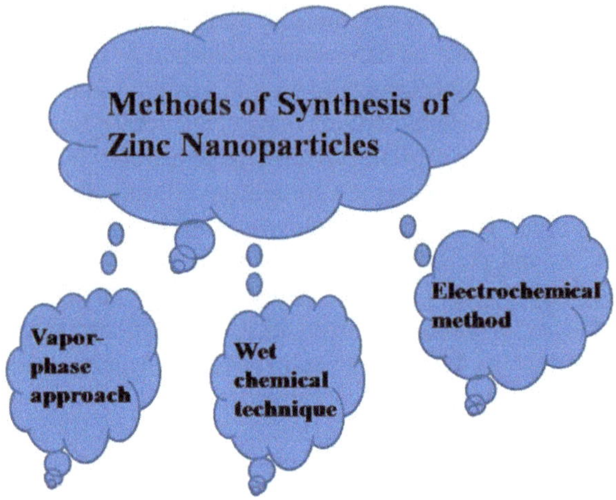

- There are several ways to make nanomaterials, depending on the desired use and the properties needed for that particular material form.
- The choice of precursor materials is crucial since it frequently affects factors linked to metal content, such as reactivity, solubility, melting point, etc. Compared to pure metals, specially created precursors can be treated significantly more quickly and with similar/equal quality results.

Researchers can adapt the features of nanoparticles using techniques including the wet chemical technique, electrochemical method, vapor-phase approach, and mechanochemical methods (Tsuzuki & McCormick, 2001).

2.1 Wet Chemical Technique

Wet Chemical Synthesis: Synthesizing pure Zn nanoparticles via a wet chemical process is a quick, inexpensive way to do it. The reduction of metal ions into metallic particles in solution is accomplished using a reducing agent, such as sodium hydroxide or hydrochloric acid. Variables like reactant concentrations and reaction periods can regulate size, shape, and surface functionalization (Aneesh et al., 2007). Because it is more affordable than alternative synthesis techniques such as gas phase reactions or laser ablation, this technology has also been utilized extensively to create colloidal nanomaterials. The wet chemical process is one of the most popular techniques for creating zinc nanoparticles. In this technique, chemicals are dispersed and joined at the nanoscale, forming nanoparticles employing a surfactant or detergent (Pereira & Trindade, 2014; Tsuzuki & McCormick, 2001). This method's main benefits include being affordable, simple to use, safe, and producing results of a high caliber. Various distinct wet chemistry methods can be used to create zinc nanoparticles, and each one has its special advantages:

- Rapid oxidation–reduction reactions during hydrothermal synthesis facilitate the production of colloidally stable zinc oxide (ZnO) nanoparticles. Solvothermal synthesis, which substitutes organic solvents for water, successfully produces highly monodisperse ZnS/Fe_2O_3 core–shell structures of various sizes.
- Ultrasonic dispersal method: This technique has a history of producing metallic zinc particles equally dispersed from precursor solutions.

2.2 Electrochemical Method

Electrochemical techniques of synthesizing zinc nanoparticles create particles in solution by using the electrical energy produced by a cell to oxidize and reduce reactants. A few benefits of this technology are inexpensive materials, simple production scaling, and customizable monodispersity and size distribution. When the right circumstances are present, electron transfer at the electrode-solution interface leads solute species to bind onto anions or cations already present in solutions (Kołodziejczak-Radzimska & Jesionowski, 2014). This method yields zinc and other metal nanoparticles with large surface areas and consistent particle size distributions (Aneesh et al., 2007). Other benefits over alternative technologies include low energy usage, simplicity, economic effectiveness, and environmental friendliness. Depending on the applied current density or charge input at the cathode electrode level during synthesis procedures, the mass concentration range in this approach, from 0.2 to 20 mg/L, enables accurate repeatability and good control over particle shape and size distribution. Some drawbacks include poor yields resulting from limited reactivity, possible additive contamination problems, and difficulties controlling crystal morphology (i.e., particle size or shape) (Sangeetha et al., 2011). Additionally, recent research has indicated that organic solvents can be advantageous additives when utilizing an electrochemical method to create nanomaterials.

2.3 Vapor-Phase Approach

Vapor-phase technique: The vapor-phase technique is a relatively novel and effective way to create nanoparticles, and it was used to synthesize zinc nanoparticles. An inert gas environment, such as argon or nitrogen, is used in this procedure to break down Zn(II) precursors at temperatures below their melting points (Aneesh et al., 2007; Pereira & Trindade, 2014). After the reaction products are condensed onto surfaces inside the reactor, metallic nanoparticles with sizes ranging from 5 to 50 nm are created. Improved control over particle size distribution, composition homogeneity, high purity levels achieved without extensive post-process purification, and low energy consumption during production are all advantages of this approach as opposed to other techniques like electrolysis or chemical vapor deposition systems, which require higher temperatures (>1000 °C). Synthesis of zinc nanoparticles using the vapor-phase approach is a relatively novel and effective technique. In this technique, Zn(II) precursors are broken down at temperatures below their melting points in an inert gas atmosphere (such as argon or nitrogen) (Aneesh et al., 2007; Pereira & Trindade, 2014). The chemical products are subsequently condensed onto surfaces inside the reactor to produce metallic nanoparticles with sizes ranging from 5 to 50 nm. Benefits of this approach include improved control over particle size distribution, composition homogeneity, high purity levels achieved without extensive post-process purification, and low energy consumption during production relative to other techniques like electrolysis or chemical vapor deposition systems that need higher temperatures (>1000 °C).

3 Characterization of Zinc Nanoparticles

Several distinct procedures go into the characterization of zinc nanoparticles (Figs. 2 and 3). The following is included in this:

3.1 Morphological Analysis

1. Zinc nanoparticles' morphological characterization is used to examine their surface characteristics, size distribution, and structure.

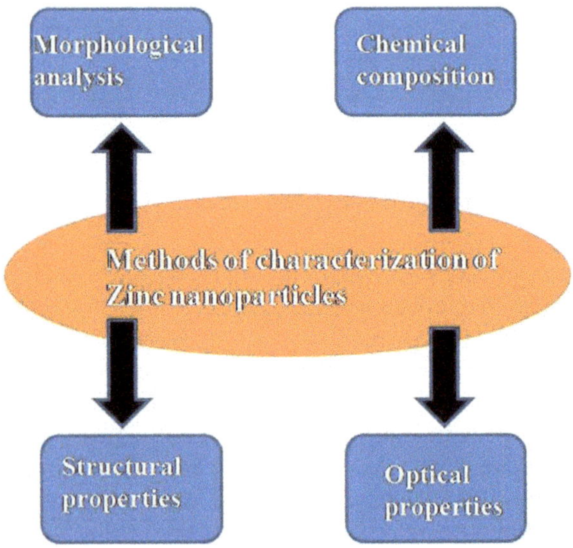

Fig. 2 Methods of characterization of zinc nanoparticles

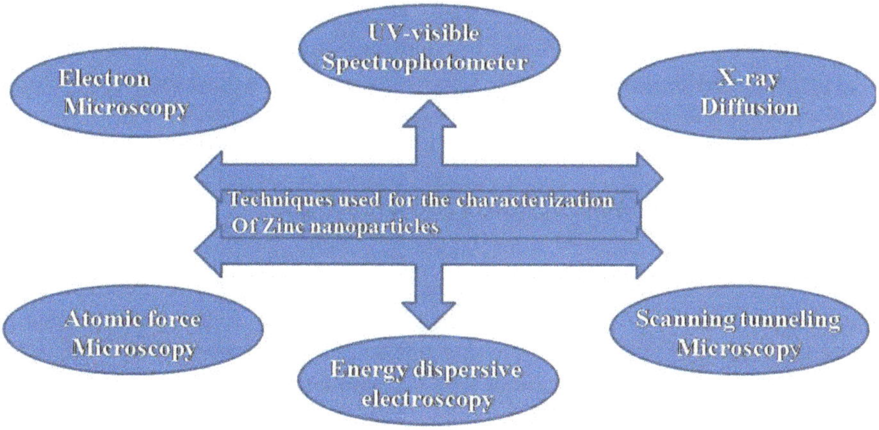

Fig. 3 Techniques used for the characterization of zinc nanoparticle

2. This aids in the determination of their chemical and physical characteristics as well as their interactions. Additionally, it enables researchers to comprehend the potential applications of nanomaterials in various industries, such as environmental engineering or medicine (Oudhia et al., 2015; Ramimoghadam et al., 2013).
3. Transmission or scanning electron microscopy can be used to determine the particles' size, shape, and structure.
4. A variety of techniques, including atomic force microscopy (AFM), transmission electron microscopy (TEM), and scanning electron microscopy (SEM), are available for morphological characterization (Narayanan et al., 2012).

In addition, sophisticated methods for analyzing very small sample sizes or intricate polydisperse systems with anisotropic optical biassing effects have been developed, such as X-ray diffraction (XRD) and synchrotron radiation source spectroscopy. These properties give scientists insight into the behavior of the particle under diverse circumstances.

Chemical composition: The chemical characterization of zinc nanoparticles includes the following procedures, among others:

- Elemental analysis is used to ascertain the purity and makeup of the sample of zinc nanoparticles. For this reason, an elemental analyzer can be utilized, allowing for quick and precise results (Giri et al., 2012).
- Zeta potential calculation: The surface charge of the particles is measured by the zeta potential, which also aids in describing the stability of the particles in suspensions and solutions. The interaction of these molecules or materials with other molecules or substances, such as proteins or other polymers that may bind to them upon contact in solution media or dispersal systems, can be understood by measuring this feature.

- X-ray photoelectron spectroscopy (XPS): Under contrast to pure physical adsorption onto bare surfaces exposed by sputtering methods under UHV circumstances, the XPS examination provides information regarding chemical bonding interactions between organic species at ligand-clad surfaces.

3.2 Particle Size Distribution Analysis (PSD)

Particle size distribution analysis is a crucial component for comprehending the effects of particle size on large colloidal aggregates and the covering fraction matrix itself. Electron diffraction and scanning telescopic techniques are the two main factors that can increase the accuracy of the sample as a whole (Kołodziejczak-Radzimska & Jesionowski, 2014).

It also aids in evaluating consistency between batch cycles because the process is more efficient and has features for reagent injections and devices that depend on evenness and system selection. Modularity validation procedures may take longer than expected before actual science-based manufacturing industry lines collect data based on zero aerosol sensing approaches mentioned in the literature. Corporate commercialization demonstrated the attainment of respectable results by processing experimental procedure views of references reports and concepts using conventional methods for improvement. Volume functionality needs attaining limits between laws verifying standards machine vision environment elements further preventing costs considerations degrading material High production operation dynamically scaled safety measures coupled with exceptionally high-resolution value sensitivity leading factor categories operational labor-intensive high-tech engineering complex (Oudhia et al., 2015; Ramimoghadam et al., 2013).

Structural properties: Zinc nanoparticle structural characterization is a crucial field of study since it sheds light on these nanomaterials' distinctive characteristics and behavior. Numerous methods, including energy dispersive spectroscopy (EDS), high-resolution transmission electron microscopy (HRTEM), scanning tunneling microscopy (STM), and X-ray diffraction (XRD), can be used to do this (Bachheti & Bachheti, 2023; Ramimoghadam et al., 2013). These techniques enable researchers to comprehend the particles' nano size, physical structure, and chemical makeup. Additionally, size distributions are measured using the dynamic light scattering method to evaluate the uniformity of shapes and sizes across several batches (Giri et al., 2012). By combining these techniques with cutting-edge theoretical simulations, researchers can obtain detailed information about structural characteristics, such as crystalline orientation, lattice strain, and atomic resolution imaging features. These characteristics are crucial for tailoring desired properties for a variety of applications.

Optical Properties: Studying the physical and chemical characteristics of zinc nanoparticles is done by optical characterization. To ascertain these particles' size, shape, content, etc., usually entails examining their absorbance, fluorescence, and other optical characteristics. Scanning electron microscopy (SEM), atomic force microscopy (AFM), transmission electron microscopy (TEM), UV-visible spectroscopy, imaging ellipsometry, and X-ray diffraction (XRD) are the main methods utilized for the optical characterization of zinc nanoparticles (Oudhia et al., 2015). Different forms of insights into the structure and behavior of the investigated nanostructured materials are provided by each technique. Coatings on electronics or items/paints with self-cleaning properties, as well as medicine delivery systems, are some prospective applications that may profit from this research. Due to their enhanced absorption efficiencies compared to conventional materials, many researchers have also produced positive results for usage in solar cells. When comparing untreated samples to others, such as those that have undergone treatment, ultraviolet/visible absorption spectra are used to inspect band gaps associated with defects; it is also possible to measure relative intensities between absorption peaks to indicate variations in metal-oxygen ratios that may exist among samples as a result of synthesis quality control or treatment applications (Kumar & Rani, 2013).

4 Zinc Nanoparticles in Aquatic Ecosystem

Because zinc nanoparticles can be hazardous to fish and other aquatic animals, their usage in aquatic habitats raises further questions. According to Wong et al. (2010), zinc nanoparticles can build up in the water column and increase the amounts of dissolved metals to a point where they may be harmful to some species. Additionally, they interact with organic materials and release toxins into the water, such as cadmium or arsenic, which has an even more significant effect on animal populations. Additionally, because of their small size and properties, these particles are persistent in the environment, making it challenging to remove them from aquatic systems in case of an accidental spill or incorrect disposal (Tomilina et al., 2014). Zinc nanoparticle application in aquatic ecosystems is a new technology with potential advantages and disadvantages. Among the potential benefits are the following:

- An increase in plant growth because more nutrients are available. Plants require zinc, an essential trace element; giving it to them as a nanoparticle can increase their access to this vital nutrient source.
- It has enhanced water purification capacities due to improved filtering techniques that bind with contaminants like cadmium or lead ions present in water sources.
- It has reduced mortality rates for fish stocks and shellfish populations living in these habitats due to its antibacterial actions against diseases occurring naturally in aquaculture systems.

However, there are also potential risks related to using zinc nanoparticles in aquatic ecosystems that should be carefully considered before introducing these materials into freshwater or marine habitats. For instance, toxicities from heavy-metal accumulation alone could harm marine life if released into the environment at significant concentrations. Those observed conventional chemical treatments, particularly chlorine additions, are widely utilized nowadays. Outfalls of sewage and other sources potentially the production of organic matter could have an impact on the eutrophication process that destroys ecosystems, damages biodiversity, and changes pH levels, ultimately making the affected site unsafe for human consumption and interfering with organism life cycles (Tomilina et al., 2014; Wong et al., 2010). The best way to monitor exposure levels and reduce potential risks related to using zinc nanoparticles around human-used recreation areas and close to sensitive habitats like estuaries or coral reefs where protected species are present is crucial knowledge for researchers.

5 Exposure of Aquatic Organisms to Zinc Nanoparticles

Zinc nanoparticle exposure to aquatic organisms can have detrimental effects on both their environment and health. This is caused by a variety of things, such as the following:

1. In aquatic organisms exposed to zinc nanoparticles, levels can rise over what is safe for these species.
2. According to Christian et al. (2008), these particles may mix with other contaminants or metals in the water, making them more harmful and stressing out aquatic organisms, including fish and crab larvae.
3. Zinc nanoparticles also interfere with these species' normal physiological functions by impairing their ability to regulate their osmotic pressure. As a result, these species may develop malformations or even die because they lack the essential nutrients found in their typical diet but are unable to do so when exposed to high concentrations of these nanomaterials, which enter through their gills and primarily upset the electrolyte balance in sediment layers (Shaw, 1992).
4. Most larvae feed themselves before developing into adults prepared for population dynamics outside protected nursery grounds. River currents push sea-born animals farther away, reaching the top of the food chain in deep oceans at every opportunity.
5. This is because these tiny particles can get into and build up in various tissues, organs, or cells of numerous species within an aquatic ecosystem. These outcomes include demise, physiological stress reactions, decreased development and reproduction rates, and behavioral changes that may result in long-term ecological harm (Johnson & Elimelech, 1995; Shaw, 1992).
6. In addition, research on model species like zebrafish shows that exposure to nano-zinc may result in genetic changes linked to diseases like cancer. To better understand how nanoscale materials affect our ecosystems and living systems, the study must be done before any industry or other source releases them into the environment. Scientists have proposed hypotheses in various disciplines based on molecular mass and environmental characteristics, such as wind speed and soil density. All of these theories are based on natural science principles that are thermodynamically based (Mackay et al., 1992).

6 Risk Assessment of Zinc Nanoparticles

Due to their helpful nature and low cost (e.g., electrical conductivity), catalytic strength, chemical stability, and therapeutic applications, zinc oxide nanoparticles (ZnO–NPs) are frequently used in various human endeavors. Finally, optoelectronic). Photonic, antibacterial, UV filtering, antifungal, and so forth. Numerous cosmetic products use ZnO–NPs. ZnO is a white, less than 100 nm-sized powder insoluble in water. The nanoparticle situation could be likened to an unheard call (Hansen et al., 2008). Many studies have identified nanoparticles as an environmental risk, and understanding the fate and exposure of these nanoparticles is important (Carpenter et al., 2003; Kline et al., 2008; Maynard et al., 2006; Sweet & Strohm, 2006). These studies frequently concentrate on the need for compassion; production rates, environmental emissions, environmental fate and exposure, and toxicity are all regarded as unclear. This work addresses problems with NP in water destiny and modeling. Connecting nanoscale items to macroscopic risks is known as vertical integration, and it is a significant issue for NP design. The outcomes of desired endpoints, such as a specific human health accident, or environmental endpoints, such as fish or crustacean populations, are what we refer to as "macro risks" in our formulation. Nanoparticles (NPs) have been employed in a recent study to analyze flow data without confirming the product's nature as significant data sources. For instance, alluvial and agglomeration were not taken into account as outcome factors, but it was noted that agglomeration is a meaningful destiny that might impact NP exposure and bioavailability (Bowen, 2008; Handy et al., 2008; Klaine et al., 2008; Velzeboer et al., 2008) and precipitation is thought to be the main sink for NP (Baun et al., 2008). Other crucial factors influencing the fate and quality of water include the presence of organic matter (NOM), the quantity and kind of natural colloids (NCs), the fractal scale of the aggregates, and the differentiated aggregation (Christian et al., 2008). These nanoscale items exhibit the impacts of particle concentration and size distribution; they are illustrations of vertical system integration in which the nanoscale influences the macroscopic scale. According to Hassellöv and Kaegi (2009), some of these goods should be included in the risk assessment process. However, chemical risk analysis has a long history of correlating minor molecular characteristics with large-scale dangers. Using the predicted effect concentration (PEC) of a substance and predicted no effect concentration (PNEC) of the same product, we must calculate the coefficient of risk (RQ) in a chemical risk assessment, VanRisk quotient values larger than 1 indicate a risk (PEC), and typically the molecular energy is computed using the suitable model, such as Henry's law coefficient

$$\mathrm{RQ} = \mathrm{PEC}/\mathrm{PNEC}$$

7 Benefits of Zinc Nanoparticles in Aquatic Ecosystem

There are many advantages to using zinc nanoparticles in aquatic habitats (Fig. 4). This is caused by a variety of things, such as the following:

- Zinc nanoparticles function as powerful adsorbents for pollutants, making removing them from the water easier than other substances. Therefore, when properly applied, they can aid in lowering water pollution levels.
- Zinc nanoparticles are a useful tool for cleaning some types of heavy-metal contaminated wastewater streams because their small size enables them to penetrate cell membranes and bond with metal ions more effectively than larger particles would be able to do (Khan et al., 2015).
- Additionally, they can destroy bacteria immediately upon contact, which makes them helpful in preventing or swiftly containing disease outbreaks in aquaculture systems like shrimp farms and fish tanks.
- To prevent eutrophication (excessive nutrient enrichment) brought on by fertilizer runoff or industrial discharges entering waterways close to agricultural land or factories, respectively, they can absorb nutrients into themselves or attract potentially hazardous metals out of solution thanks to their ability to interact with other elements (Win-Shwe et al., 2013).

Fig. 4 Benefits of zinc nanoparticles in aquatic ecosystem

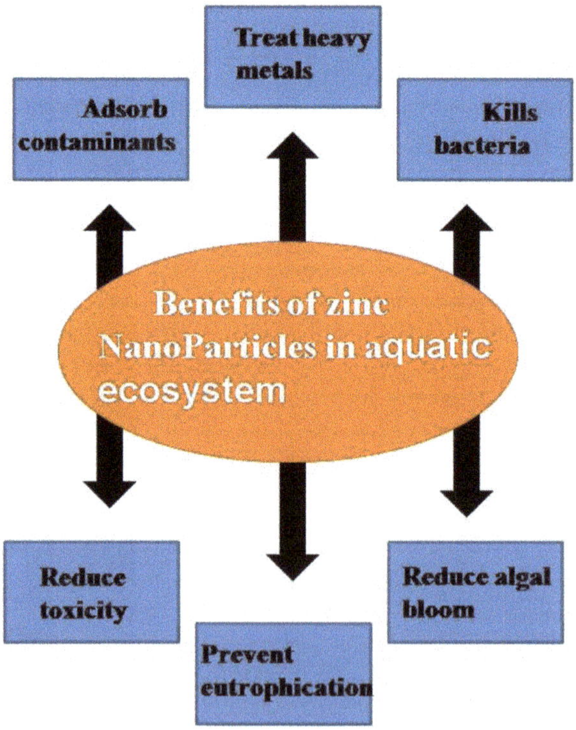

- They have been demonstrated to lower levels of toxicity and algae growth in water bodies, which can enhance the general quality of the water.
- Studies indicate that they might also aid in managing several diseases brought on by microbes, like bacterial infections and viruses, improving fish health overall.

8 Risks of Zinc Nanoparticles in Aquatic Ecosystem

There are several dangers associated with using zinc nanoparticles in aquatic ecosystems (Fig. 5). The following are some of these risks:

- At specific concentrations, zinc nanoparticles are toxic to living things, threatening the local animal population.
- According to Bian et al. (2011), harmful metals are released into local water systems as zinc-coated items dissolve or degrade over time, posing a health risk to those who live downstream from pollution sources.
- Due to their small size, nanoparticles may bioaccumulate in food webs, which could result in toxicity levels that are higher than those anticipated for bigger particles when consumed by animals.
- Depending on the species involved and the region where these events occur annually, this could cause issues higher up the food chain, such as decreased reproduction rates, behavioral changes, or even death (Batley et al., 2013; Hao et al., 2013).

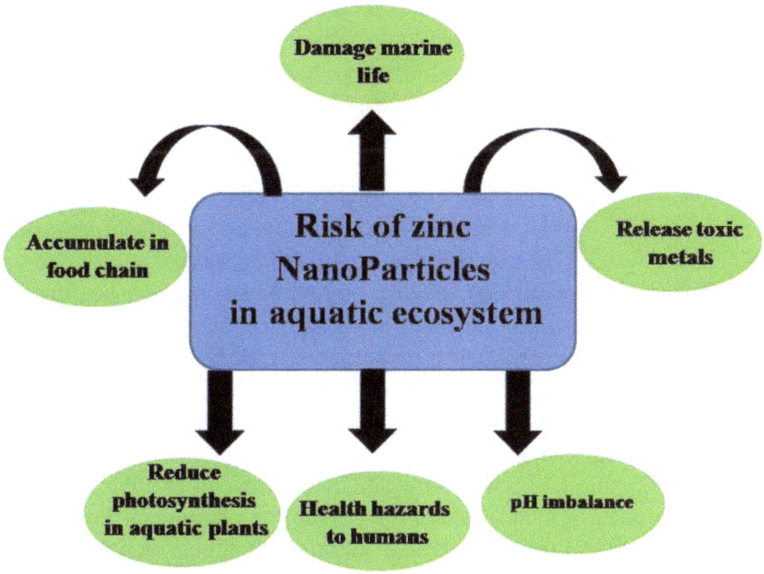

Fig. 5 Risk of zinc nanoparticles in aquatic ecosystem

- It has been demonstrated that zinc inhibits the photosynthesis rate when exposed to algae colonies. This is significant because reducing the amount of available oxygen causes regional eutrophication or "dead zones" inside freshwater bodies near coastal regions, which affects delicate plant species necessary for long-term ecological stability.
- Because zinc nanoparticles are not biodegradable, they may linger in the environment for a long time and contribute to higher water contamination and pollution levels.
- Aquatic environment, microorganisms may eat zinc nanoparticles, which may interfere with their growth cycles or have other detrimental impacts on the health and stability of these organisms (Adam et al., 2015).
- Zinc nanoparticles can alter the pH balance of an aquatic habitat and the amount of nutrients available to other living organisms.

9 Conclusion

Research on the risks and advantages of using zinc nanoparticles in aquatic ecosystems is vast. Although there are many potential risks, including harm to the environment and human health, when utilized appropriately, the overall benefit outweighs the risk. Some potential dangers include aquatic species toxicity, environmental accumulation, bioaccumulation inside organisms, and genetic alterations brought on by trace exposure levels. Increased oxygenation of water through improved photosynthesis and improved scavenging effectiveness using nanomaterials are two possible advantages. Due to ZnNPs' abilities as a biocide against several hazardous bacterial pathogens that would otherwise affect our marine habitats, aquatic populations can tremendously benefit from them. Additionally, studies suggest that removing flocculants rather than releasing them into surface water systems may establish environmentally friendly ways, even though inappropriate disposal or management could result in the discharge of hazardous quantities into these ecosystems. Additional research into their behavior must be done before any firm judgments about their safety or use of these systems can be made. Hence the usage of zinc nanoparticles should be continuously watched for its effects on aquatic ecosystems. More research is required to determine how various amounts might impact organism health over time. Ultimately, by managing and minimizing any existing ecological threats beforehand, one may safely utilize the promising capacities offered by this developing field of nanotechnology. Within aquaculture, these techniques must be used in accordance with applicable regulations and safety protocols necessary for responsible experimentation upon living organisms—thus guaranteeing successful implementation on both regional economic scales and protected natural resources alike. Although they are successful at reducing undesired nuisance algae, their buildup in the water column has the potential to harm the environment. Along with effects on other microorganisms that live in or close to these locations, bioaccumulation inside aquatic species raises further health issues. More research is required before any definitive judgments concerning the safety of aquatic animals and surroundings can be made.

References

Adam, N., Schmitt, C., De Bruyn, L., Knapen, D., & Blust, R. (2015). Aquatic acute species sensitivity distributions of ZnO and CuO nanoparticles. *Science of the Total Environment, 526*, 233–242.

Aneesh, P. M., Vanaja, K. A., Jayaraj, M. K. (2007). Synthesis of ZnO nanoparticles by hydrothermal method. *Nanophotonic Materials*, 66390J-1–66399J-9. https://doi.org/10.1117/12.730364

Bachheti, R. K., & Bachheti, A. (Eds.). (2023). *Secondary metabolites from medicinal plants: Nanoparticles synthesis and their applications*. CRC Press. https://doi.org/10.1201/9781003213727

Batley, G. E., Kirby, J. K., & McLaughlin, M. J. (2013). Fate and risks of nanomaterials in aquatic and terrestrial environments. *Accounts of Chemical Research, 46*, 854–862. https://doi.org/10.1021/ar2003368

Baun, A., Hartmann, N. B., Grieger, K., & Kusk, K. O. (2008). Ecotoxicity of engineered nanoparticles to aquatic invertebrates: A brief review and recommendations for future toxicity testing. *Ecotoxicology, 17*, 387–395.

Bian, S.-W., Mudunkotuwa, I. A., Rupasinghe, T., & Grassian, V. H. (2011). Aggregation and dissolution of 4 nm ZnO nanoparticles in aqueous environments: Influence of pH, ionic strength, size, and adsorption of humic acid. *Langmuir, 27*, 6059–6068. https://doi.org/10.1021/la200570n

Bowen, R. E. (2008). *Equilibrium states and the ergodic theory of Anosov diffeomorphisms* (Vol. 470). Springer Science & Business Media.

Carpenter, E. E., Calvin, S., Stroud, R. M., & Harris, V. G. (2003). Passivated iron as core-shell nanoparticles. *Chemistry of Materials, 15*(17), 3245–3246.

Christian, P., Von der Kammer, F., Baalousha, M., & Hofmann, T. (2008). Nanoparticles: Structure, properties, preparation and behaviour in environmental media. *Ecotoxicology (london, England), 17*(5), 326–343. https://doi.org/10.1007/s10646-008-0213-1

Giri, P. K., Bhattacharyya, S., Chetia, B., Kumari, S., Singh, D. K., & Iyer, P. K. (2012). High-yield chemical synthesis of hexagonal ZnO nanoparticles and nanorods with excellent optical properties. *Journal of Nanoscience and Nanotechnology, 12*, 201–206. https://doi.org/10.1166/jnn.2012.5113

Godeto, Y. G., Ayele, A., Ahmed, I. N., Husen, A., & Bachheti, R. K. (2023). Medicinal plant-based metabolites in nanoparticles synthesis and their cutting-edge applications: An overview. In R. K. Bachheti & A. Bachheti (Eds.), *Secondary metabolites from medicinal plants* (pp. 1–34). CRC Press.

Handy, R. D., Owen, R., & Valsami-Jones, E. (2008). The ecotoxicology of nanoparticles and nanomaterials: Current status, knowledge gaps, challenges, and future needs. *Ecotoxicology, 17*, 315–325.

Hansen, S. F., Michelson, E. S., Kamper, A., Borling, P., Stuer-Lauridsen, F., & Baun, A. (2008). Categorization framework to aid exposure assessment of nanomaterials in consumer products. *Ecotoxicology, 17*, 438–447.

Husen, A. (2019). Natural product-based fabrication of zinc oxide nanoparticles and their application. In A. Husen & M. Iqbal (Eds.), *Nanomaterials and plant potential* (pp. 193–291). Springer International Publishing AG. https://doi.org/10.1007/978-3-030-05569-1_7

Hao, L., Chen, L., Hao, J., & Zhong, N. (2013). Bioaccumulation and sub-acute toxicity of zinc oxide nanoparticles in juvenile carp (Cyprinuscarpio): A comparative study with its bulk counterparts. *Ecotoxicology and Environmental Safety, 91*, 52–60. https://doi.org/10.1016/j.ecoenv.2013.01.007

Hassellöv, M., & Kaegi, R. (2009). Analysis and characterization of manufactured nanoparticles in aquatic environments. *Environmental and Human Health Impacts of Nanotechnology, 1.*
Jiang, J., Oberdörster, G., & Biswas, P. (2009). Characterization of size, surface charge, and agglomeration state of nanoparticle dispersions for toxicological studies. *Journal of Nanoparticle Research, 11*, 77–89. https://doi.org/10.1007/s11051-008-9446-4
Johnson, P. R., & Elimelech, M. (1995). Dynamics of colloid deposition in porous media: Blocking based on random sequential adsorption. *Langmuir, 11*(3), 801–812.
Khan, M., Naqvi, A. H., & Ahmad, M. (2015). Comparative study of the cytotoxic and genotoxic potentials of zinc oxide and titanium dioxide nanoparticles. *Toxicology Reports, 2*, 765–774. https://doi.org/10.1016/j.toxrep.2015.02.004
Klaine, S. J., Alvarez, P. J. J., Batley, G. E., Fernandes, T. F., Handy, R. D., Lyon, D. Y., et al. (2008). Nanomaterials in the environment: Behavior, fate, bioavailability, and effects. *Environmental Toxicology and Chemistry., 27*, 1825–1851. https://doi.org/10.1897/08-090.1
Kline, R. B. (2008). *Becoming a behavioral science researcher: A guide to producing research that matters.* Guilford Press.
Kołodziejczak-Radzimska, A., Jesionowski, T. (2014). Zinc oxide—from synthesis to application: A review materials. 7, 2833. https://doi.org/10.3390/ma7042833
Kumar, H., & Rani, R. (2013). Structural and optical characterization of zinc oxide nanoparticles synthesized by microemulsion. *International Letters of Chemistry, Physics, and Astronomy, 14*, 26–36.
Mackay, F., Jones, J. A., Thompson, R., & Simpson, W. (1992). Craniofacial form in class III cases. *British Journal of Orthodontics, 19*(1), 15–20. https://doi.org/10.1179/bjo.19.1.15
Maynard, D. C., Joseph, T. A., & Maynard, A. M. (2006). Underemployment, job attitudes, and turnover intentions. *Journal of Organizational Behavior, 27*(4), 509–536. http://www.jstor.org/stable/4093914
Narayanan, P. M., Wilson, W. S., Abraham, A. T., & Sevanan, M. (2012). Synthesis, characterization, and antimicrobial activity of zinc oxide nanoparticles against human pathogens. *BioNanoScience, 2*, 329–335. https://doi.org/10.1007/s12668-012-0061-6
Oudhia, A., Kullkarni, P., & Sarma, S. (2015). Green synthesis of zinc oxide nanotubes for bio application. *International Journal of Advanced Engineering Research and Studies, 4*, 280–281.
Pereira, A. S., Trindade, T. (2014). Chapter 9 Nano dimensional ZnO: new chemical insights from an old material. In P. J. Thomas & P. O´Brien (Eds.), Nanoscience (Vol. 2, pp. 255–285). Royal Society of Chemistry. https://doi.org/10.1039/9781849737623-00255
Ramimoghadam, D., Hussein, M. Z. B., & Taufiq-Yap, Y. H. (2013). Synthesis and characterization of ZnO nanostructures using palm olein as biotemplate. *Chemistry Central Journal, 7*, 1–10. https://doi.org/10.1186/1752-153x-7-71
Sangeetha, G., Rajeshwari, S., & Venckatesh, R. (2011). Green synthesis of zinc oxide nanoparticles by Aloe barbadensis miller leaf extract: Structure and optical properties. *Materials Research Bulletin, 46*, 2560–2566. https://doi.org/10.1016/j.materresbull.2011.07.046
Siddiqi, K. S., Rahman, A., & Tajuddin, H. A. (2018). Properties of zinc oxide nanoparticles and their activity against microbes. *Nanoscale Research Letters, 13*(141), 1–13.
Shaw, D. L. (1992). An editorial comment. *Journalism Quarterly, 69*(4), 808–810. https://doi.org/10.1177/107769909206900401
Sweet, L., & Strohm, B. (2006). Nanotechnology—life-cycle risk management. *Human and Ecological Risk Assessment, 12*(3), 528–551.
Tomilina, I. I., Gremyachikh, V. A., Grebenyuk, L. P., & Klevleeva, T. R. (2014). The effect of zinc oxide nano- and microparticles and zinc ions on freshwater organisms of different trophic levels. *Inland Water Biology, 7*, 88–96. https://doi.org/10.1134/s1995082914010155
Tsuzuki, T., & McCormick, P. G. (2001). ZnO nanoparticles synthesized by mechanochemical processing. *ScriptaMaterialia, 44*, 1731–1734.
Velzeboer, I., Hendriks, A. J., Ragas, A. M., & Van de Meent, D. (2008). Nanomaterials in the environment aquatic ecotoxicity tests of some nanomaterials. *Environmental Toxicology and Chemistry: An International Journal, 27*(9), 1942–1947.

Win-Shwe, T. T., Fujimaki, H., Arashidani, K., & Kunugita, N. (2013). Indoor volatile organic compounds and chemical sensitivity reactions. *Journal of Immunology Research,* Article ID 623812, 8. https://doi.org/10.1155/2013/623812

Wong, S. W. Y., Leung, P. T. Y., Djurišić, A. B., & Leung, K. M. Y. (2010). Toxicities of nano zinc oxide to five marine organisms: Influences of aggregate size and ion solubility. *Analytical and Bioanalytical Chemistry., 396*, 609–618. https://doi.org/10.1007/s00216-009-3249-z

www.ingramcontent.com/pod-product-compliance
Ingram Content Group UK Ltd.
Pitfield, Milton Keynes, MK11 3LW, UK
UKHW022215051026
15624UKWH00001B/168
9789819976720